Graduate Texts in Mathematics 24

Richard B. Holmes

Geometric Functional Analysis and its Applications

Springer-Verlag New York Heidelberg Berlin

Richard B. Holmes

Purdue University
Division of Mathematical Sciences
West Lafayette, Indiana 47907

AMS Subject Classifications
Primary: 46.01, 46N05
Secondary: 46A05, 46B10, 52A05, 41A65

Library of Congress Cataloging in Publication Data

Holmes, Richard B.
Geometric functional analysis and its applications.
(Graduate texts in mathematics; v. 24)
Bibliography: p. 237
Includes index.
1. Functional analysis. I. Title. II. Series.
QA320.H63 515′.7 75–6803

Printed in the United States of America.

ISBN 0-387-90136-1 Springer-Verlag New York Heidelberg Berlin
ISBN 3-540-90136-1 Springer-Verlag Berlin Heidelberg New York

To my mother

and

the memory of my father

Preface

This book has evolved from my experience over the past decade in teaching and doing research in functional analysis and certain of its applications. These applications are to optimization theory in general and to best approximation theory in particular. The geometric nature of the subjects has greatly influenced the approach to functional analysis presented herein, especially its basis on the unifying concept of convexity. Most of the major theorems either concern or depend on properties of convex sets; the others generally pertain to conjugate spaces or compactness properties, both of which topics are important for the proper setting and resolution of optimization problems. In consequence, and in contrast to most other treatments of functional analysis, there is no discussion of spectral theory, and only the most basic and general properties of linear operators are established.

Some of the theoretical highlights of the book are the Banach space theorems associated with the names of Dixmier, Krein, James, Smulian, Bishop-Phelps, Brondsted-Rockafellar, and Bessaga-Pelczynski. Prior to these (and others) we establish to two most important principles of geometric functional analysis: the extended Krein-Milman theorem and the Hahn-Banach principle, the latter appearing in ten different but equivalent formulations (some of which are optimality criteria for convex programs). In addition, a good deal of attention is paid to properties and characterizations of conjugate spaces, especially reflexive spaces. On the other hand, the following (incomplete) list provides a sample of the type of applications discussed:

Systems of linear equations and inequalities;
Existence and uniqueness of best approximations;
Simultaneous approximation and interpolation;
Lyapunov convexity theorem;
Bang-bang principle of control theory;
Solutions of convex programs;
Moment problems;
Error estimation in numerical analysis;
Splines;
Michael selection theorem;
Complementarity problems;
Variational inequalities;
Uniqueness of Hahn-Banach extensions.

Also, "geometric" proofs of the Borsuk-Dugundji extension theorem, the Stone-Weierstrass density theorem, the Dieudonne separation theorem, and the fixed point theorems of Schauder and Fan-Kakutani are given as further applications of the theory.

Over 200 problems appear at the ends of the various chapters. Some are intended to be of a rather routine nature, such as supplying the details to a deliberately sketchy or omitted argument in the text. Many others, however, constitute significant further results, converses, or counterexamples. The problems of this type are usually non-trivial and I have taken some pains to include substantial hints. (The design of such hints is an interesting exercise for an author: he hopes to keep the student on course without completely giving everything away in the process.) In any event, readers are strongly urged to at least peruse all the problems. Otherwise, I fear, a good deal of the total value of the book may be lost.

The presentation is intended to be accessible to students whose mathematical background includes basic courses in linear algebra, measure theory, and general topology. The requisite linear algebra is reviewed in §1, while the measure theory is needed mainly for examples. Thus the most essential background is the topological one, and it is freely assumed. Hence, with the exception of a few results concerning dispersed topological spaces (such as the Cantor-Bendixson lemma) needed in §25, no purely topological theorems are proved in this book. Such exclusions are warranted, I feel, because of the availability of many excellent texts on general topology. In particular, the union of the well-known books by J. Dugundji and J. Kelley contains all the necessary topological prerequisites (along with much additional material). Actually the present book can probably be read concurrently with courses in topology and measure theory, since Chapter I, which might be considered a brief second course on linear algebra with convexity, employs no topological concepts beyond standard properties of Euclidean spaces (the single exception to this assertion being the use of Ascoli's theorem in **7C**).

This book owes a great deal to numerous mathematicians who have produced over the last few years substantial simplifications of the proofs of virtually all the major results presented herein. Indeed, most of the proofs we give have now reached a stage of such conciseness and elegance that I consider their collective availability to be an important justification for a new book on functional analysis. But as has already been indicated, my primary intent has been to produce a source of functional analytic information for workers in the broad areas of modern optimization and approximation theory. However, it is also my hope that the book may serve the needs of students who intend to specialize in the very active and exciting ongoing research in Banach space theory.

I am grateful to Professor Paul Halmos for his invitation to contribute the book to this series, and for his interest and encouragement along the way to its completion. Also my thanks go to Professors Philip Smith and Joseph Ward for reading the manuscript and providing numerous corrections. As usual, Nancy Eberle and Judy Snider provided expert clerical assistance in the preparation of the manuscript.

Table of Contents

Chapter I

Convexity in Linear Spaces

Our purpose in this first chapter is to establish the basic terminology and properties of convex sets and functions, and of the associated geometry. All concepts are "primitive", in the sense that no topological notions are involved beyond the natural (Euclidean) topology of the scalar field. The latter will always be either the real number field $\mathbb{R}$, or the complex number field $\mathbb{C}$. The most important result is the "basic separation theorem", which asserts that under certain conditions two disjoint convex sets lie on opposite sides of a hyperplane. Such a result, providing both an analytic and a geometric description of a common underlying phenomenon, is absolutely indispensible for the further development of the subject. It depends implicitly on the axiom of choice which is invoked in the form of Zorn's lemma to prove the key lemma of Stone. Several other equally fundamental results (the "support theorem", the "subdifferentiability theorem", and two extension theorems) are established as equivalent formulations of the basic separation theorem. After indicating a few applications of these ideas we conclude the chapter with an introduction to the important notion of extremal sets (in particular extreme points) of convex sets.

§1. Linear Spaces

In this section we review briefly and without proofs some elementary results from linear algebra, with which the reader is assumed to be familiar. The main purpose is to establish some terminology and notation.

A. Let X be a linear space over the real or complex number field. The zero-vector in X is always denoted by θ. If $\{x_i\}$ is a subset of X, a *linear combination* of $\{x_i\}$ is a vector $x \in X$ expressible as $x = \Sigma\lambda_i x_i$, for certain scalars λ_i, only finitely many of which are non-zero. A subset of X is a (*linear*) *subspace* if it contains every possible linear combination of its members. The *linear hull* (*span*) of a subset S of X, consists of all linear combinations of its members, and thus span(S) is the smallest subspace of X that contains S. The subset S is *linearly independent* if no vector in S lies in the linear hull of the remaining vectors in S. Finally, the subset S is a (*Hamel*) *basis* for X if S is linearly independent and span(S) $= X$.

Lemma. *S is a basis for X if and only if S is a maximal linearly independent subset of S.*

Theorem. *Any non-trivial linear space has a basis; in fact, each non-empty linearly independent subset is contained in a basis.*

B. As the preceding theorem suggests, there is no unique choice of basis possible for a linear space. Nevertheless, all is not chaos: it is a remarkable fact that all bases for a given linear space contain the same number of elements.

Theorem. *Any two bases for a linear space have the same cardinality.*

It is thus consistent to define the (*Hamel*) *dimension* $\dim(X)$ of a linear space X as the cardinal number of an arbitrary basis for X. Let us now recall that if X and Y are linear spaces over the same field then a map $T:X \to Y$ is *linear* provided that

$$T(x + z) = T(x) + T(z), \qquad x, z \in X,$$
$$T(\alpha x) = \alpha T(x), \qquad x \in X, \qquad \alpha \text{ scalar}.$$

It follows that X and Y have the same dimension exactly when they are *isomorphic*, that is, when there exists a bijective linear map between X and Y.

C. We next review some constructions which yield new linear spaces from given ones. First, let $\{X_\alpha\}$ be a family of linear spaces over the same scalar field. Then the Cartesian product $\Pi_\alpha X_\alpha$ becomes a linear space (the *product* of the spaces X_α) if addition and scalar multiplication are defined component-wise. On the other hand, let $M_1, \ldots, M_n$ be subspaces of a linear space X and suppose they are *independent* in the sense that each is disjoint from the span of the others. Then their linear hull (in X) is called the *direct sum* of the subspaces $M_1, \ldots, M_n$ and written $M_1 \oplus \cdots \oplus M_n$ or simply $\bigoplus_{i=1}^{n} M_i$. The point of this definition is that if $M = \bigoplus_{i=1}^{n} M_i$, then each $x \in M$ can be uniquely expressed as $x = \sum_{i=1}^{n} m_i$, where $m_i \in M_i$, $i = 1, \ldots, n$.

Now let M be a subspace of X. For fixed $x \in X$, the subset $x + M \equiv \{x + y : y \in M\}$ is called an *affine subspace* (*flat*) *parallel to* M. Clearly, $x_1 + M = x_2 + M$ if and only if $x_1 - x_2 \in M$, so that the affine subspaces parallel to M are exactly the equivalence classes for the equivalence relation "$\sim_M$" defined by $x_1 \sim_M x_2$ if and only if $x_1 - x_2 \in M$. Now, if we define

$$(x + M) + (y + M) = (x + y) + M,$$
$$\alpha(x + M) = \alpha x + M, \qquad \alpha \text{ scalar}$$

then the collection of all affine subspaces parallel to M becomes a linear space X/M called the *quotient space* of X by M.

Theorem. *Let M be a subspace of the linear space X. Then there exist subspaces N such that $M \oplus N = X$, and any such subspace is isomorphic to the quotient space X/M.*

Any subspace N for which $M \oplus N = X$ is called a *complementary subspace* (*complement*) of M in X. Its dimension is by definition the *codimension* of M in X. The theorem also allows us to state that symbolically

$$\operatorname{codim}_X(M) = \dim(X/M),$$

where the subscript may be dropped provided the ambient linear space X is clearly specified. In fact, this theorem seems to suggest that there is not a great need for the construct X/M, and this is so in the purely algebraic case. However, later when we must deal with Banach spaces X and closed subspaces M, we shall see that generally there will be no closed complementary subspace. In this case the quotient space X/M becomes a Banach space and serves as a valuable substitute for the missing complement.

Now let M be a subspace of X, and choose a complementary subspace $N: M \oplus N = X$. Then we can define a linear map $P: X \to M$ by $P(m + n) = m$, $m \in M$, $n \in N$. P is called the *projection* of X on M (along N). We have similarly that $I - P$ is the projection of X on N (along M), where I is the identity map on X. The existence of such projections allows us the luxury of extending linear maps defined initially on a subspace of X: if $T: M \to Y$ is linear, then $\bar{T} \equiv T \circ P$ is a linear map from X to Y that agrees with T on M. Such a map $\bar{T}$ is an *extension* of T.

D. Let X be a linear space over the scalar field $\mathbb{F}$. The set of all linear maps $\phi: X \to \mathbb{F}$ becomes a new linear space X' with linear space operations defined by

$$(\phi + \psi)(x) \equiv \phi(X) + \psi(x),$$
$$(\alpha\phi)(x) \equiv \alpha\phi(x), \qquad \alpha \in \mathbb{F}, \qquad x \in X.$$

X' is called the *algebraic conjugate* (*dual*) *space* of X and its elements are called *linear functionals* on X. Observe that if $\dim(X) = n$ (a cardinal number) then X' is isomorphic to the product of n copies of the scalar field. As we shall see many times, it is often convenient to write

$$\phi(x) = \langle x, \phi \rangle,$$

for $x \in X$, $\phi \in X'$. The reason for this is that often the vector x and/or the linear functional ϕ may be given in a notation already containing parentheses or other complications.

Since X' is a linear space in a natural fashion, we can construct its algebraic conjugate space $(X')'$, which we write simply as X''. We call X'' the *second algebraic conjugate space* of X. We then have a map $J_X: X \to X''$ defined by

$$\langle \phi, J_X(x) \rangle = \langle x, \phi \rangle, \qquad x \in X, \qquad \phi \in X'.$$

This map is clearly linear; it is called the *canonical embedding* of X into X''. This terminology is justified by the next theorem.

Theorem. *The map J_X just defined is always injective, and is surjective exactly when* $\dim(X)$ *is finite.*

Thus, under the canonical embedding J_X, the linear space X is isomorphic to a subspace of its second algebraic dual space, and this subspace is proper (not all of X'') unless X is of finite dimension. In either case, we see that if it suits our purposes, we can consider that a given linear space consists of linear functionals acting on some other linear space (namely, X').

E. The proper affine subspaces of a linear space X can be partially ordered by inclusion. Any maximal element of this partially ordered set is a *hyperplane* in X.

Lemma. *An affine subspace V in X is a hyperplane if and only if there is a non-zero $\phi \in X'$ and a scalar α such that $V = \{x \in X : \phi(x) = \alpha\} \equiv [\phi; \alpha]$.*

Thus the hyperplanes in X correspond to the *level sets* of non-zero linear functionals on X. We can alternatively say that the hyperplanes in X consist of the elements of all possible quotient spaces $X/\ker(\phi)$, where $\phi \in X'$, $\phi \neq \theta$, and $\ker(\phi) \equiv [\phi; 0]$, the *kernel* (*null-space*) of ϕ. The hyperplanes in X which contain the zero-vector are in particular seen to coincide with the subspaces of codimension one. More generally, the subspaces of codimension n (n a positive integer) are exactly the kernels of linear maps on X of rank n (that is, with n-dimensional image).

F. Suppose that X is a complex linear space. Then in particular X is a real linear space if we admit only multiplication by real scalars. This underlying real vector space X_R is called the real restriction of X. Suppose that $\phi \in X'$. Then the maps

$$x \mapsto \text{re } \phi(x),$$
$$x \mapsto \text{im } \phi(x), \qquad x \in X,$$

are clearly linear functionals on X_R, that is, they belong to X'_R. On the other hand, since $\phi(ix) = i\phi(x)$, $x \in X$, we see that

$$\text{im } \phi(x) = -\text{re } \phi(ix)$$

so that ϕ is completely determined by its real part. Similarly, if we start with $\psi \in X'_R$, and define

$$\phi(x) = \psi(x) - i\psi(ix),$$

we find that $\phi \in X'$. To sum up, the correspondence $\psi \mapsto \phi$ just defined is an isomorphism between $X'_R \equiv (X_R)'$ and $(X')_R$.

This correspondence will be important in our later work with convex sets and functions. The separation, support, subdifferentiability, etc. results all concern various inequalities involving linear functionals; it is thus necessary that these linear functionals assume only real values. Consequently, in the sequel, linear spaces will often be assumed real. The preceding remarks then allow the results under discussion to be applied to complex linear spaces also, by passage to the real restriction, the associated linear functionals being simply the real parts of the complex linear functionals.

G. We give next a primitive version of the "quotient theorem", which allows us intuitively to "divide" one linear map by another. The more substantial result involving continuity questions appears in Chapter III.

Let X, Y, Z be linear spaces and let $S: X \to Y$, $T: X \to Z$ be linear maps. We ask whether there exists a linear map $R: Y \to Z$ such that $T = R \circ S$. An obvious necessary condition for this to occur is that $\ker(S) \subset \ker(T)$; it is more useful to note that this condition is also sufficient.

Theorem. *Let the linear maps S and T be prescribed as above, and assume that* $\ker(S) \subset \ker(T)$. *Then there exists a linear map R, uniquely specified on* range(S), *such that* $T = R \circ S$.

One consequence of this theorem, important for later work on weak topologies, is the following.

Corollary. *Let X be a linear space and let* $\phi_1, \ldots, \phi_n, \psi \in X'$. *Then* $\psi \in \operatorname{span}\{\phi_1, \ldots, \phi_n\}$ *if and only if*

$$\bigcap_{i=1}^{n} \ker(\phi_i) \subset \ker(\psi).$$

H. Let M be a subspace of the linear space X. The *annihilator* M° of M consists of those linear functionals in X' that vanish at each point of M. It is clearly a subspace of X'. Similarly, if N is a subspace of X', its *pre-annihilator* $^\circ N$ consists of all vectors in X at which every functional in N vanishes. Thus:

$$M^\circ = \bigcap_{x \in M} \ker(J_X(x)),$$

$$^\circ N = J_X^{-1}(\operatorname{range}(J_X) \cap N^\circ).$$

Let $T: X \to Y$ be a linear map. The *transpose* T' is the linear map from Y' to X' defined by

$$\langle x, T'(\psi) \rangle = \langle T(x), \psi \rangle, \qquad x \in X, \qquad \psi \in Y'.$$

It may be recalled that when X and Y are (real) finite dimensional Euclidean spaces, and T is represented by a matrix (with respect to the standard unit vector bases in X and Y), then T' is represented by the transposed matrix, whence the above terminology.

Lemma. *Let* $T: X \to Y$ *be a linear map. Then* $\ker(T') = \operatorname{range}(T)^\circ$ *and* $\operatorname{range}(T') = \ker(T)^\circ$.

Thus we see that T is surjective (resp., injective) if and only if T' is injective (resp., surjective). The various constructs in the preceding sub-sections can now all be tied together in the following way. Let us say that the linear spaces X and Y are *canonically isomorphic*, written $X \cong Y$, if an isomorphism between them can be constructed without the use of bases in either space. For example, we clearly have $X \cong J_X(X)$. On the other hand, it may be recalled that none of the usual isomorphisms between a finite dimensional space and its algebraic conjugate space is canonical.

Theorem. *Let M be a subspace of the linear space X. Then*
a) $M^\circ \cong (X/M)'$;
b) $M' \cong X'/M^\circ$.

The proof of a) follows from an application of the lemma to the *quotient map* $Q_M: X \to X/M$, defined by $Q_M(x) \equiv x + M$. Since Q_M is clearly surjective, its transpose $Q'_M: (X/M)' \to X'$ is an isomorphism onto its range, which is $(\ker(Q_M))^\circ = M^\circ$. The proof of b) proceeds similarly by applying the lemma to the identity injection of M into X.

§2. Convex Sets

In this section we establish the most basic properties of convex sets in linear spaces, and prove the crucial lemma of Stone. This lemma is, in effect, the cornerstone of our entire subject, as we shall see shortly. Throughout this section, X is an arbitrary linear space.

A. Let $x, y \in X$ with $x \neq y$. The *line segment* joining x and y is the set $[x, y] = \{\alpha x + (1 - \alpha)y : 0 \leqslant \alpha \leqslant 1\}$. Similarly we put $[x, y) = [x, y]\backslash\{y\}$, and $(x, y) = [x, y)\backslash\{x\}$. If $A \subset X$, then A is *star-shaped* with respect to $p \in A$ if $[p, x] \subset A$, for all $x \in A$, and A is *convex* if it is star-shaped with respect to each of its elements. Clearly a translate of a convex set is convex, hence each affine subspace of X is convex.

Since the intersection of a family of convex sets is again convex, we can define, for any $A \subset X$, the *convex hull* of A, written co(A), to be the intersection of all convex sets in X that contain S. Thus co(A) is the smallest convex set in X that contains A. This set admits an alternative description, namely

$$\operatorname{co}(A) = \{\Sigma\alpha_i x_i : 0 \leqslant \alpha_i \leqslant 1, \Sigma\alpha_i = 1, x_i \in A\},$$

the set of all *convex combinations* of points in A. (We emphasize again that all linear combinations of vectors involve only finitely many non-zero terms.) We have, for instance, that $\operatorname{co}(\{x, y\}) = [x, y]$. More generally, if we define the *join* of two sets A and B in X to be $\cup\{[x, y] : x \in A, y \in B\}$, then

$$\operatorname{co}(A \cup B) = \operatorname{join}(\operatorname{co}(A), \operatorname{co}(B)), \tag{2.1}$$

so that if A and B are convex, then their join is convex and is, in fact, the convex hull of their union.

Let us define addition and scalar multiplication on the family $P(X)$ of non-empty subsets of X by

$$\alpha A + \beta B \equiv \{\alpha a + \beta b : a \in A, b \in B\},$$

where $A, B \subset X$ and α, β are scalars. This definition does not define a linear space structure on $P(X)$; nevertheless, it proves to be quite convenient. For instance, we can state

$$\operatorname{co}(\alpha A + \beta B) = \alpha \operatorname{co}(A) + \beta \operatorname{co}(B). \tag{2.2}$$

A set $A \subset X$ is *balanced* (*equilibrated*) if $\alpha A \subset A$ whenever $|\alpha| \leqslant 1$. The *balanced hull* of A, bal(A), is the intersection of all balanced subsets of X that contain A, and is therefore the smallest balanced set in X that contains A. Alternatively:

$$\operatorname{bal}(A) = \cup\{\alpha A : |\alpha| \leqslant 1\}.$$

Finally, a set which is both convex and balanced is called *absolutely convex*. The smallest such set containing a given set A is the *absolute convex hull* of A, written aco(A). For example, $\text{aco}(\{x\}) = [-x, x]$, if X is a real linear space. In general, we have

$$\begin{aligned}\text{aco}(A) &= \text{co}(\text{bal}(A)) \\ &= \{\Sigma\alpha_i x_i : \Sigma|\alpha_i| \leqslant 1, x_i \in A\},\end{aligned}$$

the set of all *absolute convex combinations* of points in A. In particular, we see that A is absolutely convex if and only if $a, b \in A$ and $|\alpha| + |\beta| \leqslant 1$ implies $\alpha a + \beta b \in A$.

B. We come now to the celebrated result of Stone. Two non-empty convex sets C and D in X are *complementary* if they form a partition of X, that is, $C \cap D = \varnothing$, $C \cup D = X$. An evident example of a pair of complementary convex sets occurs when X is real: choose a non-zero $\phi \in X'$ and put $C = \{x \in X : \phi(x) \geqslant 0\}$, $D = X \backslash C$.

Lemma. *Let A and B be disjoint convex subsets of X. Then there exist complementary convex sets C and D in X such that $A \subset C$, $B \subset D$.*

Proof. Let $\mathscr{C}$ be the class of all convex sets in X disjoint from B and containing A; certainly $A \in \mathscr{C}$. After partially ordering $\mathscr{C}$ by inclusion, we apply Zorn's lemma to obtain a maximal element $C \in \mathscr{C}$. It now suffices to put $D \equiv X \backslash C$ and prove that D is convex. If D were not convex, there would be $x, z \in D$ and $y \in (x, z) \cap C$. Because C is a maximal element of $\mathscr{C}$, there must be points $p, q \in C$ such that both (p, x) and (q, z) intersect B, say at points u, v, resp. (Reason by contradiction; if the last statement were false, then the following assertion (*) would hold: for all pairs $\{p, q\} \subset C$, either $(p, x) \cap B = \varnothing$ or $(q, z) \cap B = \varnothing$. Now if $(q, z) \cap B = \varnothing$, for all $q \in C$, then $C \subset \text{co}(\{z, C\})$ and C is not maximal. Consequently, there is some $\bar{q} \in C$ for which $(\bar{q}, z) \cap B \neq \varnothing$. But then, if there were a point $\bar{p} \in C$ such that $(\bar{p}, x) \cap B \neq \varnothing$, the pair $\{\bar{p}, \bar{q}\}$ would violate (*). Thus, for all $p \in C$, $(p, x) \cap B \neq \varnothing$, $C \subset \text{co}(\{x, C\})$, and C is not maximal.) Now, however, we find that $[u, v] \cap \text{co}(\{p, q, y\}) \neq \varnothing$, which contradicts the disjointness of B and C. □

C. Let A and B be subsets of X. The *core* of A *relative* to B, written $\text{cor}_B(A)$, consists of all points $a \in A$ such that for each $b \in B \backslash \{a\}$ there exists $x \in (a, b)$ for which $[a, x] \subset A$. Intuitively, it is possible to move from each $a \in \text{cor}_B(A)$ towards any point of B while staying in A. The core of A relative to X is called simply the *core* (*algebraic interior*) of A and written cor(A). Sets $A \subset X$ for which $A = \text{cor}(A)$ are called *algebraically open*, while points neither in cor(A) nor in $\text{cor}(X \backslash A)$ are called *bounding points* of A; they constitute the *algebraic boundary* of A. It is easy to see that the core of any (absolutely) convex set is again (absolutely) convex.

A second important instance of the relative core concept occurs when B is the smallest affine subspace that contains A. This subspace, aff(A) (the *affine hull* of A), can be described as $\{\Sigma\alpha_i x_i : \Sigma\alpha_i = 1, x_i \in A\}$ or, equivalently, as $x + \text{span}(A - A)$, for any fixed $x \in A$. Now the set $\text{cor}_{\text{aff}(A)}(A)$ is called the *intrinsic core* of A and written $\operatorname{icr}(A)$. In particular, when A is convex, $a \in \operatorname{icr}(A)$ if and only if for each $x \in A \backslash \{a\}$, there exists $y \in A$ such that $a \in (x, y)$; intuitively, given $a \in \operatorname{icr}(A)$, it is possible to move linearly from any point in A past a and remain in A.

In general, $\operatorname{icr}(A)$ will be empty; but in a variety of special cases we can show $\operatorname{icr}(A)$ and even $\operatorname{cor}(A)$ are not empty. For example, it should be clear that if X is a finite dimensional Euclidean space and $A \subset X$ is convex, then $\operatorname{cor}(A)$ is just the topological interior of A. But this last assertion fails in the infinite dimensional case as we shall see later, after introducing the necessary topological notions. We now work towards a sufficient condition for a convex set to have non-empty intrinsic core.

A finite set $\{x_0, x_1, \ldots, x_n\} \subset X$ is *affinely independent* (*in general position*) if the set $\{x_1 - x_0, \ldots, x_n - x_0\}$ is linearly independent. The convex hull of such a set is called an *n-simplex* with *vertices* $x_0, x_1, \ldots, x_n$. In this case, each point in the n-simplex can be uniquely expressed as a convex combination of the vertices; the coefficients in this convex combination are the *barycentric coordinates* of the point.

Lemma. *Let A be an n-simplex in X. Then* $\operatorname{icr}(A)$ *consists of all points in A each of whose barycentric coordinates is positive. In particular,* $\operatorname{icr}(A) \neq \emptyset$.

Proof. Let the vertices of A be $\{x_0, x_1, \ldots, x_n\}$. Let $a = \Sigma\alpha_i x_i$ and $b = \Sigma\beta_i x_i$ be points of A with all $\alpha_i > 0$. To show $a \in \operatorname{icr}(A)$, it is sufficient to show that $b + \lambda(a - b) \in A$ for some $\lambda > 1$. If we put $\lambda = 1 + \varepsilon$, the condition on ε becomes

$$\alpha_i + \varepsilon(\alpha_i - \beta_i) \geqslant 0, \qquad i = 0, 1, \ldots, n,$$
$$\sum_{i=0}^{n} \alpha_i + \varepsilon(\alpha_i - \beta_i) = 1.$$

Since $\sum_{i=0}^{n} (\alpha_i - \beta_i) = 1 - 1 = 0$, the second condition always holds, and since all $\alpha_i > 0$, the first condition holds for all sufficiently small positive ε. Conversely, let $a = \Sigma\alpha_i x_i$ have a zero coefficient, say $\alpha_k = 0$. Then we claim that $x_k + \lambda(a - x_k) \notin A$, for any $\lambda > 1$. For otherwise, for some $\lambda > 1$ we would have

$$x_k + \lambda(a - x_k) = \sum_{i=0}^{n} \beta_i x_i \in A.$$

It would follow that

$$a = \frac{\beta_k + \lambda - 1}{\lambda} x_k + \sum_{i \neq k} \gamma_i x_i,$$

for certain coefficients γ_i. But in this representation of a, the x_k-coefficient is clearly positive (since $\beta_k \geqslant 0$). This leads us to a contradiction, since the barycentric coordinates of a are uniquely determined, and the x_k-coefficient of a was assumed to vanish. □

The *dimension* of an affine subspace $x + M$ of X is by definition the dimension of the subspace M. The *dimension* of an arbitrary convex set A in X is the dimension of aff(A). A nice way of writing this definition symbolically is

$$\dim(A) \equiv \dim(\operatorname{span}(A - A)).$$

It follows from the preceding lemma that every non-empty finite dimensional convex set A has a non-empty intrinsic core. Indeed, if $\dim(A) = n$ (finite), then A must contain an affinely independent set $\{x_0, x_1, \ldots, x_n\}$ and hence the n-simplex $\operatorname{co}(\{x_0, x_1, \ldots, x_n\})$.

Theorem. *Let A be a convex subset of the finite dimensional linear space X. Then* $\operatorname{cor}(A) \neq \varnothing$ *if and only if* $\operatorname{aff}(A) = X$.

Proof. If $\operatorname{aff}(A) = X$, the last remark shows that $\operatorname{cor}(A) = \operatorname{icr}(A) \neq \varnothing$. Conversely, if $p \in \operatorname{cor}(A)$, and $x \in X$, there is some positive ε for which $[p, p + \varepsilon(x - p)] \subset A$. Then with $\lambda \equiv (\varepsilon - 1)/\varepsilon$, we have

$$x = \lambda p + (1 - \lambda)(p + \varepsilon(x - p)) \in \operatorname{aff}(A). \qquad \square$$

Remark. The conclusion of this theorem fails in any infinite dimensional space. More precisely, in any such space X we can find a convex set A with empty core such that $\operatorname{aff}(A) = X$. To do this we simply let A consist of all vectors in X whose coordinates wrt some given basis for X are non-negative. Clearly $A - A = X$, while $\operatorname{cor}(A) = \varnothing$.

D. Let $A \subset X$. A point $x \in X$ is *linearly accessible* from A if there exists $a \in A$, $a \neq x$, such that $(a, x) \subset A$. We write lina(A) for the set of all such x, and put $\operatorname{lin}(A) = A \cup \operatorname{lina}(A)$. For example, when A is the open unit disc in the Euclidean plane, and B is its boundary the unit circle, we have that $\operatorname{lina}(B) = \varnothing$ while $\operatorname{lin}(A) = \operatorname{lina}(A) = A \cup B$. In general, one suspects (correctly) that when X is a finite dimensional Euclidean space, and $A \subset X$ is convex then lin(A) is the topological closure of A. But we have to go a bit further to be able to prove this.

The "lin" operation can be used to characterize finite dimensional spaces. We give one such result next and another in the exercises. Let us say that a subset of A of X is *ubiquitous* if $\operatorname{lin}(A) = X$.

Theorem. *The linear space X is infinite dimensional if and only if X contains a proper convex ubiquitous subset.*

Proof. Assume first that X is finite dimensional, and let A be a convex ubiquitous set in X. Now clearly A cannot belong to any proper affine subspace of X. Hence $\operatorname{aff}(A) = X$ and thus, by **2C**, cor(A) is non-empty. Without loss of generality, we can suppose that $\theta \in \operatorname{cor}(A)$. Now, given any $x \in X$, there is some $y \in X$ such that $[y, 2x) \subset A$, and there is a positive number t such that $t(2x - y) \in A$. It is easy to see that the half-line $\{\lambda x + (1 - \lambda)t(2x - y) : \lambda \geqslant 0\}$ will intersect the segment $[y, 2x)$; but this of course means that x is a convex combination of two points in A, hence $x \in A$ also.

Conversely, assume that X is infinite dimensional. We can select a well-ordered basis for X (since any set can be well-ordered, according to Zermelo's theorem). Now we define A to be the set of all vectors in X whose last coordinate (wrt this basis) is positive. A is evidently a proper convex subset of X, and we claim that it is ubiquitous. Indeed, given any $x \in X$, we can choose a basis vector y "beyond" any of the finitely many basis vectors used to represent x. But then, if $t > 0$, we have $x + ty \in A$; in particular, $x \in \operatorname{lina}(A)$. □

E. We give one further result involving the notions of core and "lina" which will be needed shortly to establish the basic separation theorem of **4B**. It is convenient to first isolate a special case as a lemma.

Lemma. *Let A be a convex subset of the linear space X, and let $p \in \operatorname{cor}(A)$. For any $x \in A$, we have $[p, x) \subset \operatorname{cor}(A)$, and hence*

$$\operatorname{cor}(A) = \cup\{[p, x) : x \in A\}.$$

Proof. Choose any $y \in [p, x)$, say $y = tx + (1 - t)p$, where $0 < t < 1$. Then given any $z \in X$, there is some $\lambda > 0$ so that $p + \lambda z \in A$. Hence $y + (1 - t)\lambda z = (1 - t)(p + \lambda z) + tx \in A$, proving that $y \in \operatorname{cor}(A)$. Finally, given any $q \in \operatorname{cor}(A)$, $q \neq p$, there exists some $\delta > 0$ such that $x \equiv q + \delta(q - p) \in A$. It follows that $q = (\delta p + x)/(1 + \delta) \in [p, x)$. □

Theorem. *Let A be a convex subset of the linear space X, and $p \in \operatorname{cor}(A)$. Then for any $x \in \operatorname{lina}(A)$ we have $[p, x) \subset \operatorname{cor}(A)$.*

Proof. We can assume that $p = \theta$. Since $x \in \operatorname{lina}(A)$, there is some $z \in A$ such that $[z, x) \subset A$, and since $\theta \in \operatorname{cor}(A)$, there is some $\delta > 0$ such that $-\delta z \in A$. Arguing as in **2D**, given any point tx, $0 < t < 1$, the line $\{\lambda tx + (1 - \lambda)(-\delta z) : \lambda \geqslant 0\}$ will intersect the segment $[z, x)$ if δ is taken sufficiently small. Consequently, the segment $[\theta, x)$ lies in A. But now the preceding lemma allows us to conclude that in fact $[\theta, x)$ lies in $\operatorname{cor}(A)$. □

§3. Convex Functions

In this section we introduce the notion of convex function and its most important special case, the "sublinear" function. With such functions we can associate in a natural fashion certain convex sets. The geometric analysis of such sets developed in subsequent sections makes possible many non-trivial conclusions about the given functions.

A. Intuitively, a real-valued function defined on an interval is convex if its graph never "dents inward" or, more precisely, if the chord joining any two points on the graph always lies on or above the graph. In general, we say that if A is a convex set in a linear space X then a real-valued function f defined on A is *convex* on A if the subset of $X \times \mathbb{R}^1$ defined as $\{(x, t) : x \in A, f(x) \leqslant t\}$ is convex. This set is called the *epigraph* of f, written $\operatorname{epi}(f)$.

An equivalent analytic formulation of this definition is easily obtained: f is convex on A provided that

$$f(tx + (1 - t)y) \leqslant tf(x) + (1 - t)f(y),$$

for all $x, y \in A$, $0 < t < 1$. Obviously the linear functionals in X' are convex on X, and it is not hard to see that the squares of linear functionals are also convex on X. Indeed, if $\phi \in X'$ and $f \equiv \phi(\cdot)^2$, and if $x, y \in X$, then setting $\alpha = \phi(x)$, $\beta = \phi(y)$, we find for $0 < t < 1$

$$\begin{aligned} tf(x) + (1 - t)f(y) - f(tx + (1 - t)y) \\ &= t\alpha^2 + (1 - t)\beta^2 - (t\alpha + (1 - t)\beta)^2 \\ &= t(1 - t)(\alpha - \beta)^2 \geqslant 0. \end{aligned}$$

Further examples of convex functions follow from the use of elementary calculus. Let f be a continuously differentiable function defined on an open interval I. Then f is convex on I if and only if f' is a non-decreasing function on I. Consequently, if f is twice continuously differentiable on I, then f is convex on I if and only if f'' is non-negative on I. To obtain a third characterization of smooth convex functions, and to extend the preceding characterizations to higher dimensions, we consider that f is now a continuously differentiable function defined on an open convex set A in Euclidean n-space. Let $\nabla f(x)$ be its gradient at $x \in A$. The function

$$E(x, y) \equiv f(y) - f(x) - \nabla f(x) \cdot (y - x)$$

measures the discrepancy between the value of f at y and the value of the tangent approximation to f over x at y. (Here the dot denotes the usual dot product on $\mathbb{R}^n$.) Intuitively, if f is convex, this discrepancy will be non-negative at all points $x, y \in A$. To generalize the one-dimensional notion of non-decreasing derivative, let us say that the map $x \mapsto \nabla f(x)$ is *monotone* on A if

$$(\nabla f(y) - \nabla f(x)) \cdot (y - x) \geqslant 0$$

for all $x, y \in A$.

Theorem. *Let f be a continuously differentiable function defined on the open convex set A in $\mathbb{R}^n$. The following assertions are equivalent:*

a) $E(x, y) \geqslant 0$, $x, y \in A$;

b) *the map $x \mapsto \nabla f(x)$ is monotone on A;*

c) *f is convex on A.*

Proof. If $E(x, y) \geqslant 0$ throughout $A \times A$, we have

$$\begin{aligned} (\nabla f(y) - \nabla f(x)) \cdot (y - x) &= \nabla f(y) \cdot (y - x) - \nabla f(x) \cdot (y - x) \\ &\geqslant (f(y) - f(x)) - (f(y) - f(x)) = 0. \end{aligned}$$

Next, if $\nabla f(\cdot)$ defines a monotone map on A, fix $x, y \in A$ and put $g(t) = f(x + t(y - x))$. We want to see that g is convex on $[0, 1]$ or that g' is non-decreasing there. Choose $0 \leqslant \alpha < \beta \leqslant 1$. Then

$$g'(\beta) - g'(\alpha) = (\nabla f(x + \beta(y - x)) - \nabla f(x + \alpha(y - x)))\cdot(y - x)$$
$$= \frac{1}{\beta - \alpha}(\nabla f(v) - \nabla f(u))\cdot(v - u) \geqslant 0,$$

where we have put $u \equiv x + \alpha(y - x)$ and $v \equiv x + \beta(y - x)$, both in A. Thus b) implies c). Finally, let f be convex on A and fix $x, y \in A$. Define

$$h(t) = (1 - t)f(x) + tf(y) - f((1 - t)x + ty),$$

so that h is a non-negative smooth function on $[0, 1]$ and h attains its minimum at $t = 0$. Therefore, $h'(0) \geqslant 0$. Since $E(x, y) = h'(0)$, the proof is complete. □

Many further examples of convex functions will appear in due course.

B. Here we record, for future reference, some elementary properties of the class Conv(A) of all convex functions defined on a convex set A in some linear space. First, Conv(A) is closed under positive linear combinations; that is, if $\{f_1, \ldots, f_n\} \subset \text{Conv}(A)$ and $\alpha_i \geqslant 0$, $i = 1, \ldots, n$, then $\sum_1^n \alpha_i f_i \in \text{Conv}(A)$. Also, if $\{f_\alpha\} \subset \text{Conv}(A)$, and $\sup_\alpha f_\alpha(x) < \infty$ for each $x \in A$, then this supremum defines a function in Conv(A). Indeed,

$$\text{epi}(\sup_\alpha f_\alpha) = \bigcap_\alpha \text{epi}(f_\alpha).$$

The set Conv(A) is of course partially ordered by $f \leqslant g$ if and only if $f(x) \leqslant g(x)$, $x \in A$. Now let $\{f_\alpha\} \subset \text{Conv}(A)$ with each f_α non-negative on A, and suppose that the family $\{f_\alpha\}$ is "directed downwards", that is, given f_α, f_β there exists f_γ such that $f_\gamma(x) \leqslant \min\{f_\alpha(x), f_\beta(x)\}$, $x \in A$. For example, $\{f_\alpha\}$ could be a decreasing sequence. Then $\inf_\alpha f_\alpha \in \text{Conv}(A)$.

We indicate one more procedure for forming new convex functions from old. Given $f_1, \ldots, f_n \in \text{Conv}(A)$ we define their *infimal convolution* $f_1 \square \cdots \square f_n$ by

$$(f_1 \square \cdots \square f_n)(x) \equiv \inf\left\{f_1(x_1) + \cdots + f_n(x_n) : x_i \in A, \sum_1^n x_i = x\right\}.$$

This terminology is motivated by the case where $n = 2$, since we can then write

$$(f \square g)(x) = \inf\{f(y) + g(x - y) : y \in A\},$$

and be reminded of the formula for integral convolution of two functions. In practice, the functions involved in an infimal convolution will be bounded below (usually non-negative), so that the resulting function is well-defined. The convexity of the infimal convolution of convex functions is an easy consequence of the next lemma. This result is of general interest; it allows us to construct convex functions on a linear space X by prescribing their graphs in the product space $X \times \mathbb{R}^1$.

Lemma. *Let X be a linear space and K a convex set in $X \times \mathbb{R}^1$. Then the function*

$$f(x) = \inf\{t:(x, t) \in K\}$$

is convex on the projection of K on X.

The proof follows from the analytic definition of convexity in **3A**. To apply the lemma to the convexity of $f_1 \square \cdots \square f_n$ for $f_i \in \mathrm{Conv}(A)$, A convex in X, let $K = \mathrm{epi}(f_1) + \cdots + \mathrm{epi}(f_n)$. K is certainly convex in $X \times \mathbb{R}^1$ and $(x, t) \in K$ exactly when there are $x_i \in A$ and $t_i \in \mathbb{R}^1$ such that $f_i(x_i) \leqslant t_i$, $t = \sum_1^n t_i$, $x = \sum_1^n x_i$. Thus applying the procedure of the lemma yields $f_1 \square \cdots \square f_n$ which is thereby convex.

Finally, note that if $f \in \mathrm{Conv}(A)$ then the "sub-level sets" defined by $\{x \in A : f(x) \leqslant \lambda\}$ and $\{x \in A : f(x) < \lambda\}$ are convex for any real λ. However, there will be non-convex functions on A that also have this property.

C. We come now to the most important type of non-linear convex functions. Let X be a linear space. A real-valued function f on X is *positively homogeneous* if $f(tx) = tf(x)$ whenever $x \in X$ and $t \geqslant 0$. Such a function is convex if and only if $f(x + y) \leqslant f(x) + f(y)$ for all $x, y \in X$. We call such convex functions *sublinear*. In addition to the linear functions, many other examples of sublinear functions lie close at hand. Thus if $X = \mathbb{R}^n$, we can choose a number $p \geqslant 1$ and let $f(x) = \left(\sum_1^n |\xi_i|^p\right)^{1/p}$ for $x \equiv (\xi_1, \ldots, \xi_n) \in \mathbb{R}^n$. $f(x)$ is called the *p-norm* of x. Or, we can let $X = C(T)$, the linear space of all continuous real-valued functions on a compact Hausdorff space T. If Q is a closed subset of T we let $f(x) = \max\{x(t) : t \in Q\}$; this f is clearly a sublinear function on X.

Sublinear functions on linear spaces arise frequently from the following geometrical considerations. Let A be a subset of a linear space X such that $\theta \in \mathrm{cor}(A)$. Such sets A are called *absorbing*: sufficiently small positive multiples of every vector in X belong to A. We define the *gauge* (*Minkowski function*) of A by

$$\rho_A(x) \equiv \inf\{t > 0 : x \in tA\}.$$

For example, if $\phi \in X'$ and $\alpha > 0$, let A be the "slab" $\{x \in X : |\phi(x)| \leqslant \alpha\}$; then $\rho_A = |\phi(\cdot)|/\alpha$. Or, let $X = \mathbb{R}^n$ and $p \geqslant 1$; then the p-norm introduced above is the gauge defined by the *unit p-ball*

$$\left\{x = (\xi_1, \ldots, \xi_n) \in \mathbb{R}^n : \sum_1^n |\xi_i|^p \leqslant 1\right\}.$$

The primary importance of gauges in a linear space X is that they can be used to define topologies on X. This is certainly apparent in the case of the p-norms on $\mathbb{R}^n$; every one of them defines the usual Euclidean topology on $\mathbb{R}^n$ if the distance between two points in $\mathbb{R}^n$ is taken to be the p-norm of their difference. (The resulting *metric* spaces are of course not the same.)

This example leads us to the general attempt to define a metric d_A by

$$d_A(x, y) = \rho_A(x - y),$$

if ρ_A is the gauge of some given absorbing set A. Thus we are saying that two points are close if their difference lies in a small positive multiple of A. However, it is immediately apparent that more information about A is needed in order to prove that d_A is really a metric. Some of this information is given now and the topic will be continued in the next chapter.

Lemma. *Let A be an absorbing set in a linear space X.*
a) *the gauge ρ_A is positively homogeneous;*
b) *if A is convex then ρ_A is sublinear;*
c) *if A is balanced then $\rho_A(\lambda x) = |\lambda|\rho_A(x)$ for all scalars λ and all $x \in X$.*

Proof. a) Clear. b) Let $x, y \in X$ and choose $t > \rho_A(x) + \rho_A(y)$. Then there exist $\alpha > \rho_A(x)$, $\beta > \rho_A(y)$ such that $t = \alpha + \beta$. Now since A is convex, we have $z \in A$ whenever $\rho_A(z) < 1$; in particular x/α and y/β are in A. Consequently, $(x + y)/t = (x + y)/(\alpha + \beta) = (\alpha(x/\alpha) + \beta(y/\beta))/(\alpha + \beta)$ is also in A so that $\rho_A(x + y) \leqslant t$. c) Assume that $\lambda \neq 0$ and choose $t > \rho_A(x)$. Then $x \in A$ for some s, $\rho_A(x) < s \leqslant t$ and hence $\lambda x \in |\lambda| sA$ because A is balanced. Thus $\rho_A(\lambda x) \leqslant |\lambda| s$ and therefore $\rho_A(\lambda x) \leqslant |\lambda|\rho_A(x)$. The reverse inequality follows after replacing x by λx and λ by $1/\lambda$ in this argument. $\square$

D. The gauge of an absolutely convex absorbing set A is called a *semi-norm*. Thus a semi-norm ρ_A has the properties that it is sublinear and that $\rho_A(\lambda x) = |\lambda|\rho_A(x)$, for all scalars λ and vectors x. Conversely, any real-valued function ρ having these two properties is a semi-norm in the sense that there is an absolutely convex absorbing set A such that $\rho = \rho_A$. Indeed, we can take $A \equiv \{x \in X : \rho(x) \leqslant 1\}$. Since $x \in tA \Leftrightarrow \rho(x) \leqslant t$ it follows that $\rho = \rho_A$.

If $\rho = \rho_A$ is a semi-norm on X then $\ker(\rho) \equiv \{x \in X : \rho(x) = 0\}$ is a subspace of X; in fact, it is the largest subspace contained in A. When $\ker(\rho) = \{\theta\}$, we say that ρ is a *norm* on X. Thus ρ is a norm if and only if $\rho(x) = 0 \Rightarrow x = \theta$. The p-norms on $\mathbb{R}^N$ are clearly examples of norms, which justifies the use of that earlier terminology.

§4. Basic Separation Theorems

In this section we establish two elementary separation theorems for convex subsets of a linear space, making use of Stone's lemma in **2B**. Many of the major subsequent results in this book will depend in some degree on the use of an appropriate separation theorem.

A. We begin with a lemma that draws upon the results of §2. Throughout, X is a real linear space.

Lemma. *Let C and D be non-void complementary convex sets in X, and put $M \equiv \text{lin}(C) \cap \text{lin}(D)$. Then either $M = X$ or else M is a hyperplane in X.*

Proof. Since C and D are convex so are $\operatorname{lin}(C)$ and $\operatorname{lin}(D)$, and hence so is M. We claim that M is in fact an affine subspace of X. To see this, first note that $\operatorname{lin}(C) = X \backslash \operatorname{cor}(D)$ and $\operatorname{lin}(D) = X \backslash \operatorname{cor}(C)$, whence $M = (X \backslash \operatorname{cor}(C)) \cap (X \backslash \operatorname{cor}(D))$. Now let $x, y \in M$ and suppose that z is a point on the line through x and y. If $z \notin M$ then $z \in \operatorname{cor}(C) \cup \operatorname{cor}(D)$; we may suppose that $z \in \operatorname{cor}(C)$ and that $y \in (x, z)$. This entails $x \in \operatorname{lina}(C)$ and hence $y \in \operatorname{cor}(C)$ by **2E**. This contradiction proves that $z \in M$ and consequently M is an affine subspace. There is now no loss of generality in assuming that M is actually a linear subspace. Suppose that $M \neq X$; then there is a vector $p \in X \backslash M$, say $p \in \operatorname{cor}(C)$. Now $-p \in \operatorname{cor}(C) \cup \operatorname{cor}(D)$, but if $-p \in \operatorname{cor}(C)$ then $\theta \in \operatorname{cor}(C)$ also, since $\operatorname{cor}(C)$ is convex. This is not possible so it must be that $-p \in \operatorname{cor}(D)$. Now it follows that for any $x \in C$, $[-p, x] \cap M \neq \varnothing$, and, for any $y \in D$, $[p, y] \cap M \neq \varnothing$. But this means that the linear hull of p and M is all of X, since $X = C \cup D$. By definition then, M is a hyperplane. □

B. Let $H \equiv [\phi; \alpha]$ be a hyperplane in X defined by $\phi \in X'$ and the (real) scalar α. The hyperplane H determines two *half-spaces*, namely, $\{x \in X : \phi(x) \geqslant \alpha\}$ and $\{x \in X : \phi(x) \leqslant \alpha\}$. Two subsets A and B of X are *separated by* H if they lie in opposite half-spaces determined by H. This does not a priori preclude the possibility that $A \cap B \neq \varnothing$ nor that A and/or B actually lie in H. Generally, the important question is not whether A and B can be separated by a particular H, but rather by any hyperplane at all. Simple sketches suggest that an affirmative answer to this question is unlikely unless both sets are convex. Following is the "basic separation theorem".

Theorem. *Let A and B be disjoint non-empty convex sets in X. Assume that either X is finite dimensional or else that* $\operatorname{cor}(A) \cup \operatorname{cor}(B) \neq \varnothing$. *Then A and B can be separated by a hyperplane.*

Proof. By **2B** there are complementary convex sets C and D in X such that $A \subset C$ and $B \subset D$. We let $M = \operatorname{lin}(C) \cap \operatorname{lin}(D)$, as in the preceding lemma. If M is a hyperplane then it does the job of separating A and B. The lemma asserts that M can fail to be a hyperplane only if $X = \operatorname{lin}(C) = \operatorname{lin}(D)$, that is, only if both C and D are ubiquitous (**2D**). But, if X is finite dimensional, neither C nor D can be ubiquitous since they are proper (**2D** again). On the other hand, if A (resp. B) has a non-empty core, then D (resp. C) is not ubiquitous. □

We can in turn use this theorem to establish a stronger and more definitive separation principle, under the hypothesis that one of the sets to be separated has non-empty core.

Corollary. *Let A and B be non-empty convex subsets of X, and assume that* $\operatorname{cor}(A) \neq \varnothing$. *Then A and B can be separated if and only if* $\operatorname{cor}(A) \cap B = \varnothing$.

Proof. If A and B are separated by a hyperplane $[\phi; \alpha]$, then the set $\phi(\operatorname{cor}(A))$ is an open interval of reals, disjoint from the interval $\phi(B)$. Thus cor(A) and B must be disjoint. Conversely, assuming they are disjoint, they can be separated by a hyperplane $[\phi; \alpha]$ (since cor(A) is convex and algebraically open (**2C**)). But clearly if $\phi(x) \leqslant \alpha$, say, for $x \in \text{cor}(A)$, then also $\phi(x) \leqslant \alpha$ for all $x \in A$ (**2E**). Thus $[\phi; \alpha]$ separates A and B. □

C. In some cases, stronger types of separation are both available and useful. Let us say that the sets A and B are *strictly* separated by a hyperplane $H \equiv [\phi; \alpha]$ if they are separated by H and both A and B are disjoint from H, and that they are *strongly* separated by H if they lie on opposite sides of the slab $\{x \in X : |\phi(x) - \alpha| \leqslant \varepsilon\}$ for some $\varepsilon > 0$. Analytically, these two conditions can be expressed as $\phi(x) < \alpha < \phi(y)$, (respectively, as $\phi(x) \leqslant \alpha - \varepsilon < \alpha + \varepsilon \leqslant \phi(y)$), for all $x \in A$, $y \in B$ (after possibly interchanging the labels "A" and "B"). Simple examples in the plane show that convex sets A and B can be strictly separated without being strongly separated.

Some types of separation can be conveniently characterized in terms of the separation of the origin θ from the difference set $A-B$.

Lemma. *The convex sets A and B can be (strongly) separated if and only if θ can be (strongly) separated from $A-B$.*

The proof is straightforward. The assertion is not true for strict separation, however. A slightly less obvious condition for strong separation will be given next, and called the "basic strong separation theorem".

Theorem. *Two disjoint convex sets A and B in X can be strongly separated if and only if there is a convex absorbing set V in X such that $(A + V) \cap B = \varnothing$.*

Proof. If such a V exists then $A + V$ has non-empty core and so can be separated from B. Thus there exists $\phi \in X'$ such that $\phi(a + v - b) \geqslant 0$ for all $a \in A$, $b \in B$, $v \in V$. Now the interval $\phi(V)$ contains a neighborhood of 0, so there is $v_0 \in V$ with $\phi(v_0) < 0$. Hence $\phi(a) \geqslant \phi(b) - \phi(v_0)$ for all $a \in A$, $b \in V$, whence $\inf\{\phi(a): a \in A\} > \sup\{\phi(b): b \in B\}$. Thus A and B are strongly separated. Conversely, assume that A and B can be strongly separated. Then there are $\phi \in X'$ and reals α, ε, with $\varepsilon > 0$, such that $\inf\{\phi(a): a \in A\} \geqslant \alpha + \varepsilon > \alpha - \varepsilon \geqslant \sup\{\phi(b): b \in B\}$. If we put $V \equiv \{x \in X : |\phi(x)| < \varepsilon\}$ we find V is convex and absorbing and that $(A + V) \cap B = \varnothing$. □

A particular consequence of this theorem is that two disjoint closed convex subsets of $\mathbb{R}^n$ can be strongly separated, provided that one of them is bounded (hence compact). The boundedness hypothesis cannot be omitted as is shown by simple examples in $\mathbb{R}^2$.

§5. Cones and Orderings

In this section, we study a special type of convex set, the "wedge". Such sets are intimately connected with the notions of ordering in linear spaces, and positivity of linear functionals. This added structure in linear space theory is important because of its occurrence in practice, for example in function spaces and operator algebras. Wedges associated with a given convex set (support and normal wedges, recession wedges) are introduced in later sections, and play important roles in certain applications.

A. A *wedge* P in a real linear space X is a convex set closed under multiplication by non-negative scalars. Any such set defines a reflexive and transitive partial ordering on X by

$$x \leqslant y \Leftrightarrow y - x \in P.$$

This ordering has the further properties that $x \leqslant y$ entails $x + z \leqslant y + z$ for any $z \in X$, and $\lambda x \leqslant \lambda y$ whenever $\lambda \geqslant 0$. For short, we call such a partial ordering a *vector ordering* and X so equipped an *ordered linear space*. Conversely, if we start with an ordered linear space $(X, \leqslant)$ and put $P \equiv \{x \in X : x \geqslant \theta\}$, then P is a wedge in X (the *positive wedge*) which induces the given vector ordering.

A wedge P is a *cone* if $P \cap (-P) = \{\theta\}$; in this case θ is called the *vertex* of P. Since $P \cap (-P)$ is the largest subspace contained in P, this condition is equivalent to the assertion that P contains no non-trivial subspace. It is further easy to see that a wedge is a cone exactly when the induced vector ordering is anti-symmetric, in the sense that $x \leqslant y$, $y \leqslant x \Leftrightarrow x = y$.

The span of a wedge P is simply $P - P$. When $P - P = X$, the wedge is said to be *reproducing*, and X is *positively generated* by P. It is not hard to show that this situation obtains in particular whenever $\text{cor}(P) \neq \varnothing$. In terms of the associated vector ordering on X, we can state that X is positively generated by P if and only if the ordering *directs* X, in the sense that any two elements of X have an upper bound. Precisely, this means that given $x, y \in X$, there exists $z \in X$ such that $x \leqslant z$ and $y \leqslant z$.

The simplest examples of ordered linear spaces are function spaces with the natural pointwise vector ordering. If X is a linear space of functions defined on a set T, and the linear space operations are the usual pointwise ones, then it is natural to let $P = \{x \in X : x(t) \geqslant 0,\ t \in T\}$. The induced vector ordering is then defined by

$$x \leqslant y \Leftrightarrow x(t) \leqslant y(t), \qquad t \in T.$$

Let us now further specialize to the case where $X = C[0, 1]$, the space of all (real-valued) continuous functions on the interval $[0, 1]$. Clearly the pointwise vector ordering on X directs X and so the cone of non-negative functions is reproducing. On the other hand, let us consider in X the cone Q of all non-negative and non-decreasing functions in X. Now we have that $Q - Q$ is the subspace of all functions in X that are of bounded variation on $[0, 1]$. Consequently, Q is not reproducing in X.

Another interesting cone is the set $\text{Conv}(X)$ (**3B**) in the linear space of all real-valued functions on X.

B. Let X be an ordered linear space with positive wedge P. A linear functional $f \in X'$ is *positive* if $f(x) \geqslant 0$ whenever $x \in P$. Clearly a positive linear functional f is *monotone* in the sense that $x \leqslant y \Rightarrow f(x) \leqslant f(y)$. The set of all positive linear functionals forms a wedge P^+ in X' called the *dual wedge*; the induced vector ordering on X' is the *dual ordering*, and the subspace $P^+ - P^+$ is the *order dual* of X. The dual wedge is actually a cone exactly when P is reproducing.

It is not a priori clear whether or not there are any non-zero positive linear functionals on a given ordered linear space, and indeed there may be none. We now use the separation theory of §4 to give a useful sufficient condition for $P^+ \neq \{\theta\}$.

Theorem. *If the wedge P is a proper subset of X and has non-empty core, then P^+ contains non-zero elements.*

Proof. We choose an $x \in X \backslash P$ and apply **4B** to separate x and P by a hyperplane $[\phi; \alpha]$, say $\phi(x) \leqslant \alpha \leqslant \phi(y)$, $y \in P$. Now any linear functional that is bounded below on a wedge must be non-negative there. Thus $\phi \in P^+$ and $\phi \neq \theta$. □

C. We consider briefly some conditions sufficient to guarantee that a wedge P in a linear space X is actually a cone. A linear functional $\phi \in P^+$ is *strictly positive* if $x \in P$ $(x \neq \theta) \Rightarrow \phi(x) > 0$. A *base* for P is a non-empty convex subset B of P with $\theta \notin P$ such that every $x \in P$ $(x \neq \theta)$ has a unique representation of the form λb, where $b \in B$ and $\lambda > 0$. If $\phi \in P^+$ is strictly positive and we set $B \equiv [\phi; 1] \cap P$ then B is a base for P. The converse assertion is equally valid: given a base B for P, there is by Zorn's lemma a maximal element H in the class of affine subspaces which contains B but not θ. H is seen to be a hyperplane defined by a strictly positive linear functional.

Theorem. *Consider the following properties that a wedge P in X may possess*:

a) *P is a cone*;
b) *P has a base*;
c) $\text{cor}(P^+) \neq \varnothing$.

Then c) $\Rightarrow$ b) $\Rightarrow$ a); *if X is some Euclidean space, and P is closed in X, then all three properties are equivalent.*

Proof. It is clear that the existence of a base for P implies that P is a cone, so that b) $\Rightarrow$ a). Now assume that $\phi \in \text{cor}(P^+)$; it will suffice to show that ϕ is strictly positive. If not, there exists $x \in P(x \neq \theta)$ such that $\phi(x) = 0$. But since $x \neq \theta$, there must be some $\psi \in X'$ for which $\psi(x) < 0$. As $\phi \in \text{cor}(P^+)$, there is $\lambda > 0$ such that $\phi + \lambda\psi \in P^+$; however $\phi(x) + \lambda\psi(x) = \lambda\psi(x) < 0$, a contradiction. Thus c) $\Rightarrow$ b). Finally, assume that $X = \mathbb{R}^n$ for some n, and that P is closed in X. We show a) $\Rightarrow$ c). Now according to **2C**, $\text{cor}(P^+) \neq \varnothing \Leftrightarrow P^+$ is reproducing. If P^+ is not reproducing then its linear hull $P^+ - P^+$ is a proper subspace of $\mathbb{R}^n$ (here we are tacitly utilizing the usual self-duality of $\mathbb{R}^n$ with itself: $(\mathbb{R}^n)' = \mathbb{R}^n$). There is thus a non-zero linear functional $\Phi \in (\mathbb{R}^n)'' = \mathbb{R}^n$ such that Φ vanishes on $P^+ - P^+$ (**1C**).

The proof is concluded by showing that $\pm\Phi \in P$, so that P is not a cone. If, for example, $\Phi \in P$, there is a Euclidean ball V centered at θ in $\mathbb{R}^n$ such that $(\Phi + V) \cap P = \varnothing$; this follows because P is assumed closed. But now by **4C** we can strongly separate Φ and P. As in **5B**, the separating hyperplane must be defined by an element $\phi \in P^+$ with $\langle \phi, \Phi \rangle < 0$; this however is a contradiction since Φ vanishes on P^+. □

Without further hypotheses, the other conceivable implications between a), b), and c) are not valid.

§6. Alternate Formulations of the Separation Principle

In this section we establish four new basic principles involving convex sets and linear functionals, which, along with the basic separation theorems of §4, will be used repeatedly in the sequel. Of special interest here is that these new principles are in fact only different manifestations of our earlier separation principle **4B**: they are all equivalent to it and hence to each other. (In **6B** it is further noted that the existence theorem of **5B** is also equivalent to the basic separation theorem.)

A. We begin with the extension principles. In **1C** it was noted that, rather trivially, a linear map defined on a subspace of a linear space admits a (linear) extension to the whole space. For the time being, all linear maps to be extended will be linear functionals, defined on a proper subspace M of a linear space X. What will make our extension theorems interesting (and useful) is the presence of various "side-conditions" which must be preserved by the extension. If f and g are real-valued functions with common domain D, we shall write $f \leqslant g$ in case $f(x) \leqslant g(x)$ for every $x \in D$. Our first result is the "Hahn-Banach theorem".

Theorem. *Let $g \in \text{Conv}(X)$ where X is a real linear space, and suppose that $\phi \in M'$ satisfies $\phi \leqslant g|M$. Then there exists an extension $\bar{\phi} \in X'$ of ϕ such that $\bar{\phi} \leqslant g$.*

Proof. Let A be the epigraph (**3A**) of g and B the graph of ϕ in the space $Y \equiv X \times \mathbb{R}^1$. By hypothesis, $B \equiv \{(x, \phi(x)) : x \in M\}$ is a subspace of Y disjoint from the convex set A. Now A is algebraically open. To see this, choose $(x_0, t_0) \in A$ and $(x, t) \in Y$. Then for $0 \leqslant \lambda \leqslant 1$,

$$\begin{aligned} g(x_0 + \lambda x) - (t_0 + \lambda t) \\ &= g(\lambda(x_0 + x) + (1 - \lambda)x_0) - t_0 - \lambda t \\ &\leqslant \lambda g(x_0 + x) + (1 - \lambda)g(x_0) - t_0 - \lambda t \\ &= \lambda(g(x_0 + x) - t_0 - t) - (1 - \lambda)(t_0 - g(x_0)). \end{aligned}$$

Since the second term here is positive, the entire expression will be negative for sufficiently small λ, proving that $(x_0, t_0) \in \text{cor}(A)$. Thus we can separate A and B by a hyperplane $[\Phi; \alpha] \subset Y$. Since the linear functional Φ is bounded on the subspace B, $\alpha = 0$; we assume that Φ is non-negative (necessarily positive, in fact) on A. Since $(\theta, t) \in A$ for sufficiently large t, $c \equiv \Phi(\theta, 1) > 0$. Now to define the desired extension $\bar{\phi} \in X'$ we note that $\Phi(x, 0) + \Phi(\theta, t) = \Phi(x, t)$ whenever $(x, t) \in A$. That is, setting $\bar{\phi} = (-1/c)\Phi(\cdot, 0)$, we see that $g(x) < t$ implies $\bar{\phi}(x) < t$ also, so that $\bar{\phi} \leqslant g$ on X. And since $\phi(m, \phi(m)) = 0$ for $m \in M$, we see that $\bar{\phi}(m) = \phi(m)$, $m \in M$, so that $\bar{\phi}$ is the desired extension of ϕ. □

We indicate one direct and important consequence of the Hahn-Banach theorem; its derivation is outlined in exercise 1.21.

Corollary. *Let ρ be a semi-norm* (**3D**) *on the linear space X, and M a subspace of X. If $\phi \in M'$ satisfies $|\phi(\cdot)| \leqslant \rho|M$, then there is an extension $\bar{\phi} \in X'$ of ϕ such that $|\bar{\phi}(\cdot)| \leqslant \rho$.*

B. Our second extension principle concerns positive linear functionals. Let X be an ordered linear space with positive wedge P (**5A**), and let M be a subspace of X. M will be considered as an ordered linear space under the vector ordering induced by the wedge $P \cap M$. The next result, the "Krein-Rutman theorem", provides a sufficient condition for a positive linear functional (**5B**) on M to admit a positive extension to all of X.

Theorem. *With M, P, X as just defined, assume that $P \cap M$ contains a core point of P. Then any positive linear functional ϕ on M admits a positive extension to all of X.*

Proof. It will suffice to construct a positive extension on the span of P and M; we can then extend to all of X in the trivial manner of **1C**. For x in this span we define

$$g(x) = \inf\{\phi(y) : y \geqslant x, y \in M\}.$$

Now g is convex (actually sublinear; the proof is quite analogous to that of the lemma in **3C**), and we have $\phi \leqslant g|M$ on account of the monotonicity of ϕ on M. Thus we can apply the Hahn-Banach theorem (**6A**) and obtain an extension $\bar{\phi}$ (to the span of P and M) of ϕ so that $\bar{\phi} \leqslant g$. To see that this $\bar{\phi}$ is positive, choose $y_0 \in P \cap M$ and $x \in P$; we shall show that $\bar{\phi}(-x) \leqslant 0$. Now for all $t \geqslant 0$, $y_0 + tx \in P$. Thus $y_0/t \in M$ and $y_0/t \geqslant -x$, so that $\bar{\phi}(-x) \leqslant g(-x) \leqslant \phi(y_0/t) = \phi(y_0)/t$; to conclude, let $t \to +\infty$. □

In order to show that both the preceding extension theorems are equivalent to the basic separation theorem, it clearly suffices to prove that the latter is a consequence of the Krein-Rutman theorem. In turn, recalling **4C**, it suffices to show that if A is a convex set in a linear space X with non-empty core, and $\theta \notin A$, then we can separate θ from A by a hyperplane; or, in other words, we can find a non-zero linear functional in X' that assumes only non-negative values on A. Let us define $P = \{tA : t \geqslant 0\}$. Then P is a wedge (actually a cone) in X and $\operatorname{cor}(P) \neq \varnothing$. It now follows from **5B** that P^+ contains a non-zero element, which is what we wanted. Although the proof of **5B** utilized the basic separation theorem, it is clear that **5B** is also a simple consequence of the Krein-Rutman theorem.

C. Let $H \equiv [\phi; \alpha]$ be a hyperplane and A a convex set in the real linear space X. We say that H *supports* A if A lies in one of the two half-spaces (**4B**) determined by H and $A \cap H \neq \varnothing$. A point in A that lies in some such supporting hyperplane is called a *support point* of A; a support point of A is *proper* if it lies in a supporting hyperplane which does not completely contain A. There is a more general notion of supporting affine subspace (not necessarily a hyperplane) which is introduced in exercise 1.37.

The next result, the "support theorem", completely identifies the proper support points of convex sets with non-empty intrinsic core (**2C**).

Theorem. *Let A be a convex subset of a real linear space X such that* $\mathrm{icr}(A) \neq \varnothing$. *If* $x \notin \mathrm{icr}(A)$, *there exists* $\phi \in X'$ *such that* $\phi(x) > \phi(y)$, *for all* $y \in \mathrm{icr}(A)$.

Proof. We may assume that the origin θ belongs to $\mathrm{icr}(A)$. Let $M = \mathrm{span}(A)$. If $x \notin M$, we can certainly construct $\phi \in M^\circ$ with $\phi(x) > 0$. If $x \in M$, the basic separation theorem allows us to construct $\phi_0 \in M'$ such that $\phi_0(x) \geqslant \phi_0(y)$ for all $y \in \mathrm{icr}(A)$. It is clear from the linearity of ϕ_0 and the definition of core that equality can never hold here. Now any extension ϕ of ϕ_0 to all of X will serve our purpose. □

Corollary. *The proper support points of a convex set A with* $\mathrm{icr}(A) \neq \varnothing$ *are exactly those in* $A \backslash \mathrm{icr}(A)$. *In particular, if* $\mathrm{cor}(A) \neq \varnothing$, *the proper support points of A are the bounding points* (**2C**) *of A that belong to A.*

Since all finite dimensional convex sets have non-empty intrinsic core (**2C**), their support points are fully located by this corollary. Naturally, the situation is a little more complicated in the general infinite-dimensional case. Let us consider, for example, the case of the real linear space $\ell^p(d)$, where $1 \leqslant p < \infty$ and d is a cardinal number, finite or infinite. This is the usual space of real-valued functions on a set S of cardinality d which are p-th power integrable wrt the counting measure on S (the counting measure is by definition defined on all subsets of S; its value at a particular subset is the cardinality of this subset if finite, and otherwise is $+\infty$). Less formally, if $x: S \to \mathbb{R}$ and we identify x with the "d-tuple" of its values, $x = (x(s): s \in S)$, then $x \in \ell^p(d)$ if and only if $\sum_{s \in S} |x(s)|^p < \infty$. Now $\ell^p(d)$ is clearly ordered by the natural pointwise vector ordering (**5A**), and the positive wedge $P \equiv \{x \in \ell^p(d): x(s) \geqslant 0,\ s \in S\}$ is a reproducing cone in $\ell^p(d)$. However, this wedge has no core when $d \geqslant \aleph_0$ and hence no intrinsic core, so that the support theorem does not apply.

Since no hyperplane can contain P, each support point of P (if there are any) must be proper. In the case where $d > \aleph_0$, we claim that every point in P is a support point. This is so because each such point must vanish at some point in S. The characteristic function of this point in S then gives rise to a linear functional on $\ell^p(d)$ that defines a supporting hyperplane to P through the given point in P. For contrast, consider now the case where $d = \aleph_0$ and $S \equiv \{1, 2, \ldots\}$. If $x = (\xi_i) \in P$ and some $\xi_i = 0$ then the preceding argument shows that x is a support point of P. But now it is possible that no $\xi_i = 0$ and in this case x is not a support point of P.

Thus we see that in the absence of core, a particular bounding point of a convex set may or may not be a support point. More surprising, perhaps, is the possibility that a given convex set may have no support points at all. An example illustrating such a "supportless" convex set is given in exercise 1.20.

It is clear from **4C** that the present support theorem implies the basic separation theorem.

D. Let f be a convex function defined on a convex set A in some real linear space X. A linear functional $\phi \in X'$ is a *subgradient* of f at a point $x_0 \in A$ if

$$\phi(x - x_0) \leqslant f(x) - f(x_0), \qquad x \in A.$$

This definition is motivated by the result in **3A** for the case where $X = \mathbb{R}^n$, and f is differentiable at x_0. In this case, the gradient vector $\nabla f(x_0)$ was shown to satisfy the above condition (when viewed as a linear functional on $\mathbb{R}^n$ in the usual way). Thus a subgradient is a particular kind of substitute for the gradient of a convex function, in case the latter does not exist (or is not defined).

Consider, for example, the case where $A = X = \mathbb{R}^1$ and f, although necessarily continuous on $\mathbb{R}^1$ (since it is convex), is not differentiable at some x_0. In this case, as is well known, f has a left hand derivative $f'_-(x_0)$ and a right hand derivative $f'_+(x_0)$ at the point x_0, and $f'_-(x_0) \leqslant f'_+(x_0)$. Now we claim that any number t, $f'_-(x_0) \leqslant f'_+(x_0)$, defines a subgradient of f at x_0. This is so because the difference quotients whose limits define these one-sided derivatives converge monotonically:

$$\frac{f(x) - f(x_0)}{x - x_0} \downarrow f'_+(x_0), \qquad x \downarrow x_0$$

and

$$\frac{f(x) - f(x_0)}{x - x_0} \uparrow f'_-(x_0), \qquad x \uparrow x_0.$$

Thus

$$f'_+(x_0)(x - x_0) \leqslant f(x) - f(x_0), \qquad x_0 < x$$

and

$$f'_-(x_0)(x - x_0) \leqslant f(x) - f(x_0), \qquad x < x_0.$$

Other examples of subgradients are given in the exercises and in later sections.

Let us consider next the geometrical interpretation of subgradients. First we recall that when X is a real linear space, $(X \times \mathbb{R}^1)'$ is isomorphic to $X' \times \mathbb{R}^1$. Indeed, such an isomorphism occurs by associating $(\phi, s) \in X' \times \mathbb{R}^1$ with $\psi \in (X \times \mathbb{R}^1)'$, where

$$\psi(x, t) \equiv \phi(x) + st, \qquad x \in X, \qquad t \in \mathbb{R}^1.$$

Now the basic geometric interpretation to follow is that subgradients correspond to certain supporting hyperplanes of the set epi(f) (**3A**) in $X \times \mathbb{R}^1$.

Lemma. *Let A be a convex subset of X and let $f \in$ Conv(A).*

a) *$\phi \in X'$ is a subgradient of f at $x_0 \in A$ if and only if the graph of the affine function $h(x) \equiv f(x_0) + \phi(x - x_0)$ is a supporting hyperplane to* epi(f) *at the point $(x_0, f(x_0))$.*

b) *Conversely, assume that $\psi \in (X \times \mathbb{R}^1)'$ and that $H = [\psi; \alpha]$ is a supporting hyperplane to* epi(f) *at $(x_0, f(x_0))$; say $\alpha = \inf \psi(\text{epi}(f))$. Let ψ correspond to $(\phi, s) \in X' \times \mathbb{R}^1$ as above. Then, if $s \neq 0$ (intuitively, if H is "non-vertical"), we have $s > 0$ and $-\phi/s$ is a subgradient of f at x_0.*

Proof. a) By definition, ϕ is a subgradient of f at x_0 if and only if $h|A \leqslant f$. If we define $\psi \in (X \times \mathbb{R}^1)'$ by $\psi(x, t) \equiv -\phi(x) + t$, and let $\alpha = f(x_0) - \phi(x_0)$, then the inequality $h|A \leqslant f$ is equivalent to $\inf \psi(\text{epi}(f)) = \psi(x_0, f(x_0)) = \alpha$. Thus the hyperplane $[\psi; \alpha]$ supports epi(f) at $(x_0, f(x_0))$; it is clear that graph(h) $= [\phi; \alpha]$.

b) We have $\phi(x_0) + sf(x_0) \leqslant \phi(x) + st$, for all $x \in A$ and all $t \geqslant f(x)$. From this the two assertions of b) are evident. □

If there exists a subgradient ϕ of f at x_0 we say that f is *subdifferentiable* at x_0. The set of all such ϕ is the *subdifferential* of f at x_0, written $\partial f(x_0)$; it is clearly a convex subset of X'. Since the subdifferentiability of f at a given point depends, as we have just seen, on a support property of epi(f), we might suspect from the results of the previous section that in general $\partial f(x_0)$ will be empty. This is certainly the case as simple examples show. An existence theorem is thus required; the following "subdifferentiability theorem" fills this order.

Theorem. *Let A be a convex subset of the real linear space X and $f \in$* Conv(A). *Then f is subdifferentiable at all points in* icr(A).

Proof. Let $x_0 \in \text{icr}(A)$, $M = \text{span}(A - A)$ (M is the subspace parallel to aff(A)), and $B = A - x_0$. Define $g \in \text{Conv}(B)$ by $g(x) = f(x + x_0)$. Then any subgradient in $\partial g(\theta)$ will, upon extension from M' to X', also belong to $\partial f(x_0)$. In other words, there is no loss of generality in assuming that $\theta = x_0 \in \text{cor}(A)$; it is further harmless to take $f(\theta) = 0$. But now, in $X \times \mathbb{R}^1$, any point of the form (θ, t_0), $t_0 > 0$, belongs to cor(epi(f)). To see this, pick $(x, t) \in X \times \mathbb{R}^1$; we must show that $(\theta, t_0) + \lambda(x, t) \in \text{epi}(f)$ for sufficiently small $\lambda > 0$, or that $f(\lambda x) \leqslant t_0 + \lambda t$ for small λ. But the convex function $g(\lambda) \equiv f(\lambda x)$ defined on $(0, \infty)$ satisfies

$$g(\lambda)/\lambda \downarrow g'_+(0), \qquad \lambda \downarrow 0,$$

so that certainly

$$f(\lambda x)/\lambda \equiv g(\lambda)/\lambda \leqslant t_0/\lambda + t$$

for small λ. Now since cor(epi(f)) $\neq \emptyset$, by **6C** the bounding point $(\theta, 0)$ is a support point of epi(f). The corresponding hyperplane cannot be "vertical", since $\theta \in \text{cor}(A)$. Thus, by part b) of the preceding lemma, there is a subgradient of f at θ. □

To complete our circle of equivalent formulations of the basic separation principle, let us show that the subdifferentiability theorem entails this principle. From **4B** and **4C** we see that it is sufficient to prove that an algebraically open convex set A in X can be separated from any point $x_0 \notin A$. As usual, after a translation, we may assume that $\theta \in A$. Thus A is absorbing, its gauge ρ_A belongs to Conv(X) (**3C**), and $\rho_A(x_0) \geqslant 1$. By the subdifferentiability theorem, there exists $\phi \in \partial\rho_A(x_0):\phi(x - x_0) \leqslant \rho_A(x) - \rho_A(x_0)$, $x \in X$. Letting $x = \theta$ and $x = 2x_0$, and recalling that ρ_A is positively homogeneous, we see that

$$\phi(x_0) = \rho_A(x_0) \equiv \alpha$$
$$\phi(x) \leqslant \rho_A(x), \qquad x \in X.$$

Consequently, the hyperplane $[\phi; \alpha]$ separates x_0 and A (since $x \in A$ implies $\rho_A(x) \leqslant 1$ so that $\phi(x) \leqslant \rho_A(x) \leqslant 1 \leqslant \alpha$).

E. In summary, we have now established the mutual equivalence of six propositions, each of which asserts the existence of a linear functional with certain properties. These propositions are

1) the basic separation theorem (**4B**);
2) the existence of positive functionals (**5B**);
3) the Hahn-Banach theorem (**6A**);
4) the Krein-Rutman theorem (**6B**);
5) the support theorem (**6C**);
6) the subdifferentiability theorem (**6D**).

An important meta-principle is suggested by these results: if one wishes to establish the existence of a solution to a given problem, and one has some control over the choice of the linear space in which the solution is to be sought, then it will generally behoove one to choose the ambient linear space to be a conjugate space if possible. This is of course automatic in the finite dimensional case (**1D**), but does represent a restriction in the general case. We shall see many applications of this idea in subsequent sections.

§7. Some Applications

In this section we give a few elementary applications of the preceding existence theorems. Most of these results will play a role in later work. More substantial applications require the topological considerations to be developed in the next chapter. Throughout this section, X denotes a real linear space.

A. We first consider a criterion ("Helly's condition") for the consistency of a finite system of linear equations, subject to a convex constraint. The most important special cases of this result are obtained by letting the set A below be the *unit ball* of a semi-norm ρ, that is, the set $\{x \in X : \rho(x) \leqslant 1\}$ (when ρ is identically zero, this definition yields simply $A = X$).

Theorem. *Let A be an absolutely convex subset of X. Let $\{\phi_1, \ldots, \phi_n\} \subset X'$ and $\{c_1, \ldots, c_n\} \subset \mathbb{R}$. Then, a necessary and sufficient condition that for* *every $\delta > 0$ there exists $x_\delta \in (1 + \delta)A$ satisfying*

$$\phi_1(x_\delta) = c_1,$$
$$\vdots$$
$$\phi_n(x_\delta) = c_n,$$

is that for every set $\{\alpha_1, \ldots, \alpha_n\} \subset \mathbb{R}$,

$$\left|\sum_{i=1}^n \alpha_i c_i\right| \leqslant \sup\left\{\left|\sum_{i=1}^n \alpha_i \phi_i(x)\right| : x \in A\right\}.$$

Proof. The stated condition is clearly necessary for the consistency of the given system. Let us prove its sufficiency. Suppose that for some $\delta > 0$ whenever $x \in (1 + \delta)A$, we have $\phi_i(x) \neq c_i$ for some i. If we define a linear map $T: X \to \mathbb{R}^n$ by

$$T(x) = (\phi_1(x), \ldots, \phi_n(x)),$$

our assumption becomes

$$c \equiv (c_1, \ldots, c_n) \notin T((1 + \delta)A).$$

By **4B** these two sets can be separated: there is a non-zero linear functional λ on $\mathbb{R}^n$ such that

$$\begin{aligned}\lambda(c) \geqslant \sup\{\lambda(v): v \in T((1 + \delta)A)\} &= \sup\{|\lambda(v)|: v \in T((1 + \delta)A)\}\\ &= \sup\{|\lambda(T(x))|: x \in (1 + \delta)A\}.\end{aligned}$$

(The absolute values are permissible because A is a balanced set.) Now if λ is given by $\lambda(v) = \sum_1^n \alpha_i v_i$, for $v = (v_1, \ldots, v_n) \in \mathbb{R}^n$, we obtain

$$\begin{aligned}\sum_{i=1}^n \alpha_i c_i \geqslant \sup\left\{\left|\sum_{i=1}^n \alpha_i \phi_i(x)\right| : x \in (1 + \delta)A\right\}\\ = (1 + \delta)\sup\left\{\left|\sum_{i=1}^n \alpha_i \phi_i(x)\right| : x \in A\right\},\end{aligned}$$

in contradiction to Helly's condition. □

B. Next, we consider a criterion ("Fan's condition") for the consistency of a finite system of linear inequalities. Such systems are of considerable importance in the theory of linear programming and related optimization models.

Theorem. *Let $\{\phi_1, \ldots, \phi_n\} \subset X'$ and $\{c_1, \ldots, c_n\} \subset \mathbb{R}$. A necessary and sufficient condition that there exists $x \in X$ satisfying*

$$\phi_1(x) \geqslant c_1,$$
$$\vdots$$
$$\phi_n(x) \geqslant c_n,$$

is that for every set $\{\alpha_1, \ldots, \alpha_n\}$ *of non-negative numbers for which*

$$\sum_{i=1}^{n} \alpha_i \phi_i = \theta,$$

it follows that

$$\sum_{i=1}^{n} \alpha_i c_i \leqslant 0.$$

Proof. Again the necessity of the condition is clear, and we proceed to establish its sufficiency. Since a more general result will be established later, we merely outline the main steps and invite the reader to fill in the details. Let $T: X \to \mathbb{R}^n$ and c be as in the previous section, and let P be the usual positive wedge (**5A**) in $\mathbb{R}^n$. If the given system of inequalities is inconsistent then, in $\mathbb{R}^n$, the affine subspace $T(X) - c$ is disjoint from P. Let $\{b_1, \ldots, b_k\}$ be a basis for the annihilator (**1H**) of the subspace $T(X)$, and define a linear map $S: \mathbb{R}^n \to \mathbb{R}^k$

$$S(v) = Bv,$$

where B is the $k \times n$ matrix whose rows are the vectors $b_1, \ldots, b_k$. Then $S(P)$ is a closed wedge in $\mathbb{R}^k$ and, since our inequality system is inconsistent, $-S(c) \notin S(P)$. Hence, by **4C**, we can strongly separate the point $-S(c)$ from the wedge $S(P)$ by a hyperplane H in $\mathbb{R}^k$. H is a level set of a linear functional λ defined by a vector u in $\mathbb{R}^k$. We set

$$\alpha = S'(\lambda) \equiv uB \equiv (\alpha_1, \ldots, \alpha_n),$$

where S' is the transpose (**1H**) of S. The numbers $\alpha_1, \ldots, \alpha_n$ satisfy $\sum_1^n \alpha_i c_i > 0$ and $\sum_1^n \alpha_i \phi_i = \theta$, and consequently Fan's condition is violated. □

C. To illustrate the remark made in **6E** we consider one more type of system of linear inequalities. Now, however, we admit more complex systems than were covered above: infinitely many inequalities are allowed, together with an accompanying non-linear constraint. The problem will be formulated in a conjugate space, as recommended in **6E**.

We will need a result from general topology concerning compactness in function spaces. Let Y be a discrete topological space and Z a metrizable space (we are primarily interested in the special case $Z = \mathbb{R}$.) Let G be a subset of the product space Z^Y endowed with its product topology. Conditions for the compactness of G in Z^Y are contained in the following result, a special case of the "Ascoli theorem".

Lemma. *The closed set* G *is compact in* Z^Y *if (and only if)*

a) G *is equicontinuous; and*

b) *for each* $y \in Y$, $\{f(y): f \in G\}$ *has compact closure in* Z.

Now let g be a sublinear function (for example, a gauge ρ_A) defined on our real linear space X. Let J be an arbitrary index set. Given sets $\{x_j: j \in J\} \subset X$

and $\{c_j : j \in J\} \subset \mathbb{R}$ we consider the problem: find $\phi \in X'$ such that

(7.1)
$$\phi(x_j) \geqslant c_j, \qquad j \in J, \qquad g \geqslant \phi.$$

We have the following criterion (the "Mazur-Orlicz condition") for the consistency of this system.

Theorem. *The system* (7.1) *has a solution* $\phi \in X'$ *if and only if for every finite set* $\{j_1, \ldots, j_n\} \subset J$ *and every set* $\{\alpha_1, \ldots, \alpha_n\}$ *of non-negative numbers we have*

(7.2)
$$\sum_{k=1}^{n} \alpha_k c_{j_k} \leqslant g\left(\sum_{k=1}^{n} \alpha_k x_{j_k}\right).$$

Proof. As usual we need only be concerned with the sufficiency. Let us first show that for each finite set $\{j_1, \ldots, j_n\} \subset J$ the system

(7.3)
$$\phi(x_{j_k}) \geqslant c_{j_k}, \qquad k = 1, \ldots, n, \qquad g \geqslant \phi$$

has a solution ϕ. Let $c = (c_{j_1}, \ldots, c_{j_n}) \in \mathbb{R}^n$ and let P be usual positive wedge there. The set $B \equiv \{\phi(x_{j_1}), \ldots, \phi(x_{j_n}) : \phi \in X', \phi \leqslant g\}$ is a compact convex set in $\mathbb{R}^n$ (the compactness of B follows from the compactness of the set $G \equiv \{\phi \in X' : \phi \leqslant g\}$ in $\mathbb{R}^X$, which in turn is a consequence of the Ascoli theorem). Now if the system (7.3) had no solution we would have $B \cap (P + c) = \varnothing$, and consequently these two sets could be strictly separated by a hyperplane. Thus there would be numbers $\alpha_1, \ldots, \alpha_n$ and β such that

$$\sum_{k=1}^{n} \alpha_k \phi(x_{j_k}) < \beta, \qquad \phi \in G,$$

and

$$\sum_{k=1}^{n} \alpha_k (p_k + c_{j_k}) > \beta, \qquad \text{if } p_k \geqslant 0.$$

The first inequality here implies that $g\left(\sum_1^n \alpha_k x_{j_k}\right) \leqslant \beta$ and the second that $\sum_1^n \alpha_k c_{j_k} > \beta$ and also that each $\alpha_k \geqslant 0$. This is a contradiction of condition (7.2).

At this point we have proved that for each finite subset $K \subset J$, the set

$$G_K \equiv \{\phi \in G : \phi(x_k) \geqslant c_k, k \in K\}$$

is non-empty. These sets G_K are closed subsets of G and, again from what we have just shown, they have the finite intersection property. Hence, since G is compact, all the sets G_K have a non-empty intersection; any element of this intersection is clearly a solution of (7.1). □

D. Let g be a real-valued function defined on X. The *directional (Gateaux) derivative* of g at x_o in the direction x is

$$g'(x_o; x) \equiv \lim_{t \downarrow o} \frac{g(x_o + tx) - g(x_o)}{t}. \tag{7.4}$$

Replacing t by $-t$ in (7.4) we see that

$$\lim_{t \uparrow o} \frac{g(x_o + tx) - g(x_o)}{t} = -g'(x_o; -x).$$

As a preliminary to our next application, and to later work, we study this notion in the case where g is convex.

Lemma. *Let $g \in \operatorname{Conv}(X)$. For any $x_o, x \in X$, the function*

$$t \mapsto \frac{g(x_o + tx) - g(x_o)}{t} \tag{7.5}$$

is non-decreasing for $t > 0$.

Proof. Observe first that if $h \in \operatorname{Conv}(X)$ satisfies $h(\theta) = 0$, then $f(t) \equiv h(tx)/t$ is non-decreasing for $t > 0$. Because, if $0 < s \leqslant t$,

$$h(sx) \leqslant \frac{s}{t} h(tx) + \frac{t - s}{t} h(\theta),$$

so that $f(s) \leqslant f(t)$. Now apply this argument to the function $h(y) \equiv g(x_o + y) - g(x_o)$. □

Theorem. *Let $g \in \operatorname{Conv}(X)$. Given any $x_o \in X$, the directional derivative $g'(x_o; x)$ exists for all $x \in X$ and is a sublinear function of x.*

Proof. Given $x \in X$, we can establish the existence of $g'(x_o; x)$ by showing that the difference quotient (7.5) is bounded below for $t > 0$ and then applying the lemma. In the convexity inequality

$$g(su + (1 - s)v) \leqslant sg(u) + (1 - s)g(v) \tag{7.6}$$

let us replace u by $x_o + tx$, v by $x_o - x$, and s by $1/(1 + t)$. This yields

$$g(x_o) = g\left(\frac{1}{1 + t}(x_o + tx) + \frac{t}{1 + t}(x_o - x)\right)$$
$$\leqslant \frac{1}{1 + t} g(x_o + tx) + \frac{t}{1 + t} g(x_o - x),$$

whence

$$g(x_o) - g(x_o - x) \leqslant \frac{g(x_o + tx) - g(x_o)}{t}, \qquad t > 0.$$

Now the function $g'(x_o; \cdot)$ is clearly positively homogeneous (whether or not g is convex). To establish its sublinearity when $g \in \operatorname{Conv}(X)$, we return to inequality (7.6) and replace u by $x_o + 2tx$ and v by $x_o + 2ty$ for $x, y \in X$. Setting $s = \frac{1}{2}$, we obtain

$$g(x_o + t(x + y)) \leqslant \tfrac{1}{2}(g(x_o + 2tx) + g(x_o + 2ty)),$$

and so

$$\frac{g(x_o + t(x + y)) - g(x_o)}{t} \leqslant \frac{g(x_o + 2tx) - g(x_o)}{2t} + \frac{g(x_o + 2ty) - g(x_o)}{2t}.$$

Thus, when $t \downarrow 0$, we see that

$$g'(x_o; x + y) \leqslant g'(x_o; x) + g'(x_o; y). \qquad \square$$

Corollary. *Let $g \in \mathrm{Conv}(X)$ and $x_o \in X$. Then $-g'(x_o; -x) \leqslant g'(x_o; x)$, for all $x \in X$. Consequently, if $\phi \equiv g'(x_o; \cdot)$ is linear (that is, if $\phi \in X'$) then*

$$\textbf{(7.7)} \qquad \phi(x) = \lim_{t \to o} \frac{g(x_o + tx) - g(x_o)}{t}, \qquad x \in X;$$

that is, the two-sided limit as $t \to 0$ exists for all $x \in X$. Conversely, if this two-sided limit exists for all $x \in X$, then the functional ϕ defined by (7.7) is linear.

When the two sided limit in (7.7) exists for all $x \in X$, the resulting $\phi \in X'$ is called the *gradient* of g at x_o, and is written $\phi \equiv \nabla g(x_o)$. By way of illustration it is interesting to mention that when $g \in \mathrm{Conv}(A)$, where A is an open convex set in $\mathbb{R}^n$, then g has a gradient at almost every point in A and the map $x \mapsto \nabla g(x)$ is continuous on its domain in A. The proofs of these facts are not trivial and will be omitted, as the results play no role in the sequel.

E. It was observed in **6D** that when $f \in \mathrm{Conv}(\mathbb{R})$ fails to be differentiable at $x_o \in R$ then $\partial f(x_o) = [f'_-(x_o), f'_+(x_o)]$. Guided by this special situation, we consider its analogue in a more general setting, and draw some interesting conclusions relating the notions of gradient, sub-gradient, and directional derivative.

First of all, the results of **7D** allows us to assert that the subgradients of $g \in \mathrm{Conv}(X)$ at a point $x_o \in X$ are exactly the linear minorants of the directional derivative at x_o. That is,

$$\partial g(x_o) = \{\psi \in X' : \psi \leqslant g'(x_o; \cdot)\}.$$

Since ψ is linear we can re-write this formula as

$$\textbf{(7.8)} \qquad \partial g(x_o) = \{\psi \in X' : -g'(x_o; -x) \leqslant \psi(x) \leqslant g'(x_o; x), x \in X\}.$$

Theorem. *Let $g \in \mathrm{Conv}(X)$ and $x_o \in X$.*

a) *For any $x \in X$, the two-sided limit in (7.7) exists and has the value α if and only if the function $\psi \mapsto \psi(x)$ is constantly equal to α for all $\psi \in \partial g(x_o)$.*

b) *The gradient $\nabla g(x_o)$ exists in X' if and only if $\partial g(x_o)$ consists of a single element, namely $\nabla g(x_o)$.*

Proof. a) is clear from (7.8) and the fact that the limit in (7.7) exists if and only if $g'(x_o; -x) = -g'(x_o; x)$. To establish b), assume first that $\nabla g(x_o)$ exists in X'. Then given any $\psi \in \partial g(x_o)$ we see from (7.8) that

$$\psi \leqslant g'(x_o; \cdot) \equiv \nabla g(x_o),$$

so that $\psi = \nabla g(x_o)$ and hence $\partial g(x_o) = \{\nabla g(x_o)\}$. Conversely, if the gradient $\nabla g(x_o)$ fails to exist, it is because $-g'(x_o; -\bar{x}) < g'(x_o; \bar{x})$ for some $\bar{x} \in X$. Let $M = \text{span}\{\bar{x}\}$ and choose any α in the interval $[-g'(x_o; \bar{x}), g'(x_o; -\bar{x})]$. We define a functional $\bar{\psi} \in M'$ by setting $\bar{\psi}(t\bar{x}) = \alpha t$, for $t \in \mathbb{R}$. Then by our choice of α, $\bar{\psi}(x) \leqslant g'(x_o; x)$ for all $x \in M$. Now the Hahn-Banach theorem (**6A**) provides us with an extension ψ of $\bar{\psi}$ for which $\psi \leqslant g'(x_o; \cdot)$. We obtain distinct such ψ's by varying α in the indicated interval and by (7.8) all the ψ's belong to $\partial g(x_o)$. □

F. Let A be a convex absorbing set in X. It is of interest to apply the preceding results about general convex functions to the study of the gauge ρ_A of A. This will yield the insight that the linear functionals defining supporting hyperplanes to A at some bounding point in A are exactly the subgradients of ρ_A at that point. Given the geometric interpretation (**6D**) of subgradients and the fact the ρ_A is sublinear, this relationship should not be completely unexpected.

We say that the map $\tau_A: X \times X \to \mathbb{R}$ defined by

$$\tau_A(x, y) = \rho_A'(x; y).$$

is the *tangent function* of A. From **7D** it is clear that the tangent function obeys the following rules:

a) $\tau_A(x, \cdot)$ is sublinear on X;
b) $\tau_A(x, y) \leqslant \rho_A(y)$;
c) $\tau_A(x, tx) = t\rho_A(x)$, $t \in \mathbb{R}$; and
d) $\tau_A(\alpha x, \cdot) = \tau_A(x, \cdot)$, $\alpha > 0$.

Theorem. *Let A be a convex absorbing set in X with gauge ρ_A. Given $x_o \in X$ with $\rho_A(x_o) > 0$, the following assertions are equivalent for $\phi \in X'$:*
a) $\phi \in \partial\rho_A(x_o)$;
b) $\phi \leqslant \tau_A(x_o, \cdot)$;
c) $\phi(x_o) = \rho_A(x_o)$ *and* $\sup\{\phi(x): x \in A\} = 1$.

Proof. The equivalence of a) and b) is a consequence of equation (7.8). To see the equivalence of a) and c), we recall that

$$\phi \in \partial\rho_A(x_o) \Leftrightarrow \phi \leqslant \rho_A \quad \text{and} \quad \phi(x_o) = \rho_A(x_o).$$

(These implications depend only on the sublinearity of ρ_A.) Since also it is clear that

$$\phi \leqslant \rho_A \Leftrightarrow \sup\{\phi(x): x \in A\} = 1,$$

the proof is complete. □

By virtue of the support theorem (**6C**) we know that every bounding point x_o of A belonging to A is a (proper) support point of A. The theorem above tells us that $\tau_A(x_o, \cdot) \neq \theta$ in this case, and furthermore, that there is a unique hyperplane of support at x_o exactly when $\tau_A(x_o, \cdot)$ is linear. (If this functional is linear, then the unique supporting hyperplane to A at x_o is $[\tau_A(x_o, \cdot); 1] \equiv [\nabla\rho_A(x_o); 1]$.) When these conditions for uniqueness are satisfied we say that x_o is a *smooth point* of A, or that A is *smooth* at x_o. This terminology is chosen to suggest that (intuitively) the surface of A does not come together "sharply" at x_o. We have shown that smoothness of A at its bounding point x_o is equivalent to the existence of $\nabla\rho_A(x_o)$ in X'.

To illustrate these ideas, let $X = \mathbb{R}^n$, let $p \geqslant 1$, and let A be the unit p-ball (**3C**) in $\mathbb{R}^n$. We know that ρ_A is then the p-norm on $\mathbb{R}^n$:

$$\rho_A(x) = \left(\sum_{i=1}^{n} |\xi_i|^p\right)^{1/p}, \qquad x = (\xi_1, \ldots, \xi_n) \in \mathbb{R}^n.$$

By direct differentiation we compute that, for $x \neq \theta$ and $p > 1$,

$$\textbf{(7.9)} \qquad \tau_A(x, y) = \frac{\sum_{i=1}^{n} \eta_i |\xi_i|^{p-1} \operatorname{sgn} \xi_i}{\rho_A(x)^{p-1}}, \qquad y = (\eta_1, \ldots, \eta_n) \in \mathbb{R}^n.$$

Here the sigmum function sgn ξ is defined for real or complex ξ, by

$$\operatorname{sgn} \xi = \begin{cases} 0 & \text{if } \xi = 0 \\ \dfrac{\xi}{|\xi|} & \text{if } \xi \neq 0. \end{cases}$$

Suppose that x is a bounding point of A, so that $\rho_A(x) = 1$. Then equation (7.9) shows that the tangent function is linear in y. Consequently, the unit p-ball is smooth at all its bounding points and, for such points x,

$$\textbf{(7.10)} \qquad \nabla\rho_A(x) = (|\xi_1|^{p-1} \operatorname{sgn} \xi_1, \ldots, |\xi_n|^{p-1} \operatorname{sgn} \xi_n).$$

Now consider the situation when $p = 1$. A simple sketch (when $n = 2$ or 3) suggests, and (7.9) confirms, that $\tau_A(x, \cdot)$ is still linear provided no $\xi_i = 0$, that is, provided that x lies in no coordinate hyperplane in $\mathbb{R}^n$. Thus the unit 1-ball is smooth at such points and formula (7.10) remains valid. On the other hand, let us suppose that some components of x are zero; say $\xi_i = 0$ for $i \in I_o \subsetneqq \{1, 2, \ldots, n\}$. Then we compute that

$$\textbf{(7.11)} \qquad \tau_A(x, y) = \sum_{i \notin I_o} \eta_i \operatorname{sgn} \xi_i + \sum_{i \in I_o} |\eta_i|.$$

From (7.11) we see that $\tau_A(x, \cdot)$ is not linear and, in fact, that $-\tau_A(x, -y) < \tau_A(x, y)$ whenever $\eta_i \neq 0$ for some $i \in I_o$. It follows that the unit 1-ball is not smooth at any such x. In fact, we see that any hyperplane of the form $[\phi; 1]$ supports the unit 1-ball at x if ϕ is determined by $(\zeta_1, \ldots, \zeta_n)$ and

$$\begin{aligned} \zeta_i &= \operatorname{sgn} \xi_i, \qquad & i &\notin I_o \\ |\zeta_i| &\leqslant 1, & i &\in I_o. \end{aligned}$$

§8. Extremal Sets

In this section we introduce the last of our "primitive" linear space concepts: extremal subsets and points of convex sets. The fundamental idea here is that a given convex set can be "reconstructed" from knowledge of certain bounding subsets by use of the operation of taking convex combinations (and perhaps also closures, as we shall see later). There is a faint analogy with the reconstruction of a linear space from the elements of a basis and the operation of taking linear combinations, although the more complicated behavior of general convex sets permits further classifications of extremal sets and points.

A. Let E be a subset of a convex set A in the real linear space X. E is a *semi-extremal subset* of A if $A\backslash E$ is convex, and E is an *extremal subset* of A if $x, y \in A$ and $tx + (1 - t)y \in E$ for some t $(0 < t < 1)$ entails $x, y \in E$. We often write "E is A-semi-extremal" or "E is A-extremal". It is clear that each extremal subset of A is semi-extremal; the simplest examples in $\mathbb{R}^2$ show that the converse is generally false. However, when $E = \{x_o\}$ is a singleton subset of A, the two notions do coincide; when this happens, x_o is said to be an *extreme point* of A and we write $x_o \in \text{ext}(A)$. Thus the extreme points of A are just those points which can be removed from A so as to leave a convex set. Any such point is necessarily a bounding point of A.

The prototypical example is an n-simplex (**2C**): it is (by definition) the convex hull of its vertices which are the extreme points in this case. More generally, the convex hull of any subset of the vertices is an extremal subset of the n-simplex. Other possibilities can occur: on the one hand, every bounding point of the unit p-ball $(p > 1)$ in $\mathbb{R}^n$ is an extreme point, and there are no other (proper) extremal subsets; on the other hand, an affine subspace of positive dimension contains no (proper) extremal subsets at all. Examples of A-semi-extremal subsets are obtained as the intersection of A with any half-space (**4B**) in X, or more generally, as the intersection of any A-extremal set with a half-space. Any subset of ext(A) is A-semi-extremal.

The following lemma collects a variety of elementary but useful properties of (semi-) extremal sets; its proof is left as an exercise. It should be noted that the assertions below involving A-extremal sets do not require the convexity of A.

Lemma. *Let A be a convex subset of X.*

a) *The union of a family of (semi-) extremal subsets of A is A-(semi-) extremal;*

b) *The intersection of any (nested) family of A-(semi-) extremal sets is A-(semi-) extremal.*

c) *Let $E \subset B \subset A$ with B an extremal subset of A. If E is B-(semi-) extremal, then E is also A-(semi-) extremal.*

d) *If E is A-extremal then* $\text{ext}(E) = \text{ext}(A) \cap E$.

B. Let us now consider how the extremal subsets or extreme points of a given convex set A can be used to describe the set. Here we shall only consider the case where A is of finite dimension (**2C**). Thus we may as well assume that $A \subset \mathbb{R}^n$ for some n; we shall also assume that A is closed.

Lemma. *Each closed convex subset A of $\mathbb{R}^n$ contains an A-extremal affine subspace, and any two such affine subspaces are parallel.*

Proof. To prove the existence of such extremal flats in A we proceed by induction on the dimension of A. We may assume that existence has been established for sets of dimension less than $\dim(A)$, which we take to equal n. We may also assume that A has a bounding point p, for otherwise $A = \mathbb{R}^n$ and A is an extremal flat in itself. Now if H is a hyperplane supporting A at p (**6C**), the set $A \cap H$ contains an extremal flat K by the induction hypothesis. However, since $A \cap H$ is necessarily an extremal subset of A, it follows from the preceding lemma that K is also A-extremal.

Now suppose that $K_1 = p_1 + L_1$ and $K_2 = p_2 + L_2$ are two A-extremal flats parallel to the subspaces L_1 and L_2 (**1C**). We want to see that $L_1 = L_2$. If L_1 is not contained in L_2 then

(8.1) $$K_2 \subsetneqq L_1 + K_2 \subset A.$$

(The second inclusion of (8.1) can be shown as follows: let $l_1 \in L_1$ and $p_2 + l_2 \in K_2$; then for $t \geqslant 1$, $p_1 + tl_1 \in A$ and hence

(8.2) $$\left(1 - \frac{1}{t}\right)(p_2 + l_2) + \frac{1}{t}(p_1 + tl_1) \in A.$$

As $t \to +\infty$, the left side of (8.2) converges to $l_1 + p_2 + l_2$ and this must belong to A since A is closed.) Now (8.1) contradicts the assumption that K_2 is A-extremal, so that we must have $L_1 \subset L_2$. Analogously, $L_2 \subset L_1$, whence $L_1 = L_2$. □

It follows that the extremal affine subspaces of A are all parallel to a particular subspace L_A called the *lineality space* of A. The dimension of L_A is the *lineality* of A and an affine subspace of A is A-extremal exactly when it is of maximal dimension (with respect to all the affine subspaces of A), this dimension being just the lineality of A. It is easy to see that

(8.3) $$L_A = \{x \in \mathbb{R}^n : x + A = A\}.$$

A is said to be *line-free* exactly when $L_A = \{\theta\}$. We now have sufficient information to state the basic existence theorem for extreme points.

Theorem. *The closed convex set A in $\mathbb{R}^n$ has an extreme point if and only if A is line-free.*

Proof. If A is not line-free then there is a non-zero x satisfying (8.3) so that no point of A can be extreme. On the other hand, if A is line-free, the only flats contained in A are of zero dimension, hence points. The lemma now guarantees the existence of an extreme point. □

This result allows us to obtain a preliminary decomposition of the closed convex set A. Let $L_A^\perp$ be the orthogonal complement of the subspace L_A, that is, the set of all vectors in $\mathbb{R}^n$ that are orthogonal to L_A. Then we can write

(8.4) $$A = L_A + (A \cap L_A^\perp);$$

the "section" $A \cap L_A^\perp$ of A is clearly line-free and hence has an extreme point. It is not hard to show that this is the only way to express A as the orthogonal sum of a subspace and a closed line-free convex set.

C. To obtain a more complete decomposition of the closed convex set A in $\mathbb{R}^n$, we introduce the *recession cone*[1] *(asymptotic cone)* C_B of a convex set B in a real linear space X:

$$C_B \equiv \{x \in X : x + B \subset B\}.$$

Note the analogy with formula (8.3); clearly $L_B \subset C_B$, when $B \subset \mathbb{R}^n$; indeed, $L_B = C_B \cap (-C_B)$ (**5A**). We shall want to consider the set C_B especially in the case where B is line-free ($L_B = \{\theta\}$); in terms of our original convex set A under investigation, we shall be interested in C_{B_A}, where $B_A \equiv A \cap L_A^\perp$.

Lemma. *Let B be a convex subset of the real linear space X.*
a) *The recession cone C_B is a wedge in X;*
b) $C_B = \{x \in X : b + tx \in B$ *for all* $t \geqslant 0$ *and all* $b \in B\}$;
c) *if* $X = \mathbb{R}^n$ *and B is closed then C_B is closed and* $C_B = \{x \in \mathbb{R}^n : x = \lim_n t_n x_n$, *where* $x_n \in B$ *and* $t_n \downarrow 0\}$.

Proof. a) Let $x \in C_B$ so that $x + B \subset B$. Then $2x + B = x + (x + B) \subset x + B \subset B$, and more generally, $nx + B \subset B$ for every positive integer n. Since B is convex, this means that $tx + B \subset B$ for all $t \geqslant 0$, that is, $tx \in C_B$, $t \geqslant 0$. Next, if $x, y \in C_B$ and $0 < t < 1$, we have

$$\begin{aligned}((1 - t)x + ty) + B &= (1 - t)(x + B) + t(y + B)\\ &\subset (1 - t)B + tB = B,\end{aligned}$$

using the convexity of B. This proves that C_B is a wedge.

b) The inclusion from right to left here is trivial, and the reverse inclusion follows from the proof of a).

c) From part b) we see that given any $b \in B$

(8.5) $$C_B = \cap\{t(B - b) : t > 0\}$$

(whether or not $X = \mathbb{R}^n$). Since B is closed, (8.5) exhibits C_B as an intersection of closed sets, so that C_B is closed. Next, let $x \in C_B$; for any fixed

[1] The term "recession cone" is used in conformity with established terminology. To be consistent we should say "recession wedge", since this set is generally not a cone as defined in **5A**.

$b \in B$, $b + nx \equiv x_n \in B$ and hence $x = \lim_n t_n x_n$, where $x_n \in B$ and $t_n \downarrow 0$. We claim that $b_o + tx \in B$ for any fixed $b_o \in B$ and any $t > 0$. If this were not the case then for some $t_o > 0$, $b_o + t_o x \notin B$ and we could apply the strong separation theorem to find $\phi \in \mathbb{R}^{n\prime}$ and $\alpha \in \mathbb{R}$ such that

$$\sup\{\phi(b): b \in B\} \leqslant \alpha < \phi(b_o + t_o x). \tag{8.6}$$

However, $\phi(x) = \lim_n \phi(t_n x_n)$ and $\phi(t_n x_n) \leqslant t_n \alpha$, so that $\phi(x) \leqslant 0$; this entails $\phi(b_o + t_o x) \leqslant \phi(b_o) \leqslant \alpha$, in contradiction to (8.6). □

The formula in part b) provides the motivation for the term "recession cone". Note that the wedge C_B is a cone exactly when B is line-free. A procedure for computing both L_A and C_A for a given closed convex set $A \subset \mathbb{R}^n$ is indicated in exercise 1.35.

We come now to the main decomposition formula for a closed convex set $A \subset \mathbb{R}^n$, the "Klee-Minkowski-Hirsch-Hoffman-Goldman-Tucker theorem". Associated with A we have its lineality space L_A and the corresponding line-free section $B_A \equiv A \cap L_A^\perp$.

Theorem. *Let A be a closed convex subset of $\mathbb{R}^n$. Then*

$$A = L_A + C_{B_A} + \mathrm{co}(\mathrm{ext}(B_A)). \tag{8.7}$$

Proof. It will suffice (in view of (8.4)) to show that

$$B \subset C_B + \mathrm{co}(\mathrm{ext}(B)) \tag{8.8}$$

for any closed line-free convex set $B \subset \mathbb{R}^n$. We proceed (as in (**8B**)) via induction on the dimension of B and assume that (8.8) is valid for subsets B of dimension $< n$. Let p be an arbitrary point in B and let L be any line containing p. The set $B \cap L$ is then either a closed half-line or a compact line segment, since B is line-free. In the former case we can write

$$B \cap L = \{x + \lambda y : \lambda \geqslant 0\}.$$

The end-point x of this half-line is a bounding point of B and hence is contained in a hyperplane H of support to B. Applying the induction hypothesis to $B \cap H$, we have

$$\begin{aligned} x &\in C_{B\cap H} + \mathrm{co}(\mathrm{ext}(B \subset H)) \\ &\subset C_B + \mathrm{co}(\mathrm{ext}(B)), \end{aligned}$$

using (**8A**). Since y must lie in C_B and since $p = x + \bar{\lambda} y$ for some $\bar{\lambda} \geqslant 0$, it follows that

$$\begin{aligned} p = x + \bar{\lambda} y &\in (\mathrm{co}(\mathrm{ext}(B)) + C_B) + C_B \\ &\subset \mathrm{co}(\mathrm{ext}(B)) + C_B. \end{aligned}$$

In the other case, $B \cap L$ is a compact line segment and by analogous reasoning both its end-points belong to $C_B + \mathrm{co}(\mathrm{ext}(B))$; since this latter set is convex, it contains the entire line segment and, in particular, the point p. □

D. Some important consequences of the preceding theorem will now be given. The first of these, "Minkowski's theorem", follows from **8C** and the observation that a convex set $A \subset \mathbb{R}^n$ is bounded if and only if C_A (hence L_A) $= \{\theta\}$.

Corollary. *A compact convex set in $\mathbb{R}^n$ is the convex hull of its extreme points.*

Consider next the case where our closed convex set $A \subset \mathbb{R}^n$ is unbounded but line-free: $L_A = \{\theta\}$. An *extreme ray* of A is an A-extremal half-line; we write the set of all extreme rays of A as $\operatorname{rext}(A)$. The idea now is that A can be recovered (via convex combinations) from its extreme points and its extreme rays.

Lemma. *Let C be a closed (convex) cone in $\mathbb{R}^n$. Then $C = \operatorname{co}(\operatorname{rext}(C))$.*

Proof. From **5C** we know that our cone C has a base K given by $K = [\phi; 1] \cap C$, where ϕ is a linear functional that is strictly positive on $C \backslash \{\theta\}$. The base K is clearly closed and we claim also that it is bounded. For otherwise, there would exist a sequence $\{x_n\} \subset K$ with $\|x_n\|_2 \to +\infty$ (here $\|x_n\|_2$ is the 2-norm (**3C**) of x_n). Let z be a limit point of the sequence $x_n/\|x_n\|_2$; since $\|z\|_2 = 1$, $z \neq \theta$, and since C is closed, $z \in C$, whence $\phi(z) > 0$. But also

$$\phi(z) = \lim_n \phi(x_n/\|x_n\|_2) = \lim_n (1/\|x_n\|_2) = 0,$$

a contradiction; thus K must be bounded hence compact. Now an easy argument shows that if p is an extreme point of K then the ray $\{tp : t \geqslant 0\}$ is an extremal ray of C. The conclusion of the lemma is thus seen to be a consequence of Minkowski's theorem. ☐

The following result, "Klee's theorem", provides a substantial generalization of Minkowski's theorem; its proof follows directly from the lemma and **8C**.

Theorem. *A closed line-free convex set A in $\mathbb{R}^n$ is the convex hull of its extreme rays and extreme points:*

$$A = \operatorname{co}(\operatorname{ext}(A) \cup \operatorname{rext}(A)).$$

E. An important application of these structure theorems is to problems of optimization. An *optimization* (or *variational*) *problem* occurs when we are given a pair (A, f) (a *variational pair*) consisting of a set A and a real-valued function f on A; f is called the *objective* (or *cost*) *function*. The problem is to determine the number $\inf\{f(x) : x \in A\}$, called the *value* of the optimization problem, and a point in A (if any) where f attains its infimum; any such point is a *solution* of the problem. It is traditional to also refer to such an optimization problem as an *abstract mathematical program*. It should be noted that by the simple expedient of changing the sign of the objective function, problems originally requiring the maximization of some function can be converted into the present format.

Let us consider a special case of the preceding: a *finite dimensional concave program* (A, f), where A is a convex set in $\mathbb{R}^n$ for some n and f is a *concave* function on A, that is, $-f \in \operatorname{Conv}(A)$. In practice, such a problem might arise as a *linear program*, that is, $f(x) = \phi(x) + c$, where ϕ is a linear functional on $\mathbb{R}^n$ and c is a constant, or, after a change of sign, as the problem of maximizing the utility of some risk-seeking investor.

Theorem. *Let (A, f) be a finite dimensional concave program as just defined; assume that A is line-free and closed, and that f is lower semicontinuous on A. If a solution exists in A, then there is a solution in* $\operatorname{ext}(A)$. *Conversely, any solution to the program* $(\operatorname{ext}(A), f)$ *is also a solution to the original program (A, f), provided that f is bounded below on A.*

Proof. The subset F of A where f attains its infimum over A is non-empty (by hypothesis) and is in addition closed and convex. Therefore, by **8B**, F has an extreme point. Since F is also easily seen to be A-extremal it follows from **8A** that this extreme point belongs to $\operatorname{ext}(A)$. For the converse, suppose that $\inf\{f(x): x \in \operatorname{ext}(A)\}$ is attained at $p \in \operatorname{ext}(A)$. If $x = \Sigma t_i e_i \in \operatorname{co}(\operatorname{ext}(A))$, then

$$f(x) = f(\Sigma t_i e_i) \geqslant \Sigma t_i f(e_i) \geqslant \Sigma t_i f(p) = f(p).$$

Now from **8C**, $A = C_A + \operatorname{co}(\operatorname{ext}(A))$, since A is line-free (so that $L_A = \{\theta\}$). Choose any $c \in C_A$ and any $t > 1$. Then for $x \in \operatorname{co}(\operatorname{ext}(A))$ we have $tc + x \in A$ and

$$c + x = \frac{t-1}{t} x + \frac{1}{t}(tc + x),$$

so that

$$\begin{aligned} f(c + x) &\geqslant \frac{t-1}{t} f(x) + \frac{1}{t} f(tc + x) \\ &\geqslant \frac{t-1}{t} f(p) + \frac{1}{t} \inf\{f(y): y \in A\}. \end{aligned}$$

Letting $t \to \infty$, we conclude that $f(c + x) \geqslant f(p)$. □

The effect of this theorem is of course to reduce the search for solutions of the finite dimensional concave program (A, f) to the extreme points of A. In particular, when f is bounded below on A, and $\operatorname{ext}(A)$ is a finite set, then we are assured of the existence of a solution in $\operatorname{ext}(A)$. For example, if A is the strip $\{(x_1, x_2) \in \mathbb{R}^2 : r \leqslant x_1, a \leqslant x_2 \leqslant b\}$ and the concave function f is bounded below on A, then at least one of the points (r, a), (r, b) is a solution of the concave program (A, f).

F. The preceding results have dealt with finite dimensional convex sets and their extremal properties. To conclude this section we now want to give a famous example concerning the extreme points of certain infinite dimensional convex sets.

Let A be an algebra of real or complex-valued functions defined on some set Ω; we assume that A contains the function e identically equal to one on Ω. A is of course an ordered linear space with the natural pointwise vector ordering (**5A**); we let P be the positive wedge in A and P^+ the dual wedge in A'. In A' we consider the convex sets $K_0 \equiv \{\phi \in P^+ : \phi(e) \leqslant 1\}$ and $K_1 \equiv \{\phi \in K_0 : \phi(e) = 1\}$. The question now is: what are the extreme points of K_0 and K_1?

Let $A_R = \{f \in A : f \text{ is real-valued}\}$ and suppose that A_R consists only of bounded functions on Ω. Then we can derive a necessary algebraic condition for an element of K_0 or K_1 to be extreme. Note that since K_1 is K_0-extremal, it suffices to work only with K_0; indeed, $\text{ext}(K_0) = \text{ext}(K_1) \cup \{\theta\}$.

Theorem. *An extreme point of K_0 must be an algebra homomorphism of A.*

Proof. Let $\phi \in \text{ext}(K_0)$; if $\phi = \theta$ we are done. Otherwise we must show that

$$\textbf{(8.9)} \qquad \phi(fg) = \phi(f)\phi(g)$$

for all $f, g \in A$. Let us first prove this when $g = e$. Let $\psi = (1 - \phi(e))\phi$. Since $\phi \in K_0$, $\psi \in P^+$ whence $\phi + \psi \in P^+$; similarly, $\phi - \psi \in P^+$. Further,

$$\begin{aligned}\langle e, \phi + \psi\rangle &= \phi(e) + \phi(e)(1 - \phi(e))\\ &\leqslant \phi(e) + (1 - \phi(e)) = 1,\end{aligned}$$

and $\langle e, \phi - \psi\rangle = \phi(e)^2 \leqslant 1$. Therefore, $\phi \pm \psi \in K_0$ and so $\psi = \theta$, since $\phi \in \text{ext}(K_0)$. Next we prove (8.9) assuming that g is real-valued. For this we may also assume that $\theta \leqslant g \leqslant e$; because, since g is bounded (by hypothesis), there are $s, t > 0$ such that $\theta \leqslant sg + te \leqslant e$, and, if it is true that $\langle f(sg + te), \phi\rangle = \phi(f)\phi(sg + te)$, then $s\phi(fg) + t\phi(f) = s\phi(t)\phi(g) + t\phi(f)\phi(e)$. Now, with $\theta \leqslant g \leqslant e$, let $\psi(f) = \phi(fg) - \phi(f)\phi(g)$; then $\langle e, \phi + \psi\rangle = \phi(e) \leqslant 1$. Further, if $f \in P$,

$$\begin{aligned}\langle f, \phi + \psi\rangle &= \phi(f) + \phi(fg) - \phi(f)\phi(g)\\ &= \phi(f)(1 - \phi(g)) + \phi(fg) \geqslant 0,\end{aligned}$$

and similarly,

$$\begin{aligned}\langle f, \phi - \psi\rangle &= \phi(f) - \phi(fg) + \phi(f)\phi(g)\\ &= \langle f(e - g), \phi\rangle + \phi(f)\phi(g) \geqslant 0.\end{aligned}$$

Again we have shown that $\phi \pm \psi \in K_0$ and hence $\psi = \theta$. Finally, for arbitrary $f, g \in A$, define $\psi(g) = \phi(fg) - \phi(f)\phi(g)$. If $g \in P$ then by what we have just shown, $\psi(g) = 0$; thus once more $\phi \pm \psi \in K_0$ and $\psi = \theta$, completing the proof of (8.9). ∎

The algebra A was not assumed to be self-adjoint (that is, to contain $\bar{f}$ whenever it contains f). However, if $\text{ext}(K_0) \neq \varnothing$, then the self-adjointness of A follows; thus the theorem really concerns algebras of bounded functions (exercise 1.39).

Let us also remark that the same proof applies to a more general situation. Namely, let $A\#$ be a second algebra of functions defined on some set $\Omega\#$ with positive wedge $P\#$ and containing the identically-one function $e\#$. Let $K_0(A, A\#)$ consist of all positive linear maps $T: A \to A\#$ such that

$T(e) \leqslant e\#$ (T is *positive* if $T(P) \subset P\#$). Similarly we let $K_1(A, A\#) = \{T \in K_0(A, A\#): T(e) = e\#\}$. Then the preceding theorem is true for $K_0(A, A\#)$ and $K_1(A, A\#)$.

Before treating the converse of the theorem, let us note a useful fact: if $f \in A_R$ and $\phi \in K_0$ then

(8.10) $$\phi(f^2) \leqslant \phi(f)^2.$$

The proof of this follows upon consideration of the discriminant of the non-negative quadratic form $t \mapsto \langle (tf + e)^2, \phi \rangle$.

Now assume that our algebra A is self-adjoint and that ϕ is a homomorphism of A belonging to K_0. Then we can prove that $\phi \in \text{ext}(K_0)$. To do so, suppose that $\phi = \frac{1}{2}(\phi_1 + \phi_2)$ where $\phi_1, \phi_2 \in K_0$. Now, if $f \in A_R$,

$$\begin{aligned}\tfrac{1}{2}\phi_1(f)^2 + \tfrac{1}{2}\phi_2(f)^2 &\leqslant \tfrac{1}{2}(\phi_1(f^2) + \phi_2(f^2)) \\ &= \phi(f^2) = \phi(f)^2 = \tfrac{1}{4}\phi_1(f)^2 + \tfrac{1}{2}\phi_1(f)\phi_2(f) + \tfrac{1}{4}\phi_2(f)^2,\end{aligned}$$

where the first inequality is a consequence of (8.10). This argument shows that $(\phi_1(f) - \phi_2(f))^2 \leqslant 0$, whence ϕ_1 and ϕ_2 agree on A_R. Since A is self-adjoint this means that $\phi_1 = \phi_2$ and so $\phi \in \text{ext}(K_0)$ as claimed. Again, this proof generalizes to the case $K_0(A, A\#)$.

To sum up, in the case where A is a self-adjoint algebra of bounded functions, we have obtained both an algebraic characterization of the extreme points of the sets K_0 and K_1 and a geometrical interpretation of certain algebra homomorphisms of A (for many common algebras A, every homomorphism of A belongs to K_0; this is true in particular when $A = C(\Omega)$, the space of all continuous functions on the compact Hausdorff space Ω). Now at present we don't know whether extreme points or (non-zero) homomorphisms exist, but we have at least arrived at the point where knowledge of one has implications for the other. Existence proofs for either extreme points or homomorphisms involve the Axiom of Choice, usually in the guise of Zorn's lemma. We might, for example, try to utilize the latter to produce a proper maximal ideal M in A; we would then expect M to be the kernel of a homomorphism. However, in conformity with our geometric approach, we will adopt the opposite course and try to develop methods for proving the existence of extreme points for certain infinite dimensional convex sets. The eventual results bear some analogy to the finite dimensional case treated earlier in this section, but the methods are quite different. Interestingly enough, these methods require topological considerations (interesting because the notion of extreme point is purely algebraic). We turn to such considerations next.

Exercises

1.1. a) Show that the sequence of monomials $\{t^n : n = 0, 1, 2, \ldots\}$ is a linearly independent subset of the linear space of all real-valued functions defined on the interval $[0, 1]$.

b) Show that the family of complex exponentials $\{e^{i\lambda t}: -\infty < \lambda < \infty\}$ is a linearly independent subset of the linear space of all continuous bounded complex-valued functions defined on $\mathbb{R}$.

c) What is the dimension of the space in b)?

1.2. Let n be an arbitrary cardinal number. Construct a (real or complex) linear space of dimension n.

1.3. a) Let X be the linear subspace of exercise 1.1a), let $0 \leqslant t_1 < t_2 < \cdots < t_m \leqslant 1$, and let $\{c_1, c_2, \ldots, c_m\} \subset \mathbb{R}$. Show that the set $\{x \in X: x(t_i) = c_i, i = 1, \ldots, m\}$ is an affine subspace of X.

b) Let X be an arbitrary linear space. If $\phi_1, \ldots, \phi_m \in X'$, show that $\bigcap_1^m \ker \phi_i$ is a subspace of codimension $\leqslant m$.

1.4. a) Let X be an infinite dimensional linear space over the field $\mathbb{F}$. Prove that $\dim(X') = 2^{\dim(X)}$. (First show that $\dim(X') \geqslant 2^{\aleph_0}$ by considering a linear independent sequence $\{e_n\} \subset X$ and the set $\{\phi_\lambda : \lambda \in \mathbb{F}\} \subset X'$ defined by $\phi_\lambda(e_n) = \lambda^n$. Then verify

$$2^{\dim X} = \operatorname{card}(X') = 2^{\aleph_0} \dim(X') = \max(2^{\aleph_0}, \dim(X')) = \dim(X').)$$

b) Prove that two linear spaces over $\mathbb{F}$ are isomorphic if (and only if) their algebraic conjugate spaces are isomorphic. (In the case where the spaces are of infinite dimension use part a) and the generalized continuum hypothesis.)

1.5. Let $T: X \to Y$ be a linear map between linear spaces X and Y.

a) Show that $T'' \equiv (T')': X'' \to Y''$ is an extension of T.

b) If $X = Y$, show that T is always the transpose of some linear map exactly when X is finite dimensional.

1.6. Let X and Y be linear spaces. A map $T: X \to Y$ is *affine* if the map $x \to T(x) - T(\theta)$ is linear. Show that if T is affine the image $T(A)$ of a convex set $A \subset X$ is convex, and the inverse image $T^{-1}(B)$ of a convex set $B \subset Y$ is convex.

1.7. Let A be an absolutely convex set in a linear space. Show that $\operatorname{span}(A) = \bigcup_1^\infty nA$ and that $\operatorname{cor}(A)$ is again absolutely convex.

1.8. Let $A \subset \mathbb{R}^n$, for some n.

a) If A is convex show that the core of A is the (topological) interior of A, and that $\operatorname{lin}(A)$ is the closure of A.

b) Show by example in $\mathbb{R}^2$ that when A is not convex, there can be points in $\operatorname{cor}(A)$ which are not interior points of A.

c) If A is open show that $\operatorname{co}(A)$ is also open.

d) Find an example of a closed A such that $\operatorname{co}(A)$ is open (yet not all of $\mathbb{R}^n$).

e) Show that each $x \in \operatorname{co}(A)$ lies in some m-simplex with vertices in A and $m \leqslant n$. ("Caratheodory's theorem". Express x as $\sum_0^m \alpha_i x_i$

where $x_i \in A$, $\alpha_i \geqslant 0$, and $\sum_{o}^{m} \alpha_i = 1$ and assume that this representation of x involves the minimum possible number of points in A. It is to be shown that $m \leqslant n$. Proceed by contradiction, assuming that $m \geqslant n + 1$; the points $x_0, x_1, \ldots, x_m$ are then not in general position.)

f) Use the result of e) to show that co(A) is compact whenever A is compact.

1.9. Taking into account exercise 1.8a) show that a linear space X has finite dimension if and only if for all convex $A \subset X$ we have lin(lin(A)) = lin(A). (If X is infinite dimensional and $x \in X$, define $s(x)$ to be the sum of the coefficients involved in the expression of x relative to a fixed basis for X, and define $n(x)$ to be the number of these coefficients that are not zero. Let A be the set of x with non-negative coefficients such that $n(x) > 0$ and $n(x)s(x) > 1$. Then A is convex but $\theta \in \text{lin}(\text{lin}(A))\backslash A$.)

1.10. A real-valued function f defined on a linear space X is *quasi-convex* if its sublevel sets $\{x \in X : f(x) \leqslant \lambda\}$ are convex for each real λ. Show that f is quasi-convex if and only if $f(tx + (1 - t)y) \leqslant \max\{f(x), f(y)\}$ for $x, y \in X$ and $0 \leqslant t \leqslant 1$. Thus every convex function is quasi-convex but the converse fails even in $\mathbb{R}^1$.

1.11. Prove the lemma in **3C**.

1.12. Let A be a convex absorbing set in a linear space X. The gauge ρ_A determines A analytically as follows:

a) cor(A) = $\{x \in A : \rho_A(x) < 1\}$;

b) the algebraic boundary of A is $\{x \in X : \rho_A(x) = 1\}$;

c) if $\{x : \rho_A(x) < 1\} \subset B \subset \{x : \rho_A(x) \leqslant 1\}$, then $\rho_A = \rho_B$.

It follows that an absorbing non-convex set may still have a sublinear gauge.

1.13. Let $A_1, \ldots, A_n$ be convex absorbing sets in the linear space X. Express the gauge of $\bigcap_1^n A_i$ in terms of the gauges $\rho_{A_1}, \ldots, \rho_{A_n}$.

1.14. Establish the following variant of the basic separation theorem: let A and B be convex subsets of a linear space such that both icr(A) and icr(B) are non-empty; then A and B can be separated by a hyperplane if and only if icr(A) $\cap$ icr(B) = $\varnothing$ (we exclude the trivial case that $A \cup B$ already lies in a hyperplane).

1.15. Show by example in $\mathbb{R}^2$ that the lemma in **4C** is not valid for strict separation.

1.16. Let $A \subset \mathbb{R}^n$. The *polar* of A is the convex set $A^o \equiv \{x \in \mathbb{R}^n : (a, x) \leqslant 1, a \in A\}$, where $(\cdot, \cdot)$ is the usual inner product on $\mathbb{R}^n$. If now A is convex and absorbing with gauge ρ_A, show that

$$\rho_A(x) = \sup\{(x, y) : y \in A^o\}, \qquad x \in \mathbb{R}^n.$$

1.17. Sometimes it is of interest to know when two disjoint convex sets can be (strongly) separated by a given hyperplane. The simplest case is the following: we are given points $a, b, p_1, p_2, \ldots, p_n$ in $\mathbb{R}^n$ with the p_i in general position. Let H be the hyperplane $\operatorname{aff}(\{p_1, \ldots, p_n\})$. Assuming that neither a nor b lies in H, show that H strongly separates a and b if and only if the determinants $\det(\tilde{a}, \tilde{p}_1, \ldots, \tilde{p}_n)$ and $\det(\tilde{b}, \tilde{p}_1, \ldots, \tilde{p}_n)$ have opposite signs, where for $x = (\xi_1, \ldots, \xi_n)$, $\tilde{x} \equiv$ the column vector $(\xi_1, \ldots, \xi_n, 1)^T$. (Consider the condition for the line segment (a, b) to intersect H.)

1.18. Show that a wedge P in a real linear space is reproducing if and only if the dual wedge P^+ is a cone. Show that this happens in particular when P has non-empty core.

1.19. a) Let X be the space $C(T)$ (resp. $C_o(T)$) of continuous real-valued functions (resp. that vanish at infinity) on the compact (resp. locally compact) Hausdorff space T. Determine in each case the core of the positive wedge in X (the natural pointwise vector ordering is assumed.)

b) Let X be the space of real $n \times n$ symmetric matrices and let P be the wedge of positive semi-definite matrices in X. Show that P is reproducing and determine its core.

1.20. Let X be the linear space of (the usual equivalence classes of) real-valued measurable functions on the interval $[0, 1]$, and let P be the wedge of ae non-negative functions in X. Show that $P^+ = \{\theta\}$. It follows that every non-trivial linear functional in X' maps P onto all of $\mathbb{R}$; consequently P cannot be separated from any set in X, and in particular, P contains no support points. (Suppose $\phi \in P^+$ but $\phi \neq \theta$. If χ_E is the characteristic function of the measurable set $E \subset [0, 1]$, then $\langle \chi_{[0,1]}, \phi \rangle \equiv \alpha > 0$ for otherwise ϕ would annull all bounded functions in X and hence ϕ would be the zero functional. Then one of $\langle \chi_{[0,\frac{1}{2})}, \phi \rangle$ and $\langle \chi_{[\frac{1}{2},1)}, \phi \rangle$ is at least $\alpha/2$, say the former, so that $\langle 4\chi_{[0,\frac{1}{2})}, \phi \rangle \geqslant 2\alpha$. Repeating this argument, an increasing sequence of functions in X can be constructed such that ϕ cannot be defined on the (measurable) limit of this sequence.)

1.21. Let A be a convex absorbing subset of the real linear space X, M a subspace of X, and $\phi \in M'$. If $\sup\{\phi(x) : x \in A \cap M\} \leqslant 1$, then there is an extension $\bar{\phi}$ of ϕ in X' such that $\sup\{\bar{\phi}(x) : x \in A\} \leqslant 1$. (Note that the inequality on ϕ is equivalent to $\phi \leqslant \rho_A | M$.) Use this result in the case where A is absolutely convex to prove the corollary in **6A**.

1.22. Where is the hypothesis that $\operatorname{cor}(P) \cap (P \cap M) \neq \varnothing$ used in the proof of the theorem in **6B**?

1.23. Let $X = \ell^p(d)$ where $1 \leqslant p < \infty$ and d is a cardinal number $\geqslant \aleph_o$ (**6C**). Show that the positive wedge P in X has no core, and give the details for the assertions made in the text regarding the support points of P.

1.24. Let M be a subspace of the linear space X, $\phi \in M'$, and let f, g be two convex functions on X. Then there is an extension $\bar{\phi}$ of ϕ in X' such that $-g \leqslant \bar{\phi} \leqslant f$ if and only if for all pairs $x, y \in X$ with $x - y \in M$ we have $\phi(x - y) \leqslant f(x) + g(y)$. In particular, the case $M = \{\theta\}$ gives a condition for the existence of a linear functional in X' interposed pointwise between $-g$ and f. (Let $g^{-}(x) \equiv g(-x)$ and set $h = f \,\square\, g^{-}$ (**3B**); then the condition $\phi(x - y) \leqslant f(x) + g(y)$ for $x - y \in M$ is equivalent to $\phi \leqslant h|M$.)

1.25. a) Let f be the p-norm (**3B**) on $\mathbb{R}^n$ $(1 \leqslant p < \infty)$. Find all subgradients of f at the origin. Same problem if $f(x) \equiv (Qx, x)$, where Q is a symmetric positive semi-definite $n \times n$ matrix.

b) Let f be a continuous convex function defined on $\mathbb{R}^n$ (continuity is actually automatic as we shall learn later). Identifying $\mathbb{R}^n$ with $\mathbb{R}^{n'}$ in the usual way, show that $\partial f(x_o)$ is a non-empty compact convex set in $\mathbb{R}^n$, for all $x_o \in \mathbb{R}^n$.

c) Let f be the convex function on the interval $[-1, 1]$ defined by $f(x) = -\sqrt{1 - x^2}$. Show that f is not subdifferentiable at the points ± 1.

1.26. Let X be a real linear space.

a) The equality system

$$\begin{aligned} \phi_1(x) &= c_1 \\ &\vdots \\ \phi_m(x) &= c_m \end{aligned}$$

for given $\phi_i \in X'$ and $c_i \in \mathbb{R}$ is consistent if and only if for any set $\{\alpha_1, \ldots, \alpha_m\} \subset R$, $\sum_1^m \alpha_i c_i = 0$ whenever $\sum_1^m \alpha_i \phi_i = \theta$. Write the matrix version of this assertion when $X = R^n$.

b) In **7C**, prove that there is $\phi \in X'$ satisfying

$$\begin{aligned} \phi(x_i) &= c_j, \qquad j \in J, \\ g &\geqslant \phi \end{aligned}$$

if and only if the Mazur-Orlicz condition (7.2) holds with no sign restriction on $\alpha_1, \ldots, \alpha_n$.

1.27. Suppose that the inequality system in **7B** is inconsistent. Show that there is some $\varepsilon > 0$ such that for every choice of $\delta_1, \ldots, \delta_n$ with each $\delta_i \geqslant -\varepsilon$ the system

$$\begin{aligned} \phi_1(x) &\geqslant c_1 + \delta_1, \\ &\vdots \\ \phi_n(x) &\geqslant c_n + \delta_n \end{aligned}$$

is inconsistent.

1.28. Let X be a real ordered linear space whose positive wedge has non-empty core. Given an index set J and sets $\{x_j : j \in J\} \subset X$, $\{c_j : j \in J\} \subset \mathbb{R}$, suppose that for some $j_o \in J$, $x_{j_o} \in \operatorname{cor}(P)$ and $c_{j_o} > 0$. The "moment problem" is to find $\phi \in P^+$ such that $\phi(x_j) = c_j$. Show that this problem is consistent if and only if for every finite set $\{j_1, \ldots, j_n\} \subset J$ such that $\sum_1^n \alpha_k c_{j_k} = 0$ for $\{\alpha_1, \ldots, \alpha_n\} \subset \mathbb{R}$, the vector $\sum_1^n \alpha_k x_{j_k}$ is not in cor(P). Interpret this result in the case where $X = C([0, 1])$ and for $j = 0, 1, 2, \ldots, x_j(t) \equiv t^j$.

1.29. In the course of proving the theorem in **7C** we needed to know that $\sup\{\psi(x): \psi \in G\} = g(x)$, for various $x \in X$. Prove this and show that the "sup" is actually a "max".

1.30. Let $f \in \text{Conv}(\mathbb{R}^n)$. If the n (two-sided) partial derivatives of f exist at $x_o \in \mathbb{R}^n$ then f has a gradient at x_0.

1.31. Given a convex absorbing set A in a real linear space X, $x_o \in A$ with $\rho_A(x_o) = 1$, $x \in X$, and $\alpha \in \mathbb{R}$ such that $-\tau_A(x_o, -x) \leqslant \alpha \leqslant \tau_A(x_o, x)$, show that there exists $\phi \in X'$ with $\phi(x) = \alpha$ such that the hyperplane $[\phi; 1]$ supports A at x_o.

1.32. Let A be the "unit max-ball" in $\mathbb{R}^n$, that is, $A = \{(\xi_1, \ldots, \xi_n) \in \mathbb{R}^n: |\xi_i| \leqslant 1, i = 1, \ldots, n\}$. Compute ρ_A, determine the smooth points of A, and find a formula for $\nabla\rho_A$ at such points.

1.33. Let A be a convex subset of the linear space X.

a) If $p \in X$, then $p \in \text{ext}(A)$ if and only if the condition $p \pm x \in A$ for $x \in X$ implies $x = \theta$.

b) If F is a finite subset of A and $x \in \text{ext}(A) \cap \text{co}(F)$, then $x \in F$.

c) The intersection of an A-extremal set with a half-space in X is A-semi-extremal.

d) If A is a wedge in X then it is actually a cone if and only if $\theta \in \text{ext}(A)$.

1.34. Let A be a compact convex set in $\mathbb{R}^n$ and let $E \subset A$.

a) Then E is semi-extremal if and only if for all compact sets $B \subset A\backslash E$ we have $\text{co}(B) \subset A\backslash E$.

b) If E is a closed semi-extremal subset of A, then E contains an extreme point of A. (This assertion is certainly true if E is convex; in general consider a maximal convex subset of E.)

1.35. Let A be a closed convex subset of $\mathbb{R}^n$. To compute the lineality space L_A and the recession cone C_A we express A as $\{x \in \mathbb{R}^n: \phi_j(x) \geqslant c_j, j \in J\}$ for suitable linear functionals ϕ_j, reals c_j and (countable) index set J. (Use **4C**.) Then

$$L_A = \{x \in \mathbb{R}^n: \phi_j(x) = 0, j \in J\}$$

and

$$C_A = \{x \in \mathbb{R}^n: \phi_j(x) \geqslant 0, j \in J\}.$$

1.36. Let B be a closed convex set in $\mathbb{R}^n$ with $C_B = \{\theta\}$. Show that B must be bounded.

1.37. A basis $\beta = \{u_1, \ldots, u_n\}$ for $\mathbb{R}^n$ is *orthonormal* if $(u_i, u_j) = \delta_{ij}$ for $1 \leqslant i, j \leqslant n$. The associated *lexicographic order* $<_\beta$ in $\mathbb{R}^n$ is then defined by $x <_\beta y$ if there is $k \leqslant n$ such that $\xi_i = \eta_i$ for $i \leqslant k - 1$ and $\xi_k < \eta_k$

(here $x \equiv (\xi_1, \ldots, \xi_n)$ and $y \equiv (\eta_1, \ldots, \eta_n)$). Now given a compact set $A \subset \mathbb{R}^n$ we define

$$A_{u_1} = \{x \in A : (u_1, x) = \max_{y \in A} (u_1, y)\},$$
$$A_{u_1 \cdots u_k} = \{x \in A_{u_1 \cdots u_{k-1}} : (u_k, x) = \max_{y \in A_{u_1 \cdots u_{k-1}}} (u_k, y)\},$$

inductively for $k = 2, \ldots, n$. Show that $\dim(A_{u_1 \cdots u_k}) \leqslant n - k$ so that $A_{u_1 \cdots u_n}$ is a singleton set, whose unique element is denoted by $e(A, \beta)$. Show that $e(A, \beta)$ is the lexicographic maximum of A with respect to the basis β, that $e(A, \beta)$ is an extreme point of A, and that every extreme point of A arises in this manner for some (not necessarily unique) basis β.

1.38. Let X be a real linear space, A a convex subset of X, and V an affine subspace of X. We say that V *supports* A if $A \cap V$ is a non-empty extremal subset of A. (This reduces to the definition in **6C** when V is a hyperplane.)
   a) Let E be a convex A-extremal set. Show that the affine subspace $\text{aff}(E)$ supports A.
   b) If X is partially ordered with positive wedge P, a subspace $M \subset X$ is an *order ideal* if the *order interval* $\{x : y \leqslant x \leqslant z\}$ lies in M whenever $y, z \in M$. Show that M is an order ideal if and only if M supports P.

1.39. Let A be a convex subset of the real linear space X.
   a) Show that $x \in \text{icr}(A)$ if and only if x lies in no proper A-extremal set.
   b) For $x \in A$, the *A-extremal hull* $E(x)$ of x is the intersection of all extremal sets containing x and the *A-facet* $F(x)$ of x is the largest convex subset of A containing x in its intrinsic core. Prove that $E(x) = F(x)$.

1.40. Let A and $K_o \subset A'$ be defined as in **8F**. Suppose that $\text{ext}(K_o)$ is non-void. Show that the algebra A must be self-adjoint, that is, $f \in A$ implies $\bar{f} \in A$ where $\bar{f}(\omega) \equiv \overline{f(\omega)}$, $\omega \in \Omega$. (A criterion for the self-adjointness of A is that $A = A_R + iA_R$.)

1.41. Let X be a linear space and $\{\phi_1, \ldots, \phi_n\}$ a linearly independent subset of X'. Prove the existence of a subset $\{x_1, \ldots, x_n\}$ of X such that

$$\phi_i(x_j) = \delta_{ij}$$

for $1 \leqslant i, j \leqslant n$.

Chapter II

Convexity in Linear Topological Spaces

We have made good progress in developing the algebraic aspects of our subject but the needs and applications of functional analysis require more powerful methods based on topological concepts. Thus, as our next step, we consider the result of imposing on a given linear space a "compatible topology". This is hardly a novel idea; indeed, several excellent books already exist which are devoted to a detailed investigation of the many ramifications of this notion. However, our treatment is less ambitious and more pragmatic, being shaped primarily by the necessities of our intended applications. These necessities require an understanding of the properties of topologies defined by one or more semi-norms on a linear space. They also require a well-rounded duality theory and it is interesting to discover that the maximal class of linear topologies which yields the requisite duality theory is precisely the class of topologies defined by a family of semi-norms.

§9. Linear Topological Spaces

In this section we give the definition and fundamental properties of "linear topologies". This notion is too general for our purposes and it will shortly be specialized by the introduction of a geometrical constraint on the basic neighborhoods.

A. Let X be a linear space over the (real or complex) scalar field $\mathbb{F}$. We recall that a *topology* on X is a family $\mathscr{T}$ of subsets of X, closed under the formation of finite intersections and arbitrary unions, and containing in particular the empty set $\varnothing$ and the whole space X. $\mathscr{T}$ is a *linear topology* on X if it is *compatible* with the linear space structure on X; that is, if both the linear space operations

$$\begin{aligned} &(x, y) \mapsto x + y, \qquad && x, y \in X \\ &(\alpha, x) \mapsto \alpha x, \qquad && \alpha \in \mathbb{F}, \quad x \in X \end{aligned} \tag{9.1}$$

are continuous on their respective domains $X \times X$ and $\mathbb{F} \times X$. Here these product spaces are given the usual product topologies determined by $\mathscr{T}$ and the natural topology on $\mathbb{F}$. In this case the pair $(X, \mathscr{T})$ is a *linear topological space*. However, once $\mathscr{T}$ is clearly understood, it is convenient to just say that X is a linear topological space.

For any non-zero scalar $\alpha_0 \in \mathbb{F}$ and vector $x_0 \in X$ the map $x \mapsto x_0 + \alpha_0 x$ is a homeomorphism of X with itself, so that, in particular, a linear topological space is homogeneous. Consequently, the topological structure of X about any point is determined by a base of neighborhoods about the origin θ. For if $\mathscr{U}$ is a base of θ-neighborhoods, then the sets $x + U$ $(U \in \mathscr{U})$ constitute a base of x-neighborhoods. For short, we say that a base of θ-neighborhoods is a *local base* in X. The next result summarizes the fundamental working properties of a local base.

Lemma. *Let X be a linear topological space and $\mathscr{U}$ a local base in X. Then*

a) *every $U \in \mathscr{U}$ is absorbing;*
b) *if $U \in \mathscr{U}$ there exists a balanced θ-neighborhood V such that $V + V \subset U$;*
c) *if $A \subset X$ then $\bar{A}$, the closure of A in X, equals $\cap\{A + U : U \in \mathscr{U}\}$;*
d) *the topology on X is Hausdorff if and only if $\cap\{U : U \in \mathscr{U}\} = \{\theta\}$.*

Proof. Parts a) and b) follow directly from the assumed continuity of the linear space operations (9.1). Thus, given $x \in X$ and $U \in \mathscr{U}$, we have that $0x = \theta$, so there must exist an interval $(-\delta, \delta)$ for which $tx \in U$ if $-\delta < t < \delta$; this proves a). Next, since $\theta + \theta = \theta$ and addition is continuous, there is certainly some $W \in \mathscr{U}$ for which $W + W \subset U$. To complete the proof of b) it will suffice to find a balanced θ-neighborhood $V \subset W$. But the map $(\alpha, x) \mapsto \alpha x$ is continuous at $(0, \theta)$ so there is $\delta > 0$ and $N \in \mathscr{U}$ such that $\alpha N \subset W$ if $|\alpha| < \delta$. Now we can put $V = \cup\{\alpha N : |\alpha| < \delta\}$ and this V meets the requirements of b).

c) Let $A \subset X$ and let $B = \cap\{A + U : U \in \mathscr{U}\}$. If $x \in \bar{A}$ and $U \in \mathscr{U}$ then, choosing V as in b), we see that the x-neighborhood $x + V$ intersects A and so $x \in A - V = A + V \subset A + U$. Thus $x \in B$ and we have shown that $\bar{A} \subset B$. However, if $x \in B$ a completely similar argument shows that every x-neighborhood intersects A and so $x \in \bar{A}$. Therefore, $B = \bar{A}$.

d) If the topology is in fact Hausdorff and we choose any $x \neq \theta$ in X, there is some θ-neighborhood V such that $x \notin V$; consequently there is some $U \in \mathscr{U}$ such that $x \notin U$. This proves that $x \notin \cap\{U : U \in \mathscr{U}\}$. Conversely, if this intersection contains only the zero vector θ and if we choose $x \neq y$ in X, then there is some $U \in \mathscr{U}$ for which $x - y \notin U$. Selecting V as in b) we then see that $x + V$ and $y + V$ are disjoint neighborhoods of x and y, thus proving that our topology is Hausdorff. □

Theorem. *A linear topological space X has a local base consisting of closed balanced sets.*

Proof. First we note that the closure of any balanced set $A \subset X$ is again balanced. That is, if $|\lambda| \leqslant 1$ then $\lambda\bar{A} \subset \bar{A}$. To see this, choose any such λ and any $x \in \bar{A}$; if N is any neighborhood of λx we wish to show that N intersects A. By continuity there is an x-neighborhood V such that $\lambda V \subset N$. Since $x \in \bar{A}$, there exists $a \in A \cap V$. Consequently, $\lambda a \in A \cap N$, and so $\bar{A}$ is balanced if A is.

Now let W be any θ-neighborhood in X. Applying the lemma we can choose θ-neighborhoods U and V such that U is balanced and $U \subset V \subset \bar{V} \subset W$. Then $\bar{U} \subset \bar{V} \subset W$ and $\bar{U}$ is balanced. □

At this point we have obtained a few elementary properties of linear topologies. Many more connections between the algebraic and topological structure are collected together in exercise 2.1; they will be used freely as we go on. Let us also take note of a kind of converse to some of these facts. Suppose that in a linear space X we are given a family $\mathscr{U}$ of balanced absorbing sets such that whenever $U \in \mathscr{U}$ there exists $V \in \mathscr{U}$ with $V + V \subset U$ and whenever $U, V \in \mathscr{U}$ there exists $W \in \mathscr{U}$ with $W \subset U \cap V$. Then $\mathscr{U}$ is a local base for a unique linear topology $\mathscr{T}$ on X. Indeed, $\mathscr{T}$ consists of those sets $V \subset X$ such that, for every $x \in V$, there exists $U \in \mathscr{U}$ with $x + U \subset V$.

Clearly, real or complex Euclidean n-space is a linear topological space for every n. Other more substantial (infinite dimensional) examples will follow shortly, as we learn systematic procedures for constructing linear topologies.

B. Most of our subsequent interest in linear topological spaces will tend to emphasize the effects of the topology on the linear structure. This may be expected in view of our previous developments in Chapter I. However, momentarily it is of interest to adopt the opposing view: what kinds of topological spaces are obtained via the imposition of a linear topology on a given linear space? Our main point is that such topological spaces must be very "smooth".

We have already noted that a linear topological space X must be *homogeneous*, that is, given $x_1, x_2 \in X$, there exists a homeomorphism $h: X \to X$ such that $h(x_1) = x_2$. (Indeed, we can take h to be the translation $x \mapsto x + (x_2 - x_1)$.) Further, if the linear topology on X is Hausdorff then X must be regular (even completely regular; see exercise 2.6 for an important special case.) This follows from **9A** using that X contains a local base of closed sets.

Let us recall that a topological space X is *contractible* if the identity map on X is *nullhomotopic*. This means intuitively that X can be continuously shrunk to a point, and precisely that there is a continuous function (a homotopy) $h:[0, 1] \times X \to X$ such that $h(0, x) = x$ $(x \in X)$ and $h(1, x)$ is constant. Similarly, X is said to be *locally contractible* if every point $x \in X$ has a neighborhood base consisting of sets contractible to x. These conditions entail very strong connectivity properties of X. Thus if X is (locally) contractible then X is (locally) path connected and (locally) simply connected. It is clear that a linear topological space is both contractible and locally contractible, since the map $(\alpha, x) \mapsto \alpha x$ is continuous by definition.

Finally, some linear topological spaces of considerable importance in our subject (conjugate spaces with the "weak-star" topology; defined later) turn out to be expressible as a countable union of compact subsets. Such spaces then have the Lindelöf covering property (every covering by open subsets admits a countable subcovering). It is known that any regular Lindelöf space must be *paracompact* (every open covering has an open neighborhood-finite refinement), and in particular normal. The paracompactness property is a weak substitute for metrizability, which may or may not be available depending on the nature of a particular linear topology.

C. Although we have as yet established no results of any substance concerning linear topological spaces, it is nevertheless already possible to give an interesting application and we digress briefly to do so. The result to be established concerns simultaneous approximation and interpolation in abstract setting; upon suitable specialization it yields a variety of refinements of known classical approximation principles such as the Stone-Weierstrass theorem.

The main difficulty in the proof may be localized to the following lemma. We consider a convex subset A of a real linear topological space X.

Lemma. *Let ϕ be a continuous linear functional in X'. If A is dense in X then $A \cap \ker(\phi)$ is dense in $\ker(\phi)$.*

Proof. The hyperplane $H \equiv \ker(\phi)$ is closed (since ϕ is continuous) and the half-spaces $H^+ \equiv \{x \in X : \phi(x) > 0\}$, $H^- \equiv \{x \in X : \phi(x) < 0\}$ are open. Let $\mathscr{U}$ be a local base in X consisting of balanced sets (**9A**). Fix $x \in H$ and $U \in \mathscr{U}$. Now the dense set A intersects the sets $(x + U) \cap H^{+}$ at points $p^{\pm}$. Since $\phi(p^-) < 0 < \phi(p^+)$, there exists t, $0 < t < 1$, such that $tp^- + (1 - t)p^+ \in H$. We now have

$$tp^- + (1 - t)p^+ \in A \cap H \cap (x + 2U).$$

This argument shows that $A \cap H$ intersects every neighborhood of x. □

Now we come to the main result, known as the "Singer-Yamabe theorem". The idea is that if it is possible to approximate points in X from the convex set A (that is, if A is dense in X), then it is also possible to approximate while simultaneously satisfying a number of linear interpolatory conditions.

Theorem. *Let A be a dense convex subset of X and let $\phi_1, \ldots, \phi_n$ be continuous linear functionals in X'. Given any $x_0 \in X$ and any x_0-neighborhood V, there exists a point $z \in A$ such that $z \in V$ and $\phi_i(z) = \phi_i(x_0)$, $i = 1, \ldots, n$.*

Proof. After replacing A by its translate $A - x_0$ we may assume that $x_0 = \theta$. Let $M_j = \{x \in X : \phi_1(x) = \cdots = \phi_j(x) = 0\}$. Then $M_0 \equiv X \supset M_1 \supset M_2 \supset \cdots \supset M_n$ and either $M_{j+1} = M_j$ or else M_{j+1} is a closed hyperplane in M_j, for each j. Now, $A \cap M_j$ is convex, and if it is dense in M_j then by the lemma $A \cap M_{j+1}$ is dense in M_{j+1}. Thus n repeated applications of the lemma establish the desired result. □

Let us consider a particular instance of this theorem. Let $X = C(\Omega)$, the usual space of real-valued continuous functions defined on the compact space Ω. We define a linear topology (the *topology of uniform convergence on Ω*) by taking as a local base the sets

$$U_n = \{x \in C(\Omega) : \sup_{t \in \Omega} |x(t)| \leqslant 1/n\}, \tag{9.2}$$

for $n = 1, 2, \ldots$. This topology takes its name from the fact that a sequence of continuous functions coverges in this topology exactly when it converges uniformly on Ω in the usual sense. Now if μ is a finite signed measure on the Borel sets in Ω, it defines a linear functional $\phi \in X'$ by the rule

$$\phi(x) = \int_{\Omega} x \, d\mu, \qquad x \in C(\Omega).$$

Such functionals are continuous with respect to the topology of uniform convergence on Ω since

$$|\phi(x) - \phi(y)| = \left|\int_\Omega (x - y)d\mu\right| \leqslant \sup_{t\in\Omega} |x(t) - y(t)|\,|\mu|(\Omega),$$

where $|\mu|(\Omega)$ is the total variation of μ on Ω.

Now suppose that A is a subalgebra of $C(\Omega)$ that contains the constant functions and that $\mu_1, \ldots, \mu_n$ are Borel measures on Ω as above. Then the following assertions are equivalent:

i) for each $x_0 \in C(\Omega)$ and each $\varepsilon > 0$, there exists $y \in A$ such that

$$\sup_{t\in\Omega} |x_0(t) - y(t)| < \varepsilon$$

and

$$\int_\Omega x_0\, d\mu_i = \int_\Omega y\, d\mu_i, \qquad i = 1, \ldots, n;$$

ii) A separates the points of Ω.

The equivalence here follows directly from the Singer-Yamabe theorem and the Stone-Weierstrass theorem (**22E**) which asserts that A is dense in $C(\Omega)$ (in the topology of uniform convergence) exactly when A separates the points of Ω.

Note in particular that if each μ_i is a positive Borel measure concentrated at a point $t_i \in \Omega$ then $\int_\Omega x\, d\mu_i = \alpha_i x(t_i)$, for some $\alpha_i > 0$ (for details, see **22E**). Thus if A is a separating subalgebra of $C(\Omega)$ containing the constant functions we can uniformly approximate any given function x in $C(\Omega)$ by a function in A that agrees with x at a finite number of points in Ω.

If we take instead our space X to consist of all complex-valued continuous functions on a compact space Ω, then the preceding extension of the Stone-Weierstrass theorem remains valid, provided that the algebra A is also self-adjoint (**8F**).

D. We now establish some facts about products, sums, and quotients of linear topological spaces. These facts allow us to systematically construct new spaces from given ones; they represent a continuation of the development begun in **1C**.

First, let $\{X_\alpha : \alpha \in I\}$ be a family of linear topological spaces over the same field and let X be the product space $\prod_\alpha X_\alpha$. Following the classical Tychonov construction of the product topology we can define a local base in X. To do so, let $\mathscr{U}_\alpha$ be a local base in X_α. For each finite subset $J \subset I$ we choose a neighborhood $V_\alpha \in \mathscr{U}_\alpha$ for $\alpha \in J$ and define

(9.3) $$U_J = \{x \in X : x_\alpha \in V_\alpha, \alpha \in J\}.$$

Then the collection of all such U_J is a local base for the product topology on X. It is easy to see from the characterization of Hausdorff linear topologies in **9A** that the product topology on X is Hausdorff exactly when the given linear topology on each space X_α is Hausdorff.

When $X_\alpha = Y$ for each α the resulting product space is denoted Y^I; by definition, it consists of all Y-valued functions defined on the set I. In this case the product topology is often called the *topology of pointwise convergence* on I, because a net $\{f_\delta : \delta \in D\}$ of functions in Y^I converges in the product topology to $f \in Y^I$ exactly when $\lim\{f_\delta(\alpha) : \delta \in D\} = f(\alpha)$, for every $\alpha \in I$.

Consider next the situation where a linear topological space X is the algebraic direct sum of subspaces M and N: $X = M \oplus N$. If $P: X \to M$ is the associated (linear) projection along N then P is an open map, that is, if $\mathcal{O}$ is an open subset of X then $P(\mathcal{O})$ is open in M. (Because, $\mathcal{O} + N$ is open and $P(\mathcal{O}) = P(\mathcal{O} + N) = (\mathcal{O} + N) \cap M$.) We say that X is the *topological direct sum* of M and N if the map $(m, n) \mapsto m + n$ from the product space $M \times N$ to X is a homeomorphism (it is clearly a continuous isomorphism in all cases). For this to happen it is necessary and sufficient that the projection P be continuous. In turn, for P to be continuous it is evidently necessary (but not sufficient!) that the subspaces M and N be closed in X.

Finally, let M be a subspace of the linear topological space X and let $Q_M: X \to X/M$ be the quotient map (**1H**). The *quotient topology* on X/M is the strongest topology on X/M for which Q_M is continuous. This means that a set $A \subset X/M$ is considered to be open exactly when its inverse image $Q_M^{-1}(A)$ is open in X. The quotient topology is a linear topology on X/M: indeed, a local base for it is the Q_M-image of a given local base in X. When X/M is given the quotient topology, the quotient map Q_M is both continuous and open.

Theorem. *Let M be a subspace of X. The quotient topology on X/M is Hausdorff if and only if M is closed in X.*

Proof. If X/M is Hausdorff the zero-vector in X/M is a closed set and its Q_M-inverse image must be closed. But $Q_M^{-1}(\theta) = M$. Conversely, assume that M is closed. We will show that given a non-zero vector $x + M$ in X/M there is a closed neighborhood of $x + M$ that does not contain θ, and this will prove that the quotient topology is Hausdorff. Now, since $x + M \neq \theta$, $x \notin M$, so that $X \backslash M$ is an open x-neighborhood in X. Since Q_M is an open map, $Q_M(X \backslash M)$ is an open $(x + M)$-neighborhood that does not contain θ. Since the quotient topology is a linear topology, we conclude by **9A** that there is a closed $(x + M)$-neighborhood within $Q_M(X \backslash M)$. □

E. We turn next to some finite dimensional considerations. Let us say that two topological linear spaces over the same field are *isomorphic* (*linearly homeomorphic*) if there exists an algebraic isomorphism between them which is at the same time a homeomorphism. Such spaces cannot be distinguished from one another by examination of their algebraic-topological structure.

Now let $\mathbb{F}$ be either the real field $\mathbb{R}$ or the complex field $\mathbb{C}$. For each $n = 1, 2, \ldots$, $\mathbb{F}^n$ is an n-dimensional Hausdorff topological linear space over $\mathbb{F}$ in its natural (product) topology (**9D**). We claim that it is the only such space.

Theorem. *Let X be an n-dimensional Hausdorff topological linear space over the field $\mathbb{F}$. If $\{x_1, \ldots, x_n\}$ is a basis for X then the map $T:\mathbb{F}^n \to X$ defined by*

$$T(\lambda_1, \ldots, \lambda_n) = \sum_{i=1}^{n} \lambda_i x_i$$

is an isomorphism between $\mathbb{F}^n$ and X.

Proof. T is well known to be an algebraic isomorphism and an obvious induction, based on the axioms (**9A**) for a linear topological space, shows that T is also continuous. Let B be the unit 2-ball (**3C**) in $\mathbb{F}^n$; that is, $B = \{v = (\xi_1, \ldots, \xi_n) \in \mathbb{F}^n : \|v\|_2^2 \equiv \sum_1^n |\xi_i|^2 \leqslant 1\}$. In order to show that T is an open map (and hence a homeomorphism) it suffices to show that $T(B)$ contains a θ-neighborhood in X. Now the boundary $S \equiv \{v \in \mathbb{F}^n : \|v\|_2 = 1\}$ of B is compact in $\mathbb{F}^n$ and so $T(S)$ is a compact set in X that does not contain θ. Hence $X \backslash T(S)$ is a θ-neighborhood in X and so contains a balanced θ-neighborhood U. We claim that $U \subset T(B)$. Because, if $x \notin T(B)$, then $\|T^{-1}(x)\|_2 > 1$, whence

$$x/\|T^{-1}(x)\|_2 = T(T^{-1}(x)/\|T^{-1}(x)\|_2) \in T(S),$$

and so x cannot belong to U. □

This theorem admits several corollaries two of which follow below and two of which appear as exercises. First note this implication: if M is any finite dimensional subspace of a Hausdorff linear topological space X, then M is closed in X. This is because M is topologically complete, being homeomorphic to the complete metric space $\mathbb{F}^n$.

Corollary 1. *Let M and N be closed subspaces of the Hausdorff linear topological space X with N of finite dimension. Then $M + N$ is a closed subspace of X.*

Proof. Let $Q_M : X \to X/M$ be the quotient map. The subspace $Q_M(N)$ is a finite dimensional subspace of X/M, hence closed in X/M. Consequently, its inverse image $Q_M^{-1}(Q_M(N)) = M + N$ is closed. □

Corollary 2. *Let X and Y be linear topological spaces with X Hausdorff and finite dimensional. Then any linear map $R:X \to Y$ is necessarily continuous.*

Proof. Let $T:\mathbb{F}^n \to X$ be an isomorphism. Then the map $R \circ T:\mathbb{F}^n \to Y$ must be of the form $(\lambda_1, \ldots, \lambda_n) \to \lambda_1 y_1 + \cdots + \lambda_n y_n$ for suitable vectors $y_1, \ldots, y_n \in Y$. Such a map is surely continuous (as noted in the proof of the theorem). Consequently, $R = (R \circ T) \circ T^{-1}$ is continuous. □

F. We continue our finite dimensional considerations by establishing a characterization of finite dimensional linear topological spaces. We precede this result by some new terminology which also is needed for later work.

Let X be a linear topological space over the field $\mathbb{F}$ with local base $\mathscr{U}$. A sequence $\{x_n\} \subset X$ is *bounded* if $\lambda_n x_n \to \theta$ whenever $\lambda_n \to 0$ in $\mathbb{F}$. A set

$A \subset X$ is *bounded* if every sequence in A is bounded. According to exercise 2.3 this happens exactly when A is absorbed by every neighborhood in $\mathscr{U}$ (that is, given $U \in \mathscr{U}$, $\lambda A \subset U$ for sufficiently small $|\lambda|$). A set $A \subset X$ is *totally bounded* if for every $U \in \mathscr{U}$ there is a finite subset $B \subset X$ such that $A \subset B + U$. Clearly a compact set is totally bounded and every totally bounded set is bounded. The converse of this last remark is sometimes valid; see exercise 2.3. In particular, the bounded sets in $\mathbb{F}^n$ are exactly the relatively compact sets there.

Finally, a subset of X is *fundamental* in X if its linear hull is dense in X. Evidently, X has finite dimension exactly when it contains a finite fundamental subset (**9E**). We preface the main result by an abstract form of "Riesz's lemma".

Lemma. *Let A be a bounded subset and M a closed subspace of X. If there exists $\lambda \in \mathbb{F}$, $|\lambda| < 1$, such that $A \subset M + \lambda A$, then $A \subset M$.*

Proof. For any $U \in \mathscr{U}$ there is an integer n such that $\lambda^n A \subset U$. Hence $A \subset M + \lambda^n A \subset M + U$, and so $A \subset \bar{M} = M$. □

It follows immediately that if A is both bounded and fundamental, if $|\lambda| < 1$, and if $A \subset B + \lambda A$, then the set B is fundamental.

Theorem. *The linear topological space X is finite dimensional if and only if it contains a totally bounded neighborhood.*

Proof. Suppose that some $U \in \mathscr{U}$ is totally bounded. Then U is both bounded and fundamental. For any scalar λ with $0 < |\lambda| < 1$ there is a finite set B such that $U \subset B + \lambda U$; the preceding remark now shows that B must be fundamental. □

Thus we see that a locally compact (Hausdorff) linear topological space must be finite dimensional.

§10. Locally Convex Spaces

In this section we specialize the very general notion of linear topology developed in the preceding section. The reason for doing so is that the linear topology axioms are simply too weak to yield a useful duality theory. Thus in order to be able to link up the present topological considerations with the powerful linear-geometric theory of Chapter I, we find it necessary to impose an additional but crucial geometric condition on our linear topologies, namely that the topology be determined by convex neighborhoods.

A. A linear topological space X is a *locally convex space* if it contains a local base consisting of convex θ-neighborhoods. This condition implies that any x-neighborhood ($x \in X$) contains a convex x-neighborhood. The definitions (9.2) and (9.3) of local bases for $C(\Omega)$ and for $\prod_\alpha X_\alpha$ show that these spaces are locally convex (provided in the latter case that each X_α is locally convex and the local base $\mathscr{U}_\alpha$ in X_α contains only convex sets).

A *barrel* in a linear topological space X is a closed absolutely convex absorbing subset of X. It follows from **9A** and the preceding definition that any locally convex space has a local base consisting of barrels. Note that we do not claim that every barrel is a θ-neighborhood; only that enough of them are so as to define the topology. Clearly intersections and positive multiples of barrels are again barrels. In general, an absolutely convex absorbing set A in X is a barrel exactly when its gauge ρ_A is lower semi-continuous on X.

The next lemma comprises the *geometric description* of locally convex topologies; the analytical description follows momentarily.

Lemma. *Let $\mathscr{U}$ be a family of absolutely convex absorbing sets in a linear space X.*

a) *Suppose that, given $U, V \in \mathscr{U}$, there exists $W \in \mathscr{U}$ such that $W \subset U \cap V$, and that $\alpha U \in \mathscr{U}$ whenever $\alpha \neq 0$. Then $\mathscr{U}$ is a local base for a unique locally convex linear topology on X.*

b) *Whether or not $\mathscr{U}$ satisfies the conditions of* a), *there is a weakest linear topology on X such that every set in $\mathscr{U}$ is a θ-neighborhood. A local base for this topology consists of all positive multiples of finite intersections of the members of $\mathscr{U}$; in particular, the topology is locally convex.*

Part b) evidently follows from part a). The proof of part a) is omitted, being straightforward but tedious. Let us just note that the unique topology whose local base is $\mathscr{U}$ consists of those sets $V \subset X$ such that, for each $x \in V$, there exists $U \in \mathscr{U}$ with $x + U \subset V$.

B. In **3D** it was observed that the semi-norms on a linear space X are exactly the gauges of absolutely convex absorbing subsets of X. For any semi-norm ρ on X we let $U_\rho \equiv \{x \in X : \rho(x) \leqslant 1\}$ be the *ρ-unit ball* in X.

Lemma. *Let X be a linear topological space and B a barrel in X. Then*

a) *the gauge ρ_B is the only semi-norm on X for which $U_\rho = B$.*

b) *ρ_B is continuous on X if and only if B is a θ-neighborhood.*

Proof. a) We know from exercise 2.4 that $B \subset U_{\rho_B}$. Suppose that $x \in U_{\rho_B}$; then for every $\varepsilon > 0$ there is $y_\varepsilon \in B$ such that $x = (1 + \varepsilon)y_\varepsilon$. Clearly $y_\varepsilon \to x$ as $\varepsilon \downarrow 0$ so that $x \in \bar{B} = B$. Thus $B = U_{\rho_B}$. Now if also $B = U_\rho$ for some semi-norm ρ then we would have $\rho(x) \leqslant 1 \Leftrightarrow \rho_B(x) \leqslant 1$. For any fixed $y \in X$ and $\varepsilon > 0$, put $x = y/(\rho(y) + \varepsilon)$ and then put $x = y/(\rho_B(y) + \varepsilon)$ to obtain $\rho_B(y) \leqslant \rho(y) + \varepsilon$ and $\rho(y) \leqslant \rho_B(y) + \varepsilon$.

b) If ρ_B is continuous then B contains the open θ-neighborhood $\{x \in X: \rho_B(x) < 1\}$. Conversely, if B is a θ-neighborhood, for each $\varepsilon > 0$, $\rho_B(\varepsilon B)$ lies in the interval $[0, \varepsilon]$. This proves that ρ_B is continuous at θ. Since

$$|\rho_B(x) - \rho_B(y)| \leqslant \rho_B(x - y),$$

it now follows that ρ_B is (uniformly) continuous on X. □

One implication of this lemma is that the unit balls of all the continuous semi-norms on a given locally convex space X constitute a local base of barrels in X. However, this would be an inordinately large local base. It is more interesting (and practical) to be able to define the topology on X with as few semi-norms as possible. The following details constitute the *analytic description* of locally convex topologies.

Let $\mathcal{N}$ be a family of continuous semi-norms on X. We say that $\mathcal{N}$ is a *base of continuous semi-norms* on X if, for any continuous semi-norm σ on X, there exist $t > 0$ and $\rho \in \mathcal{N}$ such that $\sigma \leqslant t\rho$ (equivalently, $U_\rho \subset tU_\sigma$). Thus if $\mathcal{N}$ is a base of continuous semi-norms on X, then $\{tU_\rho : \rho \in \mathcal{N}, t > 0\}$ is a local base in X.

The most common way of specifying a locally convex topology by means of semi-norms is to make use of **10A**. Suppose that we are given a family $\mathcal{N}$ of semi-norms on a linear space X. Then there is a weakest locally convex topology $\mathcal{T}$ on X for which all the semi-norms in $\mathcal{N}$ are continuous. $\mathcal{T}$ is said to be *generated* by $\mathcal{N}$ and we often call $\mathcal{T}$ the *$\mathcal{N}$-topology*. A local base for $\mathcal{T}$ consists of all positive multiples of finite intersections of ρ-unit balls for $\rho \in \mathcal{N}$ and a base of continuous semi-norms on $(X, \mathcal{T})$ is given by the collection of suprema of the finite subsets of $\mathcal{N}$.

Another perspective on the $\mathcal{N}$-topology is gained from the following considerations. Each semi-norm $\rho \in \mathcal{N}$ defines a pseudometric d_ρ on X by

$$d_\rho(x, y) = \rho(x - y).$$

That is, the relations

$$\begin{aligned} d_\rho(x, y) &\geqslant 0, \\ d_\rho(x, y) &= d_\rho(y, x), \\ d_\rho(x, y) &\leqslant d_\rho(x, z) + d_\rho(y, z) \end{aligned}$$

hold for all $x, y, z \in X$. (The pseudo-metric d_ρ interacts with the linear structure on X by virtue of being *translation invariant*, that is, the relation

$$d_\rho(x + z, y + z) = d_\rho(x, y)$$

holds for all $x, y, z \in X$.) Each pseudo-metric d_ρ defines in the usual way a pseudo-metric topology on X and the $\mathcal{N}$-topology is simply the least upper bound of these d_ρ-topologies for $\rho \in \mathcal{N}$. The point of these remarks is to suggest that a property of the $\mathcal{N}$-topology is likely to be the conjunction of the corresponding properties in all the d_ρ-topologies. This idea is made forcefully clear by the next theorem which summarizes the main operating characteristics of $\mathcal{N}$-topologies.

Theorem. *Let $\mathcal{N}$ be a family of semi-norms on the linear space X.*

a) *The $\mathcal{N}$-topology is Hausdorff if and only if for each non-zero $x \in X$ there exists $\rho \in \mathcal{N}$ for which $\rho(x) > 0$.*

b) *A net $\{x_\delta : \delta \in D\} \subset X$ converges to $x \in X$ if and only if $\lim_\delta \rho(x_\delta - x) = 0$, for every $\rho \in \mathcal{N}$.*

c) *A subset A of X is bounded in the $\mathcal{N}$-topology if and only if A has finite ρ-diameter for every $\rho \in \mathcal{N}$.*

d) *A subset A of X is totally bounded in the $\mathcal{N}$-topology if and only if for every $\varepsilon > 0$ and every $\rho \in \mathcal{N}$ there is a finite (ε, ρ)-net in A.*

Let us clarify some terminology here. The *ρ-diameter* of the set A is the number $\sup\{\rho(x - y): x, y \in A\}$. An *$(\varepsilon, \rho)$-net* in A is a subset B of A such that for every $x \in A$ there exists $y \in B$ with $\rho(x - y) < \varepsilon$. Thus, parts c) and d) may be reworded: the set $A \subset X$ is bounded (resp., totally bounded) in the $\mathcal{N}$-topology exactly when A is bounded (resp., precompact) in each d_ρ-topology ($\rho \in \mathcal{N}$). The proof of this theorem is straightforward and is left to the exercises, as is a criterion for metrizability of an $\mathcal{N}$-topology.

C. Recall (**3D**) that a norm ρ on a linear space X is a semi-norm with trivial kernel: $\rho(x) = 0$ only if $x = \theta$. Geometrically, ρ is the gauge of an absolutely convex absorbing set which contains no proper subspace. A locally convex space is *normable* if its topology is the $\mathcal{N}$-topology for $\mathcal{N}$ consisting of a single norm.

Theorem. *A locally convex space X is normable if and only if there exists a proper bounded θ-neighborhood in X.*

Proof. If the topology on X is defined by a norm ρ then the ρ-unit ball (**10B**) is such a θ-neighborhood. Conversely, if such a θ-neighborhood exists, it must contain a barrel B. The gauge ρ_B must be a norm since $\rho_B(x) = 0$ implies $nx \in B$ for $n = 1, 2, \ldots$, so that $x = \theta$ or else B would not be bounded. Since ρ_B is continuous (**10B**) the ρ_B-topology is weaker than the original topology. But since B is bounded, the sets $\left\{\frac{1}{n} B : n = 1, 2, \ldots\right\}$ form a local base in X; hence the ρ_B-topology is exactly the original topology on X. □

Corollary. *Let $\{X_\alpha : \alpha \in I\}$ be a family of normable spaces over the same field. Then the product space $\prod_\alpha X_\alpha$ is normable if and only if the index set I is finite.*

The proof results directly from the theorem and exercise 2.3b.

D. In practice a normable locally convex space is specified analytically rather than geometrically as in **10C**. That is, there is given a pair (X, ρ) consisting of a linear space X and a norm ρ on X. X is then considered to be topologized by the $\mathcal{N}$-topology with $\mathcal{N} = \{\rho\}$; the resulting locally convex space is called a *normed linear space*. Since ρ is a norm the pseudo-metric d_ρ (**10B**) is actually a metric on $X \times X$. The study of normed linear spaces and the interplay between the resulting algebraic-geometrical-topological structure is one of the major objects of this book.

Let us now consider some prototypical examples of normed linear spaces. In doing so and in subsequent work we shall adhere to the tradition of writing a norm as $\|\cdot\|$.

Example 1. (Spaces of continuous functions). Let Ω be a topological space and let $\mathbb{F}$ be either $\mathbb{R}$ or $\mathbb{C}$. Then the space $C_b(\Omega, \mathbb{F})$ is the linear space of all bounded continuous $\mathbb{F}$-valued functions f on Ω normed by the *uniform norm*

$$(10.1) \qquad \|f\|_\infty \equiv \sup\{|f(t)|: t \in \Omega\}.$$

Convergence in the associated metric is uniform convergence on Ω (as in **9C**). The case where Ω is discrete is of interest here; in this case $C_b(\Omega, \mathbb{F})$ comprises the space of all bounded $\mathbb{F}$-valued functions on Ω. □

Example 2. (Spaces of integrable functions). Let (Ω, Σ, μ) be a positive measure space and let $p \geqslant 1$. The space $L^p \equiv L^p(\Omega, \mu, \mathbb{F})$ is the linear space of all (equivalence classes of) p-th power μ-integrable functions $f: \Omega \to \mathbb{F}$ normed by the *p-norm*

$$(10.2) \qquad \|f\|_p \equiv \left(\int_\Omega |f(\cdot)|^p \, d\mu\right)^{1/p}.$$

The subadditivity of the p-norm is not entirely obvious; it is in fact equivalent to Minkowski's inequality in measure theory. Convergence in the associated metric is called *convergence in the mean of order p.* Note that this example subsumes the earlier case of the p-norm on $\mathbb{F}^n$ (**3C**). □

Example 3. (Spaces of measures). Let (Ω, Σ) be a measurable space. Then the space $\mathscr{M}(\Omega, \Sigma, \mathbb{F})$ is the linear space of all $\mathbb{F}$-valued countably additive set functions μ defined on the σ-algebra Σ, normed by

$$(10.3) \qquad \|\mu\|_v \equiv \sup\left\{\sum_{i=1}^{\infty} |\mu(A_i)| : \{A_i\} \text{ partitions } \Omega\right\}.$$

To say that $\{A_i\}$ partitions Ω means that $\{A_i\}$ is a sequence of pairwise disjoint subsets of Σ whose union is Ω. As is well known, if in the right hand side of (10.3) we replace Ω by an arbitrary set $A \in \Sigma$, then the formula defines a finite positive measure $|\mu|$ on Σ, called the *total variation* (*measure*) of μ. Thus $\|\mu\|_v \equiv |\mu|(\Omega)$. □

Example 4. (Spaces of Lipschitz functions). Let (Ω, d) be a metric space. Then the space $\text{Lip}(\Omega, d, \mathbb{F})$ is the linear space of all bounded $\mathbb{F}$-valued functions on Ω which satisfy a Lipschitz condition on Ω (in the sense that

$$(10.4) \qquad \|f\|_d \equiv \sup\{|f(s) - f(t)|/d(s, t): s, t \in \Omega, s \neq t\}$$

is finite), normed by

$$(10.5) \qquad \|f\|_L \equiv \max\{\|f\|_\infty, \|f\|_d\}.$$

Convergence in the associated metric is much stronger than uniform convergence; for an illustration, see exercise 2.9. □

Example 5. (Spaces of analytic functions). Let Ω be a subset of $\mathbb{C}$ consisting of a simple closed curve and its interior; for example, let $\Omega = \{z \in \mathbb{C}: |z| \leqslant 1\}$. Then the space $A(\Omega)$ is the linear space of all continuous complex-valued functions on Ω which are analytic in the interior of Ω, normed by the uniform norm (10.1). We may similarly define the space

$H^\infty(\Omega)$ to be the linear space of all bounded analytic functions defined on the interior of Ω, again normed by the uniform norm. □

At this point we have a large variety of examples at our disposal, because in addition to the preceding examples we can consider subspaces, quotient spaces and (finite) products. A subspace M of a normed linear space X is of course again a normed linear space with norm equal to the restriction of the given norm on X to M. Similarly, the quotient space X/M is normable, provided that M is closed in X. To see this, we recall from **9D** that a local base in X/M is given by the Q_M-image of a local base in X. Since X is normed we let $A \subset X/M$ be the Q_M-image of the unit ball in X. The set A fails to be a barrel in X/M only because it may fail to be closed; in any event, its gauge ρ_A is a norm on X/M which defines the quotient topology. We can give a formula for this quotient norm as follows:

(10.6) $$\|x + M\| \equiv \rho_A(x + M) = d(x, M)$$

where the right hand side of (10.5) is defined for all $x \in X$ by

(10.7) $$d(x, M) = \inf\{\|x - y\| : y \in M\}.$$

It was shown in **10C** that a product $X = X_1 \times \cdots \times X_n$ of normed linear spaces $X_1, \ldots, X_n$ is normable. The question remains of actually constructing a norm on X in terms of the given norms $\|\cdot\|_i$ defined on X_i. A general way of doing this is to select any monotonic norm ρ on $\mathbb{R}^n$ and then to define

(10.8) $$\|x\| = \rho(\|x_1\|_1, \ldots, \|x_n\|_n)$$

for all $x = (x_1, \ldots, x_n) \in X$. The norm ρ on $\mathbb{R}^n$ is *monotonic* provided that $\rho(u) \leqslant \rho(v)$ whenever $\theta \leqslant u \leqslant v$ in $\mathbb{R}^n$. In particular, the p-norms on $\mathbb{R}^n$ are monotonic.

One final remark should be made about our list of examples of normed linear spaces. In all cases these examples were linear spaces of scalar-valued functions with certain special properties. Much more complicated examples can be constructed by replacing the range space $\mathbb{F}$ (the scalar field) by some normed linear space X (necessarily complex for Example 5). Then we consider the preceding examples with the feature that our functions are now X-valued. We must then modify the definitions (10.1)–(10.4) by replacing the absolute values on the right hand sides by the norm in X. Thus, for example, (10.1) becomes

(10.9) $$\|f\|_\infty = \sup\{\|f(t)\| : t \in \Omega),$$

where $\|f(t)\|$ is the X-norm of the vector $f(t) \in X$.

In addition to the foregoing list of spaces of scalar or vector-valued functions, there is one other class of normed linear spaces that is very important in practice. These spaces consist of the continuous linear operators acting between two given normed linear spaces. We defer further discussion of such spaces until we have learned how they should be normed.

§11. Convexity and Topology

In this section we study the basic topological properties of convex sets in linear topological spaces, and learn conditions for the continuity of linear mappings on such spaces. The results of this section will allow us to establish sharper forms of the separation, support, and extension theorems in the context of linear topological spaces.

A. The following lemma summarizes information that we shall need concerning the topology of convex sets. We write the interior of a set A as $\text{int}(A)$, and we say that A is *solid* if its interior is non-empty.

Lemma. *Let A be a convex subset of a linear topological space X. Then $\bar{A}$ is convex. If A is also solid, then*

a) $\text{int}(A) = \text{cor}(A)$;

b) $\bar{A} = \overline{\text{int}(A)}$ *and* $\text{int}(A) = \text{int}(\bar{A})$ (*in particular*, $\text{int}(A)$ *is dense in* A);

c) $\text{lin}(A) = \bar{A}$ (*in particular*, $\text{lin}(A)$ *is convex*);

d) *the algebraic boundary of A is its topological boundary.*

Proof. The map $f: X \times X \times [0, 1] \to X$ defined by $f(x, y, t) = tx + (1 - t)y$ is continuous; hence

$$f(\bar{A} \times \bar{A} \times [0, 1]) = f(\overline{A \times A \times [0, 1]}) \subset \overline{f(A \times A \times [0, 1])} \subset \bar{A}.$$

Thus we see that $\bar{A}$ is convex.

The key observation for the rest of the proof is the fact that

$$\textbf{(11.1)} \qquad t\bar{A} + (1 - t)\text{int}(A) \subset \text{int}(A),$$

for $0 \leqslant t < 1$. To prove this, it suffices to show that the left hand side of (11.1) lies in A since it is clearly open. Let $p \in \text{int}(A)$. Then $(1 - t)(\text{int}(A) - p)$ is an open θ-neighborhood and so

$$\begin{aligned} t\bar{A} &= \overline{tA} \subset tA + (1 - t)(\text{int}(A) - p) \\ &= tA + (1 - t)\text{int}(A) - (1 - t)p \subset A - (1 - t)p. \end{aligned}$$

a) From exercise 2.1 we know that $\text{int}(A) \subset \text{cor}(A)$. Suppose that $x \in \text{cor}(A)$ and $p \in \text{int}(A)$. Then there exists $y \in A$ such that $x \in [p, y)$. Since this segment lies in $\text{int}(A)$ by (11.1), we see that $\text{cor}(A) \subset \text{int}(A)$.

b) Clearly $\overline{\text{int}(A)} \subset \bar{A}$. Suppose that $x \in \bar{A}$ and $p \in \text{int}(A)$. Then $[p, x) \subset \text{int}(A)$ by (11.1) and so $x \in \overline{\text{int}(A)}$. To prove the second assertion of part b), we see that $\text{int}(A) \subset \text{int}(\bar{A})$ is trivial. Suppose that $x \in \text{int}(\bar{A})$ and $p \in \text{int}(A)$. Then $x \in [p, y)$ for some $y \in \bar{A}$; by (11.1) again, this implies that $x \in \text{int}(A)$.

c) Clearly $\text{lin}(A) \subset \bar{A}$. Suppose that $x \in \bar{A}$ and $p \in \text{int}(A)$. Then $[p, x) \subset \text{int}(A)$ shows immediately that $x \in \text{lin}(A)$.

d) If x is a bounding point of A, then $x \in \text{lin}(A)\backslash\text{cor}(A) = \bar{A}\backslash\text{int}(A) \equiv \partial(A)$, the topological boundary of A. This reasoning is clearly reversible. □

Let A be a subset of the linear topological space X. We define the *closed convex hull* of A, $\overline{\text{co}}(A)$, to be the smallest closed convex set in X that contains A:

$$\overline{\text{co}}(A) = \cap\{K \subset X : A \subset K, K \text{ closed and convex}\}.$$

It follows from the lemma that

$$\overline{\text{co}}(A) = \overline{\text{co}(A)},$$

that is, $\overline{\text{co}}(A)$ is simply the closure of the ordinary convex hull $\text{co}(A)$.

Let A and B be two subsets of X whose closed convex hulls are compact. Then

$$\overline{\text{co}}(A \cup B) = \text{co}(\overline{\text{co}}(A) \cup \overline{\text{co}}(B))$$

is compact, since the right hand side (the join (**2A**) of $\overline{\text{co}}(A)$ and $\overline{\text{co}}(B)$) is the continuous image of the compact space $\overline{\text{co}}(A) \times \overline{\text{co}}(B) \times [0, 1]$ under the mapping $(x, y, t) \mapsto tx + (1 - t)y$. Analogously, if at least one of $\overline{\text{co}}(A)$, $\overline{\text{co}}(B)$ is compact, and α, β are scalars, then

$$\overline{\text{co}}(\alpha A + \beta B) = \alpha\, \overline{\text{co}}(A) + \beta\, \overline{\text{co}}(B).$$

Because, being closed and convex, the set $\overline{\text{co}}(\alpha A) + \overline{\text{co}}(\beta B)$ contains $\text{co}(\alpha A + \beta B)$ (making use of equation (2.2)); on the other hand,

$$\begin{aligned} \overline{\text{co}}(\alpha A + \beta B) &= \overline{\text{co}(\alpha A + \beta B)} \\ &\supset \overline{\text{co}}(\alpha A) + \overline{\text{co}}(\beta B). \end{aligned}$$

Finally, we employ the general fact

$$\overline{\text{co}}(cA) = c\, \overline{\text{co}}(A)$$

for any scalar c.

B. The following proposition, concerning the preservation of boundedness properties, requires us to work in locally convex spaces; both conclusions can fail in a general linear topological space.

Lemma. *Let A be a bounded (resp. totally bounded) subset of a locally convex space X. Then the absolutely convex hull of A is bounded (resp. totally bounded).*

Proof. Let $\mathcal{N}$ be a base of continuous semi-norms (**10B**) on X. We know (**10B** again) that A is bounded if and only if it has finite ρ-diameter for each $\rho \in \mathcal{N}$. Now if A has finite ρ-diameter for a semi-norm ρ the same is clearly true for $\text{bal}(A)$ and for $\text{co}(\text{bal}(A)) \equiv \text{aco}(A)$ (**2A**). Now suppose that A is totally bounded. Then for each $\varepsilon > 0$ and each $\rho \in \mathcal{N}$, there is a finite (ε, ρ)-net in A (**10B**). To show that $\text{bal}(A)$ is totally bounded, we select $\varepsilon > 0$, $\rho \in \mathcal{N}$, an $\left(\frac{\varepsilon}{2}, \rho\right)$-net $\{x_1, \ldots, x_n\}$ in A and an $\varepsilon/2\gamma$-net $\{\alpha_1, \ldots, \alpha_m\}$ in the set of scalars $\{\alpha : |\alpha| \leqslant 1\}$; here $\gamma \equiv \max\{\rho(x_1), \ldots, \rho(x_n)\}$. Then we see that the set $\{\alpha_i x_j : 1 \leqslant i \leqslant m, 1 \leqslant j \leqslant n\}$ is an (ε, ρ)-net in $\text{bal}(A)$. Finally we show that co(A) is totally bounded if A is. Given ε, ρ, and $\{x_1, \ldots, x_n\} \subset A$ as above, let $B = \text{co}(\{x_1, \ldots, x_n\})$. Then B is totally bounded (actually compact by exercise 1.8 and **9E**); let $\{y_1, \ldots, y_m\}$ be an $\left(\frac{\varepsilon}{2}, \rho\right)$-net in B. We claim that the y's form an (ε, ρ)-net in co(A). Because, if $x = \sum \alpha_i a_i$ is in co(A), there are points $x_{k(i)}$ in $\{x_1, \ldots, x_n\}$ such that $\rho(a_i - x_{k(i)}) < \varepsilon/2$. Consequently, setting $y = \sum \alpha_i x_{k(i)}$, we have $\rho(x - y) < \varepsilon/2$; since $y \in B$, there is some y_j with $\rho(y - y_j) < \varepsilon/2$, whence $\rho(x - y_j) < \varepsilon$. □

The chief interest in this result is that it provides a means of constructing or recognizing compact convex sets. That is, if a set A is known to be totally bounded we intuitively suspect that the sets $\overline{\text{co}}(A)$ and $\overline{\text{aco}}(A) \equiv \overline{\text{aco}(A)}$ should be compact. Whether they actually are or not depends on the completeness properties of X. We briefly consider such properties next.

C. Let X be a locally convex space. A net $\{x_\delta : \delta \in D\}$ in X is a *Cauchy net* if

$$\lim_{\delta, \delta' \in D} (x_\delta - x_{\delta'}) = \theta.$$

A subset A of X is *complete* (resp., *semi-complete*) if every Cauchy net (resp., sequence) contained in A has a limit in A. X is called *quasi-complete* if every closed bounded subset of X is complete. In particular, a quasi-complete space must be semi-complete, since any Cauchy sequence is necessarily bounded.

A complete normed linear space is called a *Banach space*. All the examples in **10D** of normed linear spaces are actually Banach spaces (exercise 2.8). In the next chapter we shall see some important examples of quasi-complete but not complete spaces.

Theorem. *Let X be a quasi-complete locally convex space. If A is a totally bounded subset of X then $\overline{\text{aco}}(A)$ (and hence $\overline{\text{co}}(A)$) is compact in X.*

Proof. The most expedient proof of this theorem is based on the notion of universal net. Recall that a *universal net* in a topological space A is a net in A with the property that for each set $B \subset A$, the net is eventually in B or else eventually in $A \backslash B$. Now in order to show that $\overline{\text{aco}}(A)$ is compact we must show that any net $\{x_\delta : \delta \in D\}$ in $\overline{\text{aco}}(A)$ has a cluster point in $\overline{\text{aco}}(A)$. Since every net has a universal subnet, we may as well suppose that $\{x_\delta : \delta \in D\}$ is already universal. We are going to show that a universal net in $\overline{\text{aco}}(A)$ must be Cauchy; once this is done we can invoke the completeness of $\overline{\text{aco}}(A)$ to conclude that this Cauchy net converges in $\overline{\text{aco}}(A)$ and so finish the proof.

Let U be an arbitrary θ-neighborhood in X and choose a balanced θ-neighborhood V such that $V + V \subset U$ (**9A**). Then since $\overline{\text{aco}}(A)$ is totally bounded (being the closure of a totally bounded set), there is a set $B = \{y_1, \ldots, y_n\} \subset X$ such that $\overline{\text{aco}}(A) \subset B + V$. Because our net is universal it must eventually lie in one of the sets $y_1 + V, \ldots, y_n + V$, say in $y_j + V$.

That is, there exists $\gamma \in D$ such that $x_\delta \in y_j + V$ whenever $\delta \geqslant \gamma$ (in the ordering of D). Consequently, if $\delta, \delta' \geqslant \gamma$ then

$$\begin{aligned} x_\delta - x_{\delta'} &\in (y_j + V) - (y_j + V) \\ &= V - V = V + V \subset U, \end{aligned}$$

and so our universal net is indeed a Cauchy net. □

It is to be noted that this argument made no use of local convexity beyond the appeal to **11B**. It follows that any complete and totally bounded subset of a linear topological space must be compact. The converse is also true, and much easier to prove.

D. Now we consider some criteria for the continuity of linear mappings between linear topological spaces. The only satisfactory general results occur when the target space is finite dimensional. We begin with a very general elementary proposition, whose straightforward proof is omitted.

Lemma. *Let $T: X \to Y$ be a linear map between linear topological spaces X and Y. The following are equivalent:*

a) *T is continuous on X;*

b) *T is continuous at θ in X;*

c) *T is uniformly continuous on X, that is, for each θ-neighborhood $V \subset Y$, there is a θ-neighborhood $U \subset X$ such that $T(x) - T(y) \in V$ if $x - y \in U$.*

Corollary. *If X and Y are locally convex then a linear map $T: X \to Y$ is continuous if and only if for every continuous semi-norm ρ on Y, there is a continuous semi-norm σ on X and a constant $\beta > 0$ such that*

$$\rho(T(x)) \leqslant \beta\sigma(x), \qquad x \in X. \tag{11.2}$$

Suppose in particular that X and Y are normed linear spaces. Then we can state that $T: X \to Y$ is continuous if and only if for some $\beta > 0$

$$\|T(x)\| \leqslant \beta\|x\|, \qquad x \in X. \tag{11.3}$$

The smallest such β is called the *norm* of T and written $\|T\|$, that is

$$\|T\| \equiv \sup\left\{\frac{\|T(x)\|}{\|x\|} : x \neq \theta\right\}. \tag{11.4}$$

Note that this quantity serves as the Lipschitz constant for T. On account of the inequality (11.3) continuous linear maps between normed linear spaces are frequently called *bounded* linear maps, since they map bounded sets into bounded sets. Let $B(X, Y)$ be the linear space of all bounded linear maps between the normed linear spaces X and Y. Then $B(X, Y)$ is again a normed linear space with norm defined by (11.4). Such spaces constitute the class of examples suggested at the end of **10D**.

We now give a useful condition for a certain important class of linear maps to be continuous. Recall that the *rank* of a linear map is the dimension of its image.

Theorem. *Let $T:X \to Y$ be a linear map of finite rank between linear topological spaces X and Y. Then T is continuous if and only if its kernel is closed in X.*

Proof. The condition is trivially necessary. Conversely, suppose that $N \equiv \ker(T)$ is closed. Then by **9D**, the quotient space X/N is a finite dimensional Hausdorff linear topological space, and so, by **9E**, any linear map defined on X/N must be continuous. In particular the map $\hat{T}:X/N \to Y$ defined by

$$\hat{T}(x + N) = T(x)$$

is continuous (and also injective). Thus $T = \hat{T} \circ Q_N$ is also continuous. □

This theorem of course applies in the case of linear functionals: a linear functional ϕ in X' is continuous if and only if its hyperplanes $[\phi; \alpha]$ are closed for all scalars α. We denote the set of all such ϕ by X^*; it is a subspace of X' but possibly trivial in the sense that its only element may be θ. We shall soon see that the local convexity of X is sufficient to insure that X^* is in fact a usefully "large" subspace of X'. First we note a corollary to the preceding theorem for linear functionals.

Corollary. *Let X be a linear topological space. A linear functional $\phi \in X'$ is discontinuous if and only if one (and hence all) of its level sets $[\phi; \alpha]$ is dense in X.*

Proof. It only needs to be observed that if ϕ is discontinuous then its kernel is not closed in X and so is properly contained in its closure. Since this closure is a subspace it must be all of X because a hyperplane is by definition (**1E**) a maximal proper flat in X. □

Observe that this corollary implies that a discontinuous linear functional defined on a real linear space X cannot be bounded above or below on any (proper) open subset of X.

E. Now it is time to reconsider the separation and support principles of sections 4 and 6. The necessity for this is due to the phenomenon of dense hyperplanes, possible in infinite dimensional spaces. Separation or support assertions involving such hyperplanes are devoid of useful geometric implication. To avoid this problem we must work exclusively with closed hyperplanes, or equivalently, with continuous linear functionals. Following is the topological version of the basic separation theorem of **4B**, to be called henceforth simply the "separation theorem".

Theorem. *Let A and B be convex subsets of the real linear topological space X, and assume that A is solid. Then A and B can be separated by a closed hyperplane if and only if* $\text{int}(A) \cap B = \varnothing$.

Proof. Since $\text{cor}(A) = \text{int}(A) \neq \varnothing$, we know from **4B** that A and B can be separated by some hyperplane exactly when $\text{int}(A) \cap B = \varnothing$. Let $[\phi; \alpha]$ be any such hyperplane. Then $[\phi; \alpha] \cap \text{int}(A) = \varnothing$ so that the level set $[\phi; \alpha]$ is not dense in X. By **11D** it follows that ϕ is continuous and that $[\phi; \alpha]$ must be closed. □

As a consequence we note a geometric criterion for the existence of non-trivial linear functionals on X, namely, the existence of a proper solid convex set in X. For if such a set exists, it can be separated from any point outside it by a closed hyperplane which is a level set of some non-zero continuous linear functional. This observation in turn yields an even more striking corollary which provides most of the motivation for the study of locally convex topologies.

Corollary 1. *Let X be a locally convex Hausdorff space. Then the space X^* separates the points of X.*

Proof. Let x and y be distinct points in X. We are to find a continuous linear functional ϕ on X such that $\phi(x) \neq \phi(y)$. Letting $z = x - y \neq \theta$, we shall find such a ϕ for which $\phi(z) \neq 0$. Let V be a convex θ-neighborhood that does not contain z. Then z and V can be separated by a closed hyperplane $[\psi; \alpha]$. Since $\psi(V)$ is an interval containing 0 in its interior, $\psi(z)$ cannot be 0. Thus ψ is (the real part (**1F**) of) a continuous linear functional ϕ and $\phi(z) \neq 0$. □

The Hausdorff restriction here is crucial; the assertion of the corollary is false for non-Hausdorff spaces. Consequently, all locally convex spaces will henceforth be assumed to be Hausdorff, unless the contrary is explicitly allowed. The space X^* will be called the *continuous dual* (*topological conjugate space*) of X. One of the major techniques in our subject is the characterization of a property of a space X by means of a "dual" property of the space X^*.

Corollary 2. *A closed solid convex subset of a real linear topological space is supported at every boundary point by a closed hyperplane.*

This is an immediate consequence of the theorem, and is the topological version of the support theorem of **6C**.

F. Here we give our final separation theorem in linear topological spaces. Its necessity is evidenced by the fact that at present we don't even know if a point can be separated from a closed convex set (disjoint from the given point). We can only be sure of this if the convex set were known to be solid. However, we can exploit the fact that the point is a compact set. What is needed is the topological version of the basic strong separation theorem of **4C**.

Theorem. *Two convex sets A and B in a locally convex space X can be strongly separated by a closed hyperplane if and only if $\theta \notin \overline{B - A}$.*

Proof. The condition is clearly necessary. Conversely, if $\theta \notin \overline{B - A}$ there is an open absolutely convex set V in X such that $V \cap (B - A) = \varnothing$, or, such that the open convex set $A + V$ is disjoint from B. By **11E** we can separate these sets by a closed hyperplane which, just as in **4C**, strongly separates A and B. □

Corollary. *Let A and B be disjoint closed convex subsets of a locally convex space with A compact. Then A and B can be strongly separated by a closed hyperplane.*

Proof. The set $A - B$ is closed and does not contain θ. □

A profound extension of this corollary is given later in **15D**. See also exercise 2.26.

G. We conclude this section with sharpened forms of the Hahn-Banach theorem (**6A**) and its corollary. There is an obvious analogous sharpening of the Krein-Rutman theorem (**6B**); see exercise 2.45.

Theorem. *Let M be a subspace of the linear topological space X and let $\phi \in M^*$.*

a) *Assume that X is real, that $g \in \mathrm{Conv}(X)$ is continuous on X, and that $\phi \leqslant g|M$. Then there is an extension $\bar{\phi} \in X^*$ of ϕ such that $\bar{\phi} \leqslant g$.*

b) *Assume that ρ is a continuous semi-norm on X and that $|\phi(\cdot)| \leqslant \rho|M$. Then there is an extension $\bar{\phi} \in X^*$ of ϕ such that $|\bar{\phi}(\cdot)| \leqslant \rho$.*

The proof is an immediate consequence of the existence of the extensions (**6A**) and the continuity criteria in **11D**.

§12. Weak Topologies

In this section we introduce a class of linear topologies which, together with the normable topologies, leads to the most important examples of locally convex spaces (at least for our purposes). The ultimate reason for the importance of these so-called "weak topologies" is the compactness criterion presented below in **12D**.

A. Let X be a linear space and Y a subset of the algebraic conjugate space X'. The $\mathcal{N}$-topology (**10B**) defined on X by means of the family of semi-norms

$$\mathcal{N} = \{|\phi(\cdot)| : \phi \in Y\}$$

is called the *weak topology on X generated by Y*, and denoted $\sigma(X, Y)$. It is the weakest topology on X in which all the functionals belonging to Y are continuous. It is clear that $\sigma(X, Y)$ is unchanged if we replace Y by $\mathrm{span}(Y)$, so we shall assume always that Y is a subspace of X'. From the general properties of $\mathcal{N}$-topologies given in **10B** we can deduce the following important facts about weak topologies. Recall that a subset $Y \subset X'$ is *total* if it separates the points of X; that is, Y is total if $\phi(x) = 0$ for all $\phi \in Y$ implies $x = \theta$.

Lemma. *Let X be a linear space and Y a subspace of X'.*

a) *The weak topology $\sigma(X, Y)$ is a Hausdorff topology if and only if Y is total.*

b) *A net $\{x_\delta : \delta \in D\}$ in X converges to x in the $\sigma(X, Y)$ topology if and only if $\lim\{\phi(x_\delta) : \delta \in D\} = \phi(x)$, for all $\phi \in Y$.*

c) *The following properties of a set $A \subset X$ are equivalent:*

i) *A is $\sigma(X, Y)$-bounded;*

ii) *A is $\sigma(X, Y)$-totally bounded;*

iii) $\sup\{|\phi(x)| : x \in A\} < \infty$, *for all $\phi \in Y$.*

d) *A $\sigma(X, Y)$-local base in X is given by the family*

$$\{x \in X : |\phi(x)| \leqslant 1, \phi \in \mathscr{F} \in \mathscr{F}\},$$

where $\mathscr{F}$ consists of the finite subsets of Y.

Important cases in practice of weak topologies occur when X is already a locally convex (Hausdorff) space and Y is its continuous dual X^*. The resulting (Hausdorff) topology $\sigma(X, X^*)$ is called simply the *weak topology* on X, and we speak accordingly of *weakly bounded* or *weakly compact* sets in X, and of *weak convergence* of nets in X. Even more important is the topology $\sigma(X^*, X) \equiv \sigma(X^*, J_X(X))$ defined on the continuous dual X^* of X by the evaluation functionals

$$\phi \mapsto \phi(x), \qquad x \in X, \qquad \phi \in X^*.$$

The map J_X here is the canonical embedding of X into X'' introduced in **1D**; we have tacitly restricted the domain of these functionals to be X^* rather than all of X'. The topology $\sigma(X^*, X)$ is called the *weak** (*weak-star*) *topology* on X^*, and we may consider *weak*-compact* sets in X^*, *weak*-convergence* of nets in X^*, etc.

Let us consider in general a weak topology $\sigma(X, Y)$ on a linear space X generated by a total subspace $Y \subset X'$. We know that each $\phi \in Y$ is $\sigma(X, Y)$-continuous on X and we ask if there are any other $\sigma(X, Y)$-continuous functionals in X'.

Theorem. *A functional $\phi \in X'$ is $\sigma(X, Y)$-continuous on X if and only if $\phi \in Y$.*

Proof. If ϕ is $\sigma(X, Y)$-continuous then ϕ is bounded (say by 1) on some $\sigma(X, Y)$-θ-neighborhood in X. By part d) of the lemma it follows that there are $\phi_1, \ldots, \phi_n \in Y$ such that $|\phi(x)| \leqslant 1$ whenever $|\phi_i(x)| \leqslant 1$, $1 \leqslant i \leqslant n$. In particular, if $x \in \cap \{\ker(\phi_i) : 1 \leqslant i \leqslant n\}$, then $|\phi(kx)| \leqslant 1$ for all k, whence $\phi(x) = 0$. Thus

$$\cap\{\ker(\phi_i) : 1 \leqslant i \leqslant n\} \subset \ker(\phi),$$

so that, by **1G**, $\phi \in \operatorname{span}\{\phi_1, \ldots, \phi_n\} \subset Y$. □

Let us consider some implications of this theorem for a locally convex space X and its weak topology $\sigma(X, X^*)$. First note that the weak topology really is weaker than the given topology on X. This follows, for example, from part b) of the lemma which implies that any convergent net in X is also weakly convergent. Now the theorem implies that every continuous linear functional on X is necessarily weakly continuous. Thus, even though the weak topology on X may be strictly weaker than the given topology, the two topologies yield exactly the same continuous linear functionals. We also see that the two topologies yield the same closed convex sets.

Corollary 1. *Let A be a convex subset of a real locally convex space X. Then A is closed if and only if it is weakly closed.*

Proof. Assume that A is closed and that $x \in X \backslash A$. By **11F** x can be strongly separated from A by a hyperplane $[\phi; \alpha]$ for some $\phi \in X^*$. Thus, by part b) of the lemma again, no net in A can converge weakly to x. That is, x does not belong to the weak closure of A. □

This corollary shows us in turn that the weak closure of any set $A \subset X$ lies in $\overline{\text{co}}(A)$, since this latter set is weakly closed. In particular, if X is metrizable (for example, if X is a normed linear space), then each point in the weak closure of A is the limit (in the original topology on X) of a sequence of convex combinations of points in A.

Corollary 2. *Let X be a normed linear space. Then the norm on X is a weakly lower semicontinuous function.*

This result follows directly from the weak closure of the unit ball (and its positive multiples) in X. The implication is that

(12.1) $$\|x\| \leqslant \lim_\delta \inf \|x_\delta\|,$$

whenever $\{x_\delta : \delta \in D\}$ is a net in X that converges weakly to x.

B. Let X be a locally convex space and let $J_X : X \to X^{*\prime}$ be the canonical embedding of X into the algebraic conjugate space of X^*:

$$\langle \phi, J_X(x) \rangle \equiv \phi(x), \qquad \phi \in X^*, \qquad x \in X.$$

Now X^* with its $\sigma(X^*, X)$ topology is a locally convex space and $J_X(X)$ is its continuous dual. (Although a simple enough consequence of **9A**, this is still an important remark. Explicitly it asserts that any weak*-continuous linear functional defined on X^* must be an *evaluation functional*, that is, it must have the form $\phi \mapsto \phi(x_o)$ for some fixed $x_o \in X$.) Thus $J_X(X)$ has its own weak* topology. With $J_X(X)$ so topologized and with X topologized with the weak topology, the map $J_X : X \to J_X(X)$ is an isomorphism. Thus every locally convex space is *weakly reflexive* in that it is canonically isomorphic to the dual of its dual, provided that the appropriate topologies are used on the spaces involved. Because of its universal validity this result is not of great usefulness in practice; it is much more important (but not always possible) to achieve such an isomorphism when X has its given topology. We shall return to this problem in Chapter III.

C. In exercise 1.16 the notion of the polar of a subset of $\mathbb{R}^n$ was introduced. We now extend this definition. Let X be a real linear topological space. The *polar* of a set $A \subset X$ is

$$A^\circ = \{\phi \in X^* : \phi(x) \leqslant 1, x \in A\}.$$

Since $\theta \in A^\circ$ always, we see that A° is a non-empty convex and weak*-closed subset of X^*. As an example, if A is a subspace of X, its polar is the intersection of X^* and its annihilator subspace (**1H**) in X'.

The *bipolar* of A is

$${}^\circ A^\circ \equiv {}^\circ(A^\circ) = \{x \in X : \phi(x) \leqslant 1, \phi \in A^\circ\}.$$

The precise relation between a set in X and its bipolar is quite important and is contained in the following "bipolar theorem".

Theorem. *Let A be a subset of a real locally convex space. Then*

(12.2)
$$^\circ A^\circ = \overline{\text{co}}(\{\theta\} \cup A).$$

Proof. It is clear that the set $\{\theta\} \cup A$ lies in $^\circ A^\circ$. Since $^\circ A^\circ$ is convex and (weakly) closed we see that $^\circ A^\circ \supset \overline{\text{co}}(\{\theta\} \cup A)$. On the other hand, if any closed half-space (**4B**) contains $\{\theta\} \cup A$, it must also contain $^\circ A^\circ$. Taking into account exercise 2.13, we obtain the reverse inclusion. □

Corollary 1. $(^\circ A^\circ)^\circ = A^\circ$.

Corollary 2. *Let A be a subset of a real locally convex space X. Then* span(A) *is weakly dense in X if and only if $\phi = \theta$ is the only functional in X^* to annul every point of A.*

Proof. Let $M = \text{span}(A)$ and let $\bar{M}^w$ be its weak closure in X. Then $\bar{M}^w = X$ if and only if $(\bar{M}^w)^\circ = X^\circ = \{\theta\}$, since $\bar{M}^w$ and X are each equal to their own bipolars. But $\bar{M}^w = {}^\circ M^\circ$ so that our condition for $\bar{M}^w = X$ becomes

$$\{\theta\} = (\bar{M}^w)^\circ = (^\circ M^\circ)^\circ = M^\circ,$$

which is clearly equivalent to the condition stated in the corollary. □

Corollary 3. *Let X be a real locally convex space. A subset of X^* is total if and only if its span is weak*-dense in X^*.*

Example. To gain some insight into the significance of weak*-density of subspaces, we consider a particular case. Let $L^1 \equiv L^1([0, 1], \mathbb{R})$ be the Banach space of real-valued Lebesgue integrable functions on the interval $[0, 1]$ (a special case of the spaces in Example 2, **10D**). Now every essentially bounded measurable function g on $[0, 1]$ defines a continuous linear functional on L^1 by

(12.3)
$$f \mapsto \int_0^1 f(t)g(t)dt, \qquad f \in L^1,$$

and the norm (**11D**) of this functional is just the essential sup norm $\|g\|_\infty$. In this fashion we obtain a norm-preserving isomorphism from $L^\infty \equiv L^\infty([0, 1], \mathbb{R})$ into L^{1*}. On the other hand, if $\phi \in L^{1*}$ then the formula

$$\mu(E) \equiv \phi(\chi_E)$$

defines an absolutely continuous measure μ on the measurable subsets $E \subset [0, 1]$ whose Radon-Nikodym derivative g belongs to L^∞ with $\|g\|_\infty = \|\phi\|$. Since

$$\int_0^1 \chi_E(t)g(t)dt = \int_0^1 \chi_E \, d\mu = \mu(E) = \phi(\chi_E),$$

we see that g defines via the formula (12.3) a continuous linear functional on L^1 that agrees with ϕ on each characteristic function, hence on each simple function, and finally (by continuity) on all of L^1. Thus we may identify

L^{1*} with the space L^∞. (The same result holds if $[0, 1]$ is replaced by any σ-finite measure space.)

Now the space $C([0, 1], \mathbb{R})$ is evidently a proper closed subspace of L^∞. We claim that it is weak*-dense, however. To see this, it is enough (by Corollary 3) to verify that the continuous functions are total over L^1. That is, we must show that if, for some $f \in L^1$,

$$\int_0^1 f(t)g(t)dt = 0$$

for all continuous g, then $f = \theta$. There are various ways to prove this last assertion. One may note, for example, that the hypothesis implies that f has an everywhere vanishing Fourier transform, so that by the uniqueness of Fourier transforms, $f = \theta$. Otherwise, it may be attacked directly (a good exercise!) by use of standard measure theoretic tools such as the regularity of Lebesgue measure and the dominated convergence theorem. □

D. We consider next a criterion (the "Alaouglu-Bourbaki theorem") for the weak*-compactness of subsets of the continuous dual X^* of a linear topological space X. It is possible to give more general (but more complicated) criteria for compactness in any $\sigma(X, Y)$ topology. However, such criteria are difficult to apply in practice, and, in any event, most practical situations can be handled by the clear and simple condition below. Also we point out that very sophisticated (and useful) compactness conditions will be given in the next chapter for complete locally convex spaces (especially, Banach spaces).

Let G be a subset of X^*. Since G consists of continuous linear functionals we can assert that G is equicontinuous if and only if there is some balanced θ-neighborhood $U \subset X$ such that $\phi(x) \leqslant 1$ for $x \in U$ and $\phi \in G$. In other words, G is equicontinuous if and only if $G \subset U^\circ$, for some balanced θ-neighborhood $U \subset X$.[(2)]

Suppose, for example, that X is a normed linear space with unit ball $U(X) \equiv \{x \in X : \|x\| \leqslant 1\}$. We have seen (**11D**) that X^* is also a normed linear space with norm defined by

$$\|\phi\| \equiv \sup\left\{\frac{|\phi(x)|}{\|x\|} : x \neq \theta\right\} = \sup\{|\phi(x)| : \|x\| = 1\}.$$

It follows that

$$U(X)^\circ = U(X^*) \equiv \{\phi \in X^* : \|\phi\| \leqslant 1\}.^{(3)}$$

Consequently, since

$$(tU)^\circ = t^{-1}U^\circ \qquad (t \neq 0)$$

---

(2) Polars such as U° have only been defined in real spaces; in case our space X here is complex we apply this condition to the real restriction X_R (**1F**).

(3) If X is complex this equation means that $U(X^*)$ consists of those functionals whose real parts belong to $U(X)^\circ$; it is justified by the fact that $\|\phi\| = \|re\ \phi\|$, $\phi \in X^*$.

always holds for polars, we see that a subset G of X^* is equicontinuous exactly when it is bounded in the norm topology on X^*.

Theorem. *Let U be a θ-neighborhood in the linear topological space X. Then U° is weak*-compact in X^*. In particular, every equicontinuous subset of X^* is relatively weak*-compact.*

Proof. Let V be a balanced θ-neighborhood contained in U; since $U^\circ \subset V^\circ$, it will suffice to show that V° is weak*-compact. Now, by part c) of the lemma in **12A**, V° is totally bounded because, for any $x \in X$, there is $\varepsilon > 0$ for which $\varepsilon x \in V$, whence

$$\sup\{|\phi(x)| : \phi \in V^\circ\} \leqslant \varepsilon^{-1} < \infty$$

On the other hand, V° is certainly weak*-complete. To see this, let $\{\phi_\delta\}$ be a weak*-Cauchy net in V°. Then $\{\phi_\delta(x)\}$ is a Cauchy net of scalars for each $x \in X$ so that $\lim_\delta \phi_\delta(x) \equiv \phi(x)$ exists and defines a linear functional $\phi \in X'$. But, since $|\phi_\delta(x)| \leqslant 1$ for each $x \in V$, we must have $|\phi(x)| \leqslant 1$ also, so that $\phi \in V^\circ$. □

Note that this result can also be considered as another application of the Ascoli theorem (**7C**).

Corollary 1. *Let X be a normed linear space. Then every bounded subset of X^* is relatively weak*-compact, and every ball in X^* is weak*-compact.* (A *ball* in a normed linear space X is a set of the form

$$\{x \in X : \|x - x_o\| \leqslant r\} \equiv x_o + rU(X),$$

for some $x_o \in X$ and $r > 0$.) The converse of this corollary is not true in general unless X is complete; this will be shown in the next chapter (see also the following sub-section).

Corollary 2. *Let $A_1, \ldots, A_n$ be closed convex θ-neighborhoods in a locally convex space. Then*

$$(A_1 \cap \cdots \cap A_n)^\circ = \operatorname{co}(A_1^\circ \cap \cdots \cap A_n^\circ). \tag{12.4}$$

Proof. This is a consequence of the general formula for the polar of the intersection of an arbitrary family of closed convex sets containing θ (exercise 2.28) and the Alaoglu-Bourbaki theorem which guarantees that each of the polars A_i° is weak*-compact. It remains only to apply exercise 2.21 to conclude that the right hand side of (12.3) is weak*-compact. □

E. Let X be a normed linear space. We remain interested in the problem of deciding whether a given weak*-compact set $G \subset X^*$ is necessarily bounded (or, equivalently, equicontinuous). We noted after Corollary 1 of **12D** that this will not generally be the case unless X is a Banach space.

Example. Let X be the subspace of $\ell^1(\aleph_o)$ (**6C**) consisting of sequences of scalars that have only a finite number of non-zero terms. Let t_n be a sequence of positive numbers such that $\lim t_n = +\infty$, and let $\phi_n \in X^*$ be defined by

$$\phi_n(x) = \xi_n, \qquad x = (\xi_1, \xi_2, \ldots) \in X.$$

Then the set $G \equiv \{\theta, t_1\phi_1, t_2\phi_2, \ldots\}$ in X^* is clearly unbounded since $\|t_n\phi_n\| = t_n$. However, the sequence $\{t_n\phi_n\}$ converges weak* to θ so that G is weak*-compact. □

We see next that such a situation cannot occur in the presence of convexity.

Theorem. *A weak*-compact convex set G in X^* is bounded.*

Proof. We can write

$$G = \cup\{G \cap nU(X^*) : n = 1, 2, \ldots\},$$

exhibiting G as a countable union of weak*-closed subsets. Since any compact topological space is a Baire space (**17A**), one of the sets $G \cap nU(X^*)$ must be solid. That is, there exists $\phi \in G$ and a weak*-θ-neighborhood $V \subset X^*$ such that

$$(\phi + V) \cup G \subset nU(X^*).$$

Since $G - G$ is weak*-bounded (actually w^*-compact) it is absorbed by V: there exists λ, $0 < \lambda < 1$, such that $\lambda(G - G) \subset V$. Because G is convex we see that

$$\begin{aligned}(1 - \lambda)\phi + \lambda G &\subset (\phi + \lambda(G - G)) \cap G \\ &\subset (\phi + V) \cap G \subset nU(X^*).\end{aligned}$$

This inclusion shows that λG lies in some ball in X^* and hence so does G. □

F. In studying new topological spaces such as linear spaces X with a weak topology $\sigma(X, Y)$ (**12A**), one is naturally concerned with their metrizability. Unfortunately, except for generally uninteresting special cases (such as finite dimensionality), the weak topologies are not definable by a metric. This assertion will be justified in the next chapter, as an application of some results about Banach spaces.

In spite of this general disappointment we shall see next that the restriction of the $\sigma(X, Y)$ topology to certain subsets of X is metrizable. Such results can be quite useful.

Theorem. *Let X be a locally convex space and Y a total subspace of X^*. Let A be a $\sigma(X, Y)$-compact subset of X and suppose there exists a countable set $G \subset Y$ which separates the points of A. Then the (relative) $\sigma(X, Y)$ topology on A is a metric topology.*

Proof. Consider the product space $\mathbb{F}^G$ where $\mathbb{F}$ is the scalar field associated with X. Since G is countable this space is metrizable as we see by recalling the construction (9.3) of a local base in $\mathbb{F}^G$ and the metrizability criterion of exercise 2.5. Hence each subset of $\mathbb{F}^G$ is metrizable and we will be done if we can show that A is homeomorphic to some such subset. To do this we define $Q: A \to \mathbb{F}^G$ by

$$Q(x) = J_X(x)|G, \qquad x \in A,$$

where J_X is, as usual, the canonical embedding of X into $X^{*\prime}$. Then Q is certainly continuous by the definitions of the relevant topologies (the relative $\sigma(X, Y)$ topology on A, the topology of pointwise convergence on G). Furthermore Q is injective because G separates the points of A. Consequently, since A is compact Q must be a homeomorphism. □

Corollary 1. *Let X be a separable locally convex space and let A be a weak*-closed equicontinuous subset of X^*. Then in its (relative) weak* topology A is a compact metric space.*

Proof. This is a consequence of the Alaoglu-Bourbaki theorem (**12D**) and the preceding result where we choose the set G to be any countable dense subset of X. □

The most important special case of this corollary occurs when X is a normed linear space; we can then completely characterize those cases where the conclusion of Corollary 1 holds.

Corollary 2. *Let X be a normed linear space. Then the following statements are equivalent.*

a) *$U(X^*)$ is weak*-metrizable;*
b) *every ball in X^* is weak*-metrizable;*
c) *X is separable.*

Proof. We need only check that a) implies c). Since every metric space is first countable there exists a countable θ-neighborhood base $\{V_n\}$ in $U(X^*)$ and hence a sequence $\{A_n\}$ of finite subsets of X such that

$$\{\phi \in U(X^*): |\phi(x)| \leqslant 1, x \in A_n\} \subset V_n, \qquad n = 1, 2, \ldots.$$

Let $A = \cup\{A_n : n = 1, 2, \ldots\}$. If $\phi \in U(X^*)$ vanishes at each point of A, then $\phi \in \cap\{V_n : n = 1, 2, \ldots\} = \{\theta\}$. It now follows from **11F** that $\overline{\text{span}}(A) = X$ and so X is separable (exercise 2.22). □

The analogue of Corollary 2 for the weak metrizability of balls in a normed linear space X is also valid, the necessary and sufficient condition being the separability of X^* (in its norm topology). However, this situation occurs less in practice than that of the corollary.

A topological space Ω is called *sequentially compact* if every sequence in Ω has a convergent subsequence (with limit in Ω). For example, every compact metric space is sequentially compact. In particular, if X is a separable normed linear space, the ball $U(X^*)$ is weak*-sequentially compact. What happens if X is not separable?

Example. Let m be the linear space of all bounded scalar sequences $x = (\xi_1, \xi_2, \ldots)$ normed by $\|x\|_\infty \equiv \sup\{|\xi_1|, |\xi_2| \ldots\}$. m is a non-separable normed linear space (in fact, a Banach space). Hence the ball $U(m^*)$ is weak*-compact but not weak*-metrizable. We claim that it is not weak*-sequentially compact. To see this, let $\phi_n \in m^*$ be defined by $\phi_n(x) = \xi_n$ for $x \in m$. Clearly $\|\phi_n\| = 1$ for all n. But $\{\phi_n\}$ has no weak*-convergent subsequence. On the other hand, $\{\phi_n\}$ has a weak*-cluster point in $U(m^*)$. That is, there exist weak*-convergent subnets of $\{\phi_n\}$ none of which is a subsequence. □

G. Let X be a normed linear space. We shall see in the next chapter that if X is complete then X^* is weak*-quasi-complete. However, as we now show, it is never the case that X^* is weak*-complete, unless X is finite dimensional.

Example. Let X have infinite dimension. According to exercise 2.14, there exists a linear functional $\phi \in X' \backslash X^*$. For each finite dimensional subspace M of X, there is, by exercise 2.2, a closed complementary subspace of M in X and consequently there is a continuous projection $P_M: X \to M$. We define $\phi_M = \phi \circ P_M$. Each ϕ_M is continuous by **11D**. Now the collection $\mathcal{M}$ of finite dimensional subspaces of X can be partially ordered by inclusion and then forms a directed set. The net $\{\phi_M : M \in \mathcal{M}\}$ is consequently a weak*-Cauchy net in X^* with no weak*-limit in X^*. □

What is going on here is the following. As a product of complete spaces the space $\mathbb{F}^X$ of all scalar-valued functions on X is complete in its product topology, the topology of pointwise convergence on X. The subspace X' of $\mathbb{F}^X$ is closed (hence complete) in this topology and the preceding argument shows that X^* is a proper dense subspace of X'.

It is also true but somewhat harder to prove that X is never weakly complete (unless it is finite dimensional, of course). We shall give a criterion (reflexivity) for X to be weakly quasi-complete in the next chapter.

§13. Extreme Points

In this section we continue the discussion of §8 concerning the extreme points of a convex set and their usefulness in describing that set. A basic difficulty (**8F**) is the very existence of an extreme point and we deal with this problem first. Then we give the infinite dimensional analogue of Minkowski's theorem (**8D**). Frequent applications of these results appear in this and later sections.

A. Throughout this section X will be a real locally convex (Hausdorff) space. Let A be a closed convex subset of X. If X is finite dimensional we were able to give a concise necessary and sufficient condition for A to have an extreme point, namely that A be line-free (**8B**). As simple examples show (exercise 2.30), this result is generally false in infinite dimensional spaces, although it is clearly always necessary for A to be line-free. Our first result, a nice application of the strong separation theorem (**11F**), shows that a sufficiently strong topological assumption (compactness) entails the existence of extreme points, whether or not our set is even convex. As noted in **8F**, this type of result is somewhat surprising because the notion of extreme point is strictly algebraic. It is at this point that an appreciation for the efficacy of topological methods in functional analysis should really begin.

Lemma. *Let A be a (non-empty) compact subset of X. Then A has an extreme point.*

Proof. The family $\mathscr{A}$ of all compact extremal subsets of A is non-empty (since $A \in \mathscr{A}$) and is partially ordered by inclusion. The intersection of any nested family in $\mathscr{A}$ is non-empty (by compactness) and A-extremal (by **8A**). Therefore, there exists a minimal element B of $\mathscr{A}$. We claim that B contains only one point which, in that case, must belong to ext(A), If not, there are distinct points $p, q \in B$. By **11E** these points can be separated by a continuous linear functional $\phi \in X^*: \phi(p) \neq \phi(q)$. But now the set

$$B \cap [\phi; \min\{\phi(x): x \in B\}]$$

is a proper compact extremal subset of B. By **8A** this set is also A-extremal and this contradicts the minimality of B. Thus B must be a singleton set. □

There is an application of this theorem to concave programming problems (**8E**). Let f be a concave function defined on X and let A be a non-empty compact subset of X. We consider the optimization problem (A, f).

Corollary. *If f is lower semicontinuous on A then f attains its minimum on A at an extreme point of A.* □

Proof. The set where f attains its minimum is a non-empty compact extremal subset of A. This set has an extreme point which, by **8A**, must also be an extreme point of A. □

Note that this corollary applies in particular to the case where $f \in X^*$.

B. The result to be given next is an extended form of the "Krein-Milman theorem", one of the most important general principles of geometric functional analysis. It provides two conditions for a subset B of a compact convex set $A \subset X$ to satisfy $\overline{\text{co}}(B) = A$. In particular the conditions are satisfied when $B = \text{ext}(A)$. The additional operation of closure is necessary now in contrast to the finite dimensional case (**8D**) because, for example, co(ext(A)) will not be closed in general (exercise 2.31).

Theorem. *Let B be a subset of the compact convex set $A \subset X$. The following conditions are equivalent:*

a) $\overline{\text{co}}(B) = A$;

b) $\inf\{\phi(x): x \in B\} = \min\{\phi(x): x \in A\}$, *for any* $\phi \in X^*$;

c) $\text{ext}(A) \subset \overline{B}$.

Proof. The equivalence of a) and b) follows directly from the strong separation theorem **11F** and the fact that

$$\inf\{\phi(x): x \in B\} = \inf\{\phi(x): x \in \overline{\text{co}}(B)\}$$

for every $\phi \in X^*$. Note that the compactness of A is not needed for this conclusion. The preceding corollary shows that c) implies b). It remains to prove c) from a). This will follow from the general fact that

(13.1) $$\text{ext}(\overline{\text{co}}(B)) \subset \bar{B},$$

a result of independent interest. To prove (13.1), let $x \in \text{ext}(\overline{\text{co}}(B))$; we must show that $(x + V) \cap B \neq \varnothing$ for any θ-neighborhood V (by **10A** it may be assumed that V is a barrel). Now B is totally bounded and hence there is a finite subset $\{x_1, \ldots, x_n\} \subset B$ such that

$$B \subset \cup\{x_i + V : i = 1, \ldots, n\}.$$

Since the sets $K_i \equiv \overline{\text{co}}((x_i + V) \cap B) \subset \overline{\text{co}}(B) = A$ are compact and convex, we have

$$\overline{\text{co}}(B) = \overline{\text{co}}(\cup\{K_i : i = 1, \ldots, n\}) = \text{co}(\cup\{K_i : i = 1, \ldots, n\}),$$

the last equality by exercise 2.21. It now follows from exercise 1.33 b) that x actually belongs to some K_i. In particular, $x = x_i + v$ from some $v \in V$, whence $x_i = x - v$ is a point in $(x + V) \cap B$. □

We now list a few corollaries of this theorem, the first two pertaining to an arbitrary compact convex set $A \subset X$ (see also exercise 2.32).

Corollary 1. *Let B be a non-empty closed semi-extremal subset of A. Then B contains an extreme point of A.*

Corollary 2. *A lower semicontinuous quasi-concave function f on A attains its minimum on A at an extreme point of A.*

Proof. A quasi-concave function is by definition the negative of a quasi-convex function (exercise 1.10). It follows that the sets $\{x \in A : \alpha < f(x)\}$ are open and convex for every real α. Thus we see that the set $B \subset A$ where f attains its minimum satisfies the conditions of Corollary 1 and so contains an extreme point of A. □

The optimization result given in Corollary 2 is known as "Bauer's minimum principle". There is of course an analogous statement pertaining to the maximum of an upper semicontinuous quasi-convex function.

Now suppose that X is a normed linear space. Then in the weak*-topology any ball in X^* is a compact convex set (**12D**) and so is the closed convex hull of its extreme points. This fact yields a geometric strengthening of Corollary 1 in **11E**.

Corollary 3. *For every normed linear space X the extreme points of $U(X^*)$ separate the points of X.*

We can paraphrase this corollary by stating that to any pair x, y of distinct points of X there corresponds a $\phi \in \text{ext}(U(X^*))$ such that $\phi(x) \neq \phi(y)$. The fact that the balls in the dual space of a normed linear space are well supplied with extreme points has an interesting implication. Namely, if a given space X has the property that $\text{ext}(U(X))$ is empty it follows that X cannot be a dual space. Examples of such spaces occur in exercise 2.30.

C. In the two preceding sections we have been successful in developing a viable extreme point theory for compact sets. It has further been noted that such results do not extend much beyond the compact situation because of the existence of closed bounded (convex) sets with no extreme points. One avenue of extension is available, however, and this is to the case of locally compact sets. This extension will result in a generalization of Klee's theorem (**8D**) to arbitrary infinite dimensional (locally convex) spaces. Two preliminary lemmas are required, the first being a bit stronger than we need but having some independent interest.

Lemma 1. *A (non-zero) cone C in a real locally convex space X is locally compact if and only if it has a compact base, in which case C is necessarily closed.*

Proof. Assume that C is locally compact. There is then a closed convex θ-neighborhood $U \subset X$ for which $C \cap U$ is compact. Let D be the intersection of C with the boundary of U. Then $\overline{\mathrm{co}}(D) \subset C \cap U$ and hence is compact. Since $\theta \in \mathrm{ext}(C)$ and $\theta \notin D$ we have by (13.1) that $\theta \notin \overline{\mathrm{co}}(D)$. Now let H be a closed hyperplane strongly separating θ and $\mathrm{co}(D)$. We then have that $B \subset C \cap H$ is a base for C and that $B \equiv C \cap U$, whence B is compact.

Conversely, suppose that B is a compact base for C. Then for all $t > 0$ the sets $K_t \equiv \{\lambda x : x \in B, 0 \leqslant \lambda \leqslant t\}$ are compact. Let H again be a closed hyperplane strongly separating θ and B; we can assume that $H = [\phi; 1]$ for some $\phi \in X^*$. Now let $x_o \in C$; we wish to find a compact x_o-neighborhood in C. Since $x_o = t_o b_o$ for suitable $t_o > 0$ and $b_o \in B$ (**5C**), we have that $x_o \in K_{2t_o}$. But if $\beta \equiv \inf\{\phi(x) : x \in B\}$ then the set $\{x \in C : \phi(x) \leqslant 2\beta t_o\} \subset K_{2t_o}$ is the desired neighborhood.

The proof that C must be closed if it has a compact base is left as an easy exercise. □

Note that the linear functional ϕ used to separate θ from the base B is a strictly positive functional (**5C**) on $X : \phi(x) > 0$ for all $x \in C$, $x \neq \theta$.

In **8C** we introduced the recession cone C_A of a convex set and in exercise 1.36 it was noted that if $C_A = \{\theta\}$ and A is closed in $\mathbb{R}^n$ then A must be compact. The same argument applies in our more general (infinite dimensional) setting, provided we hypothesize that A is locally compact: $C_A = \{\theta\}$ if and only if A is compact. We now use this observation to establish the existence of extreme points.

Lemma 2. *A (non-empty) closed, convex, locally compact, and line-free set A in X has an extreme point.*

Proof. We may assume that A is not compact. Then C_A is a non-trivial closed cone in X (closure follows from equation (8.5)). Further C_A is itself locally compact since a translate of it lies in A. Let $\phi \in X^*$ be a strictly positive linear functional and let K be the half-space $\{x \in X : \phi(x) \leqslant 0\}$. If we translate C_A and K to a point $x_o \in A$ we see that the set $A \cap (x_o + K) \equiv B$ must be compact, since otherwise, by its local compactness, it would contain a half-line outside the set $x_o + C_A$, in contradiction to the definition of

C_A. Finally, either B is contained in the hyperplane $\{x \in \bar{X} : \phi(x) = \phi(x_o)\}$ in which case $\text{ext}(B) \subset \text{ext}(A)$ (**8A**), or else B has an extreme point not in this hyperplane; but such a point must again be an extreme point of A. □

Now we can give the general version of Klee's theorem. In particular our approach here provides a new but less direct proof of the original finite dimensional result in **8D**.

Theorem. *Let A satisfy the hypotheses of Lemma 2. Then*

$$A = \overline{\text{co}}(\text{ext}(A) \cup \text{rext}(A)). \tag{13.2}$$

Proof. Let B be the right-hand side of (13.2). If B were properly contained in A we could strongly separate B from a point in $A \backslash B$ by a closed hyperplane H. By Lemma 2 there is an extreme point p of $A \cap H$ which, by definition, does not belong to $\text{ext}(A)$. There is hence a line L in X such that $p \in \text{cor}(A \cap L)$, where $A \cap L$ is either a line segment or a half-line. In the former case we claim that the end-points of $A \cap L$ are both extreme points of A, which would then imply that $p \in B$, in contradiction to the choice of p. To prove this claim, let q be an end-point of $A \cap L$. If $q \notin \text{ext}(A)$ there are distinct points $u, v \in A$ with $q \in (u, v)$. Then, if z is a point of $A \cap L$ in the half-space of H that does not contain q, it follows that $p \in \text{cor}(H \cap \text{co}(u, v, z))$ which again contradicts the choice of p. Finally, if $A \cap L$ is a half-line, we see analogously that it must in fact be an extreme ray of A; this entails $p \in B$ which is again a contradiction. □

D. Let M be a closed linear subspace of a real normed linear space X. It is frequently of interest to determine how well an element $x_o \in X \backslash M$ can be approximated by members of M. (A classical situation occurs when $X = C([a, b], \mathbb{R})$ and M consists of all polynomials of degree at most n, for some n.) By definition this closeness of approximation is given by the quantity

$$d(x_o, M) \equiv \inf\{\|x_o - y\| : y \in M\}.$$

We are going to see that the extreme points of certain convex sets in X^* play a role in the determination of this value.

Let $\phi \in U(M^\circ)$ where M° is the annihilator (or, equivalently, the polar) of M in X^*. Then for any $y \in M$,

$$|\phi(x_o)| = |\phi(x_o - y)| \leqslant \|\phi\| \, \|x_o - y\| \leqslant \|x_o - y\|$$

whence

$$|\phi(x_o)| \leqslant d(x_o, M). \tag{13.3}$$

On the other hand, we can separate M and the ball $\{x \in X : \|x_o - x\| \leqslant d(x_o, M)\}$ to obtain a functional $\psi \in M^\circ$ such that $\psi(x_o + x) \geqslant 0$ for $\|x\| \leqslant d(x_o, M)$. Thus

$$-\psi(x_o) \leqslant \inf\{\psi(x) : \|x\| \leqslant d(x_o, M)\} = -d(x_o, M)\|\psi\|,$$

so that the functional $\phi \equiv \dfrac{\psi}{\|\psi\|}$ satisfies $\phi \in U(M^\circ)$ and

$$d(x_o, M) \leqslant \phi(x_o). \tag{13.4}$$

Combining (13.3) and (13.4) we obtain

$$\begin{aligned} d(x_o, M) &= \max\{\phi(x_o) : \phi \in U(M^\circ)\} \\ &= \sup\{\phi(x_o) : \phi \in \operatorname{ext}(U(M^\circ))\}, \end{aligned} \tag{13.5}$$

where the second equality is a consequence of **13A** applied to the weak*-continuous linear function $\phi \mapsto \phi(x_o)$ on the weak*-compact set $U(M^\circ)$ (the compactness of $U(M^\circ)$ follows from **12D** and the fact that subspaces of the form M° are weak*-closed).

We now develop a technique for recognizing the extreme points of sets of the type $U(M^\circ)$. Let ρ be a continuous semi-norm on X (which may be any real locally convex space for the moment). For any $\phi \in U_\rho^\circ$ (the polar of the ρ-unit ball (**10B**)) we define a set

$$A_\phi = \{x \in X : \rho(x) - \phi(x) \leqslant 1\}.$$

Each such set is an unbounded convex θ-neighborhood in X. We give next a preliminary result for the case $M = \{\theta\}$.

Lemma. *A functional $\phi \in U_\rho^\circ$ is an extreme point of U_ρ° if and only if the difference set $A_\phi - A_\phi$ is dense in X.*

Proof. In general, by the strong separation theorem, a convex set $K \subset X$ fails to be dense in X exactly when some non-zero $\psi \in X^*$ is bounded (above) on K. Suppose first that $\phi \notin \operatorname{ext}(U_\rho^\circ)$. Then there is a non-zero $\psi \in X^*$ such that $\phi \pm \psi \in U_\rho^\circ$. Hence $|\langle x, \phi \pm \psi\rangle| \leqslant \rho(x)$ for all $x \in X$ and in particular $\psi(x) \leqslant \rho(x) - \phi(x)$. Thus ψ is bounded above by 1 on $\pm A_\phi$ and hence by 2 on $A_\phi - A_\phi$; consequently, $A_\phi - A_\phi$ is not dense in X. Conversely, if we assume that $A_\phi - A_\phi$ is not dense, there is some non-zero $\psi \in X^*$ bounded above (say by 1) on $A_\phi - A_\phi$. Since A_ϕ is a balanced set containing A_ϕ it follows that $\sup\{|\psi(x)| : x \in A_\phi\} \leqslant 1$. It remains to prove that $\phi \pm \psi \in U_\rho^\circ$ for this will show that $\phi \notin \operatorname{ext}(U_\rho^\circ)$. To do this select any $x \in U_\rho$ and set $\alpha = \rho(x) - \phi(x)$.

Case 1: $\alpha = 0$. In this case we have $tx \in A_\phi$ for every $t > 0$ whence $|\psi(x)| \leqslant 1/t$ and so $\psi(x) = 0$. Thus $\langle x, \phi \pm \psi\rangle = \phi(x) = \rho(x) \leqslant 1$.

Case 2: $\alpha > 0$. We have $\rho(x/\alpha) - \phi(x/\alpha) = \alpha/\alpha = 1$, so that $x/\alpha \in A_\phi$ and hence $|\psi(x/\alpha)| \leqslant 1$. Thus $|\psi(x)| \leqslant \alpha \equiv \rho(x) - \phi(x)$, so that again $\langle x, \phi \pm \psi\rangle \leqslant \rho(x) \leqslant 1$. ☐

We can now establish the main characterization of extreme points of sets of the type $U(M^\circ)$, known as the "Buck-Phelps theorem".

Theorem. *Let ρ be a continuous semi-norm on the real locally convex space X and let M be a linear subspace of X. A functional $\phi \in M^\circ \cap U_\rho^\circ$ is an extreme point of $M^\circ \cap U_\rho^\circ$ if and only if*

$$X = M + A_\phi - A_\phi. \tag{13.6}$$

Proof. Let σ be the semi-norm $d(\cdot, M)$, that is, $\sigma(x) \equiv \inf\{\rho(x - y): y \in M\}$. Then $U_\sigma = M + U_\rho$ and we define $B_\phi = \{x \in X: \sigma(x) - \phi(x) \leqslant 1\}$ for any $\phi \in U_\sigma^\circ$. By the lemma we know that $\phi \in \operatorname{ext}(U_\sigma^\circ)$ if and only if $B_\phi - B_\phi$ is dense in X. Now it is clear that $U_\sigma^\circ = M^\circ \cap U_\rho^\circ$. We show next that $M + A_\phi - A_\phi$ is dense in $B_\phi - B_\phi$. Since $\sigma(x + y) \leqslant \rho(x)$ for all $y \in M$, we see that $M + A_\phi \subset B_\phi$ and then that $M + A_\phi - A_\phi \subset B_\phi - B_\phi$. Next, select any $x \in B_\phi$ and set $u_n = \dfrac{n-1}{n} x$ for $n = 1, 2, \ldots$. We have $\sigma(u_n) - \phi(u_n) < 1$ so that there exists $y_n \in M$ such that $\rho(u_n + y_n) < 1 + \phi(u_n) = 1 + \phi(u_n + y_n)$; that is, $u_n + y_n \in A_\phi$. Similarly given $z \in B_\phi$ we can analogously define $v_n = \dfrac{n-1}{n} z$ and vectors $y'_n \in M$. We then have $(u_n + y_n) - (v_n + y'_n) \in A_\phi - A_\phi$ so that $u_n - v_n \in M + A_\phi - A_\phi$ and $\lim_n(u_n - v_n) = x - z$. This establishes the density of $M + A_\phi - A_\phi$ in $B_\phi - B_\phi$ as claimed above.

We now know that $\phi \in \operatorname{ext}(M^\circ \cap U_\rho^\circ)$ if and only if $M + A_\phi - A_\phi$ is dense in X. It remains to see that this density is equivalent to (13.6). However, this is a consequence of the fact that the sets A_ϕ are solid which entails that $M + A_\phi - A_\phi$ is also solid. Thus any $x \in X \setminus (M + A_\phi - A_\phi)$ could be separated from $M + A_\phi - A_\phi$ by a closed hyperplane, but this contradicts the density of $M + A_\phi - A_\phi$ in X. □

Corollary. *A functional $\phi \in M^\circ \cap U_\rho^\circ$ is an extreme point of $M^\circ \cap U_\rho^\circ$ if and only if for each $n = 1, 2, \ldots$,*

$$X = M + \frac{1}{n} A_\phi - \frac{1}{n} A_\phi.$$

Observe that

$$\frac{1}{n} A_\phi = \left\{x \in X: \rho(x) - \phi(x) \leqslant \frac{1}{n}\right\}.$$

E. Let us now give a few examples of the extreme point structure of the unit balls in certain normed linear spaces. We omit most of the details of the following assertions; filling these in should constitute an interesting exercise. The notation of **10D** is utilized when possible.

Example 1. Let X be either $C_b(\Omega, \mathbb{F})$ or else $L^\infty(\mu, \mathbb{F})$, where $\mathbb{F} \equiv \mathbb{R}$ or $\mathbb{C}$ and, in the latter case, μ is a σ-finite measure on some measure space. The extreme points of $U(C_b(\Omega, \mathbb{F}))$ are the functions with modulus one everywhere on Ω. Similarly, any μ-measurable function f with $\mu(\{t: |f(t)| \neq 1\}) = 0$ defines an extreme point of $U(L^\infty(\Omega, \mathbb{F}))$ and every extreme point is so obtained. (Interestingly enough, for all these spaces except $X = C_b(\Omega, \mathbb{R})$ we have

$$U(X) = \overline{\operatorname{co}}(\operatorname{ext}(U(X))). \tag{13.7}$$

However, when $X = C_b(\Omega, \mathbb{R})$ the validity of (13.7) is equivalent to a topological constraint on Ω, namely that Ω should be *totally disconnected*, which means that there is a base for the topology of Ω consisting of sets which are both open and closed. The proof of (13.7) depends on the equivalence of a) and b) in **13B** and on further knowledge of the continuous dual space X^*.) □

Example 2. Let Ω be the unit disc $\{z \in \mathbb{C}: |z| \leqslant 1\}$ and let $X = A(\Omega)$. As in Example 1, any function in X which is of modulus one on the boundary $\partial\Omega$ of Ω (the unit circle) is an extreme point of $U(X)$. More generally, because the functions in X are analytic on the interior of Ω, it follows that any function in $U(X)$ which has modulus one on a subset of positive (Lebesgue) measure of $\partial\Omega$ is also extreme. The complete answer is that $f \in U(X)$ is an extreme point if and only if

(13.8) $$\int_{-\pi}^{\pi} \log(1 - |f(e^{it})|)dt = -\infty.$$

(To see that this condition is necessary, assume that it fails. Select a continuous function h on $\partial\Omega$ such that $\theta \leqslant h(\cdot) \leqslant 1 - |f(\cdot)|$ and such that h is of class C^1 on each open arc of the set where $|f(\cdot)| < 1$. Then if we define

$$g(z) = \exp\left[\frac{1}{2\pi}\int_{-\pi}^{\pi} \frac{e^{it} + z}{e^{it} - z} \log h(t)dt\right],$$

we will have $g \in A(\Omega)$ and $\|f \pm g\|_\infty \leqslant 1$).

The same result holds for the space $H^\infty(\Omega)$, although some preliminary work is needed to establish the existence of boundary values on $\partial\Omega$ before the condition (13.8) can be applied. For both these spaces it is again true that formula (13.7) is valid. □

Example 3. Let $X = \text{Lip}([0, 1], d, \mathbb{F})$ where d is the usual metric on $[0, 1]$ and, as usual, $\mathbb{F} = \mathbb{R}$ or $\mathbb{C}$. Any function $f \in X$ is differentiable almost everywhere and, in fact, $\|f'\|_\infty = \|f\|_d$. If f has modulus one at each point of $[0, 1]$ then certainly $f \in \text{ext}(U(X))$ by Example 1. Also for f to belong to $\text{ext}(U(X))$ it is necessary that $\|f\|_\infty = 1$ (otherwise, we just add and subtract a suitable constant and see thereby that f is not extreme). Then we have that $f \in U(X)$ is extreme if and only if $|f'(\cdot)| = 1$ almost everywhere in the set where $|f(\cdot)| < 1$. (To see that this condition is necessary, we can proceed by contradiction. Let E be a compact subset of $\{t \in [0, 1]: |f(t)| < 1\}$ with positive measure such that $\|f'|E\|_\infty < 1$. Then we can choose $t_o \in [0, 1]$ such that the function

$$g(s) \equiv \int_0^s \chi_E(t)\{\chi_{[0, t_o]}(t) - \chi_{(t_o, 1]}(t)\}dt$$

belongs to X and $g'(\cdot)$ vanishes off E. Then for sufficiently small $\delta > 0$, $\|f \pm \delta g\|_L \leqslant 1$.) Again we remark that (13.7) is valid for this example. □

Example 4. We consider again $X = C_b(\Omega, \mathbb{F})$ but now we try to identify $\text{ext}(U(X^*))$. In this example we shall assume that Ω is a compact Hausdorff space. This is a very important case in practice. Our task is facilitated by the results of **8F**. We let P be the positive wedge in X and let $P^* \equiv P^+ \cap X^*$ be the continuous dual wedge in X^*. If we now define $K = \{\phi \in P^*: \phi(e) = 1\}$

where e is the function constantly equal to one on Ω then we know from **8F** that $\operatorname{ext}(K)$ consists of the algebra homomorphisms of X. In particular, given any point $t \in \Omega$ the *evaluation functional* δ_t defined by $\langle x, \delta_t \rangle = x(t)$ for $x \in X$ belongs to $\operatorname{ext}(K)$. Since K is clearly an extremal subset of $U(X^*)$ it follows that each evaluation functional is an extreme point of $U(X^*)$. More generally,

$$E \equiv \{\alpha\delta_t : |\alpha| = 1, t \in \Omega\} \subset \operatorname{ext}(U(X^*)).$$

Now evidently $^\circ E = U(X)$, so by the bipolar theorem (**12C**) we have

$$U(X^*) = {}^\circ E^\circ = \overline{\text{co}}^*(E),$$

where "$\overline{\text{co}}^*$" refers to weak*-closure. But since E is weak*-compact (by virtue of being a continuous image of the compact set $\{\alpha \in \mathbb{F} : |\alpha| = 1\} \times \Omega$), we see by applying (13.1) that

$$\operatorname{ext}(U(X^*)) \subset \bar{E}^* = E \subset \operatorname{ext}(U(X^*)).$$

Thus we have achieved the identification

(13.9) $$\operatorname{ext}(U(C(\Omega, \mathbb{F})^*)) = \{\alpha\delta_t : \alpha \in \mathbb{F}, |\alpha| = 1, t \in \Omega\},$$

for every compact Hausdorff space Ω. □

The discussion of this class of examples is continued in exercise 2.35 for the case of non-compact Ω. Also, in contrast with the preceding examples, $U(C(\Omega, \mathbb{F})^*)$ is not now generally equal to the norm-closure of the convex hull of its extreme points (consider, for example, the case $\Omega = [0, 1]$).

Example 5. A normed linear space $(X, \|\cdot\|)$ is *strictly normed* if $\|x + y\| = \|x\| + \|y\|$ implies that $x = ty$ for some $t \geqslant 0$ or else $y = \theta$. This constraint on the norm is easily seen to be equivalent to the geometric condition that $U(X)$ be rotund, where a convex set is *rotund* if every bounding point is an extreme point. From our present point of view such spaces are not very interesting since, for example, condition (13.7) is automatically fulfilled. Examples of strictly normed spaces are the $L^p(\Omega, \mu, \mathbb{F})$ spaces for $1 < p < \infty$; this may be shown by consideration of the condition for equally in Hölder's inequality. □

Example 6. For our final examples we consider normed linear spaces X and Y (over the same scalar field) and study some extreme points of $U(B(X, Y))$. A map $T \in B(X, Y)$ is an *isometry* if T "preserves the norm", that is, if $\|T(x)\| = \|x\|$, for all $x \in X$. It is easy to see that if Y is strictly normed then any isometry in $B(X, Y)$ is an extreme point of $U(B(X, Y))$. Now we assume that $X = Y$ and abbreviate $B(X, Y)$ to $B(X)$. Then we claim that the identity map I (where $I(x) \equiv x$ for all $x \in X$) is an extreme point of $U(B(X))$ (whether or not X is strictly normed). In fact, one can prove the much stronger assertion that I is a *vertex* of $U(B(X))$ in the sense that the set $\{\phi \in B(X)^* : \|\phi\| = 1 = \phi(I)\}$ is total over $B(X)$. In other words, the intersection of all hyperplanes of support to $U(B(X))$ that contain I

is just $\{I\}$. Examples of such functionals ϕ are the *double evaluation* functionals $\omega_{f,x} \in B(X)^*$ defined by

$$\omega_{f,x}(T) = \langle T(x), f \rangle, \qquad T \in B(X),$$

where $x \in X, f \in X^*$, $\|f\| = \|x\| = f(x) = 1$. (The proof that I is a vertex is not entirely straightforward for general spaces X but it should be clear in the special finite dimensional case where $X = \mathbb{R}^n$ or $\mathbb{C}^n$.)

Once it is known that I is a vertex of $U(B(X))$ for some X it then readily follows that any isometry $T \in B(X)$ whose range is all of X is also a vertex. This can be seen by observing that the map $S \mapsto T^{-1} \circ S$ is an isometry on $B(X)$ that sends T into I. Thus, for example, when $X = \mathbb{R}^n$ or $\mathbb{C}^n$ (with the usual Euclidean norm) then every linear map on X defined by a unitary matrix is a vertex of $U(B(X))$. Furthermore, in this case a strong converse is valid: every extreme point of $U(B(X))$ is defined by a unitary matrix, and hence is a vertex. (If $T \in \text{ext}(U(B(X)))$ is defined by $T(x) = Ax$ for some square matrix A, then we can express A as VDU^H ("singular value decomposition") where U and V are unitary, U^H is the hermitian transpose of U, and D is a diagonal matrix with diagonal entries $0 \leqslant d_1, d_2, \ldots, d_n \leqslant 1$. Because T is extreme each d_j is either 0 or 1. If some $d_j = 0$ we define a linear map S by $S(x) = \langle x, u_j \rangle v_j$ where u_j (resp. v_j) is the jth column of U (resp. V). Then $\|T \pm S\| \leqslant 1$ contradicting that T is extreme. Thus we see that A is unitary.) □

§14. Convex Functions and Optimization

In this section we resume our general discussion of convex functions which was begun in §3 and continued in **6D** and **7D-E**. Further developments depend on topological considerations reflected in continuity assumptions about the functions. Our approach constitutes a noteworthy application of the geometric theory developed in earlier sections. In particular, it is interesting to observe how the existence of various separating or supporting hyperplanes to certain convex sets entails analytical information about a given convex function or program.

A. We first discuss conditions which insure the continuity of a given convex function f defined on an open convex set A in a linear topological space X. The main point to be made is that, except in very pathological cases, f is automatically continuous on A. We have already seen a special case of this in **11D** where f was a linear function: $f \in X'$. We note that if f is continuous at a point p in A then f is certainly bounded on some neighborhood of p. It is striking that this trivial necessary condition forces f to be continuous throughout A.

Theorem. *Let A be an open convex subset of the linear topological space X. If $f \in \text{Conv}(A)$ is bounded above on a neighborhood of a point $p \in A$ then f is continuous at every point in A.*

Proof. Let us first see that f must be bounded from above on a neighborhood of any point $q \in A$. Given such a point q, there is $t > 1$ such that $p + t(q - p) \in A$ (**11A**). Now, suppose that $\alpha \equiv \sup\{f(x): x \in p + V\} < \infty$ for some balanced θ-neighborhood V; then we claim that f is bounded on the q-neighborhood $q + (1 - 1/t)V$. Indeed, if $z = q + (1 - 1/t)v$ for some $v \in V$, then

$$f(z) = f\left(q - \left(1 - \frac{1}{t}\right)p + \left(1 - \frac{1}{t}\right)(p + v)\right)$$
$$\leqslant \frac{1}{t} f(p + t(q - p)) + \left(1 - \frac{1}{t}\right)\alpha.$$

To complete the proof it will suffice to show that if f is bounded above on some neighborhood of $p \in A$ then f is continuous at p. Choose α and V as above and let $0 < \varepsilon < 1$. Then if $z \in p + \varepsilon V$ we can write

$$z = (1 - \varepsilon)p + \varepsilon(p + v)$$

for some $v \in V$, and therefore

$$f(z) \leqslant (1 - \varepsilon)f(p) + \alpha\varepsilon,$$
$$f(z) - f(p) \leqslant \varepsilon(\alpha - f(p)).$$

On the other hand, since, for any $v \in V$, we can write

$$p = \frac{1}{1 + \varepsilon}(p + \varepsilon v) + \left(1 - \frac{1}{1 + \varepsilon}\right)(p - v),$$

we have

$$f(p) \leqslant \frac{1}{1 + \varepsilon} f(p + \varepsilon v) + \frac{\varepsilon}{1 + \varepsilon} f(p - v)$$
$$\leqslant \frac{1}{1 + \varepsilon} f(z) + \frac{\alpha\varepsilon}{1 + \varepsilon}.$$

This yields

$$\varepsilon(f(p) - \alpha) \leqslant f(z) - f(p).$$

Thus we see that

$$|f(z) - f(p)| \leqslant \varepsilon(\alpha - f(p)),$$

for all $z \in p + \varepsilon V$, proving that f is continuous at p. □

Some important corollaries are now at hand. First, if A is not open but is solid then the theorem applies to the interior of A. Next, suppose that A has no interior. We may view A as a subset of its *closed affine hull* $\overline{\text{aff}}(A) \equiv \overline{\text{aff}(A)}$. Relative to $\overline{\text{aff}}(A)$ the set A may be solid. This will occur exactly when there is a point $p \in A$ and a θ-neighborhood V in the closed linear subspace $\overline{\text{span}}(A - A) \equiv \overline{\text{span}(A - A)}$ such that $p + V \subset \overline{\text{aff}}(A)$. Such points p (if any) constitute the *relative interior* of A (written rel-int(A)), and evidently the theorem still applies to these points.

Corollary 1. *If $f \in$ Conv(A) is bounded above on some neighborhood of a relative interior point of A, then f is continuous throughout* rel-int(A).

The concept of relative interior of a convex set is the appropriate substitute for the intrinsic core (**2C**) when dealing with infinite dimensional linear topological spaces. In particular, then, we have the following strong result about convex functions with finite dimensional domain.

Corollary 2. *If* $f \in \mathrm{Conv}(A)$ *where* $A \subset \mathbb{R}^n$ *for some* n, *then* f *is continuous throughout* $\mathrm{icr}(A)$.

When X is a normed linear space we can make a still stronger assertion about the continuity of convex functions with domain in X. Namely such a function must satisfy a Lipschitz condition throughout some neighborhood of each point of continuity; see exercise 2.40.

It is clear from the theorem that continuity of $f \in \mathrm{Conv}(A)$ at some point $p \in \mathrm{int}(A)$ is equivalent to upper semicontinuity of f at p. On the other hand, when f is only known to be lower semicontinuous at p the theorem need not apply. Nevertheless (exercise 3.50), it is a consequence of the Baire category theorem that when X is a Banach space and f is lower semicontinuous at every point of A, then f is continuous throughout $\mathrm{cor}(A)$ (when A is closed we actually have $\mathrm{cor}(A) = \mathrm{int}(A)$, again by the Baire theorem). If we recall that f is lower semicontinuous on A if and only if the sub-level sets $\{x \in A : f(x) \leqslant \lambda\}$ are closed for all λ, then we have the basis for the proof of the second part of the next corollary, which establishes the connection between continuity properties of convex functions and topological properties of their epigraphs (**3A**).

Corollary 3. *Let* $f \in \mathrm{Conv}(A)$ *where* A *is a convex subset of the linear topological space* X.

a) f *is continuous throughout* $\mathrm{int}(A)$ *if and only if* $\mathrm{epi}(f)$ *is solid*;

b) f *is lower semicontinuous on* A *if and only if* $\mathrm{epi}(f)$ *is a closed subset of* $X \times \mathbb{R}^1$.

B. Let $f \in \mathrm{Conv}(A)$ where A is a convex set in the real linear topological space X. Subgradients of f were introduced in **6D** and were shown to exist (in X') at each intrinsic core point of A. As usual, we would like these linear functionals to be continuous and it is natural to inquire as to what hypothesis on f will ensure this. The following result is a satisfactory answer.

Theorem. *Let* $p \in A$. *The set* $\partial f(p) \cap X^*$ *of continuous subgradients is a weak*-closed convex set. If* f *is continuous at* $p \in \mathrm{int}(A)$, *this set is also non-empty and weak*-compact.*

Proof. The first assertion is clear from the definitions. If now f is continuous at p and $\phi \in \partial f(p)$ then $\phi \leqslant f - (f(p) - \phi(p))$ on A. In particular ϕ is bounded (above) on some p-neighborhood and hence continuous by **11D**. To complete the proof it is now sufficient to show that $\partial f(p)$ is relatively weak*-compact in X^*. From **12D** we see that this will be true if there is a θ-neighborhood $V \subset X$ such that $\partial f(p) \subset V^\circ$. But there is such a neighborhood, namely $V = \{x \in X : |f(p + x) - f(p)| < 1\}$. □

From now on we shall consider only continuous subgradients for convex functions; in particular, the set $\partial f(p)$ will always be considered to belong to X^* (rather than X'). In practical terms this results in little loss of generality, since we just observed it to be the case whenever f is continuous at p.

Let us note that when X is a normed linear space and f is continuous at p then $\partial f(p)$ is a bounded subset of X^*. This is a consequence of the convexity and weak*-compactness of $\partial f(p)$ (**12E**).

C. Let f, A, and X be as in the preceding section, and let us suppose that the directional derivative $f'(p; x)$ is defined for some $p \in A$ and every $x \in X$. From **7D** we know that this will be the case in particular if $p \in \text{int}(A)$. Now if $\phi \in \partial f(p)$ it follows from formula (7.8) that the directional derivative function $f'(p; \cdot)$ is bounded below on some θ-neighborhood in X. This means that as we move linearly away from the point p the value of f cannot drop off too sharply. We show next that this condition is actually equivalent to the subdifferentiability of f at p, provided that X is locally convex.

Theorem. *Let $f \in \text{Conv}(A)$ where A is a convex subset of the real locally convex space X. For any $p \in A$ we have $\partial f(p) \neq \varnothing$ if and only if there is some θ-neighborhood $V \subset X$ such that $-\infty < \inf\{f'(p; x): x \in V\}$.*

Proof. Since X is locally convex we can assume (**10A**) that V is a barrel with gauge ρ_V. Since $f'(p; \cdot)$ is positively homogeneous we have

$$\gamma \rho_V(x) \leqslant f'(p; x), \qquad x \in X,$$

where $\gamma \equiv \inf\{f'(p; x): x \in V\}$. Now $|\gamma|\rho_V$ is a continuous seminorm on X so that its epigraph E is a solid convex set in $X \times \mathbb{R}^1$. Also, $\text{int}(-E) \cap \text{epi}(f'(p; \cdot))$ is void; for otherwise it would contain a point (x, t) and then $t < -|\gamma|\rho_V(x) = \gamma\rho_V(x) \leqslant f'(p; x) \leqslant t$, a contradiction. Consequently, we can separate these two convex sets by a closed hyperplane $[\psi; \alpha]$ where $\psi(x, t) \equiv \phi(x) + t$ for $x \in X$, $t \in \mathbb{R}^1$ (**6D**). Since $\text{epi}(f'(p; \cdot))$ is a wedge we must have $\alpha = 0$. Further, we must have $s \neq 0$ because, if $s = 0$, then it would follow that $\phi(x) < 0$ for every $x \in X$, since (x, t) always belongs to $\text{int}(-E)$ for sufficiently small negative t. But now part b) of the Lemma in **6D** allows us to conclude that $-\phi/s \in \partial f(p)$. □

When X is finite dimensional there is an even more striking implication of the failure of f to be subdifferentiable at a point $p \in A$.

Corollary. *If $\dim(X) < \infty$ and $\partial f(p) = \varnothing$ for some $p \in A$, then there there exists $x \in X$ such that $f'(p;x) = -\infty$.*

Actually, any $x \in \text{icr}(A - p)$ yields $f'(p; x) = -\infty$ when $\partial f(p) = \varnothing$. This corollary may be illustrated by the function $f(t) \equiv -\sqrt{1 - t^2}$ defined on $A \equiv [-1, 1] \subset \mathbb{R}^1$. We see that $\partial f(\pm 1) = \varnothing$ (exercise 1.25) and that $f'(\pm 1; \mp 1) = -\infty$.

The theorem also yields a new proof that $\partial f(p) \neq \varnothing$ whenever $p \in \text{int}(A)$ and f is continuous at p. For if we put $g(x) = f(p + x) - f(p)$ then g is continuous at θ and $f'(p; \cdot) \leqslant g$. This shows that $f'(p; \cdot)$ is continuous at θ and hence certainly bounded below on some θ-neighborhood.

D. We give one final and important general relation between directional derivatives and subgradients of convex functions. Let $f \in \text{Conv}(A)$ where A is a solid convex set in the real linear topological space X. Then from equation (7.8) we have, for $p \in \text{int}(A)$.

(14.1)
$$\begin{aligned} -f'(p; -x) &\leqslant \inf\{\psi(x):\psi \in \partial f(p)\} \\ &\leqslant \sup\{\psi(x):\psi \in \partial f(p)\} \leqslant f'(p; x), \end{aligned}$$

for any $x \in X$. We now see when the outside inequalities in (14.1) become equalities.

Theorem. *If X is a real locally convex space and $f \in \text{Conv}(A)$ is continuous at $p \in \text{int}(A)$, then*

(14.2)
$$f'(p; x) = \max\{\psi(x):\psi \in \partial f(p)\}$$

and

(14.3)
$$-f'(p; -x) = \min\{\psi(x):\psi \in \partial f(p)\}$$

for every $x \in X$.

Proof. By **14B** the set $\partial f(p)$ is weak*-compact so that the max and min in (14.2) and (14.3) are attained. We shall just prove (14.2) as (14.3) then follows by an analogous argument (or even by just a change in sign of x). Suppose that

(14.4)
$$\max\{\psi(x):\psi \in \partial f(p)\} < \alpha < f'(p; x)$$

for some $x \in X$ and $\alpha \in \mathbb{R}$. Arguing as in **7E** we define a linear functional $\bar{\psi}$ on $M \equiv \text{span}\{x\}$ by $\bar{\psi}(tx) = \alpha t$ for all $t \in \mathbb{R}$. Then on M, $\bar{\psi} \leqslant f'(p; \cdot)$ and hence by **11G** there is a continuous extension ψ on $\bar{\psi}$ to all of X such that $\psi \leqslant f'(p'; \cdot)$. Since $\psi(x) \equiv \alpha$ we have arrived at a contradiction to (14.4). □

It is an instructive exercise to give an alternative proof of this theorem by separating $\text{epi}(f)$ from the ray $\{(p + tx, f(p) + tf'(p; x)):t \geqslant 0\}$ in $X \times \mathbb{R}^1$. In any case we now have the exact analogue of the results in **7E** for continuous subgradients provided that we make the usual continuity hypothesis on f.

Corollary 1. *With the same hypotheses on f, p, A and X, we have that the gradient $\nabla f(p)$ exists in X^* if and only if the subdifferential $\partial f(p)$ consists of a single element, namely $\nabla f(p)$.*

The above theorem has some further more substantial corollaries. We give one now and another in the next subsection. Consider a fixed set $A \subset X$ and let $p \in A$. Then the gradient map $f \to \nabla f(p)$ is a linear map from the space of smooth functions on A into X^*. When we drop the smoothness requirement we still obtain an analogue of the gradient map by imposing convexity conditions: A is a convex set and $f \in \text{Conv}(A)$. The analogue is now the subdifferential map $f \mapsto \partial f(p)$, considered as a map from the wedge $\text{Conv}(A)$ into the weak*-closed convex subsets of X^*. We show that this map, although in no ways linear (Conv(A) is not even a linear space), still generally respects the wedge operations on Conv(A).

Corollary 2. *Let $f, g \in \mathrm{Conv}(A)$ and assume both are continuous at $p \in A$. Then for any non-negative numbers s and t we have*

$$\partial(sf + tg)(p) = s\partial f(p) + t\partial g(p).$$

Proof. Let $h = sf + tg$. Then

$$\begin{aligned}\max\{\psi(x):\psi \in \partial h(p)\} &= h'(p; x) = sf'(p; x) + tg'(p; x)\\ &= s\max\{\psi(x):\psi \in \partial f(p)\} + t\max\{\psi(x):\psi \in \partial g(p)\}\\ &= \max\{\psi(x):\psi \in s\partial f(p) + t\partial g(p)\}, \qquad x \in X.\end{aligned}$$

Now the sets $\partial h(p)$ and $s\partial f(p) + t\partial g(p)$ are both convex and weak*-compact, and the first contains the second (using **7E**, for example). They must therefore be equal (**13B**). □

Again it is an instructive exercise to give a direct proof by use of a separating hyperplane argument in $X \times \mathbb{R}^1$. This will also yield a slightly stronger version of Corollary 2, in that it will be seen that the continuity of only one of the functions f and g at p need be assumed.

E. As another application of the preceding theorem we derive a global criterion of the solvability of convex optimization problems. Suppose we are given the variational pair (A, f) consisting of a convex set $A \subset X$ (real, locally convex) and $f \in \mathrm{Conv}(A)$. The general problem then is to minimize f over A, and in particular, to decide whether a given point $p \in A$ is a solution in the sense that

$$f(p) = \min\{f(x):x \in A\}.$$

To accomplish this we introduce the set $F(p; A)$ of *feasible directions* of A at p as the set of all $x \in X$ for which some $\delta > 0$ exists (depending on x) such that $p + tx \in A$ for $0 \leqslant t < \delta$. This concept is related to our earlier notion (**8C**) of the recession cone C_A by

(14.5) $$C_A = \cap\{F(p; A):p \in A\}$$

provided that A is closed (actually, (14.5) does not even require that A be convex). Now the set $F(p; A)$ is a wedge in X; we let $F(p; A)^*$ be the continuous dual wedge, that is, the wedge of continuous linear functionals on X which assume non-negative values on $F(p; A)$. We then have the following optimality principle ("Pshenichnii's condition").

Theorem. If f is continuous at $p \in A$ then p is a solution of the convex program (A, f) if and only if

(14.6) $$\partial f(p) \cap F(p; A)^* \neq \varnothing.$$

Proof. Suppose that condition (14.6) holds and that $\phi \in \partial f(p) \cap F(p; A)^*$. Since A is convex we have $x - p \in F(p; A)$ for all $x \in A$, and so

$$0 \leqslant \phi(x - p) \leqslant f(x) - f(p), \qquad x \in A.$$

Thus p is a solution of our program. (Note that the continuity hypothesis was not needed for this implication.)

Conversely, suppose that p is a solution but that (14.6) does not hold. Then, in X^* the origin does not belong to the weak*-closed convex set $F(p; A)^* - \partial f(p)$ (the weak*-closure of this set results from the weak*-compactness of $\partial f(p)$). Applying the strong separation theorem and **14A** we obtain $x_o \in X$ such that

$$\delta \equiv \inf\{\phi(x_o) : \phi \in F(p; A)^* - \partial f(p)\} > 0,$$

or

(14.7) $$\inf\{\phi(x_o) : \phi \in F(p; A)^*\} \geqslant \delta + \max\{\phi(x_o) : \phi \in \partial f(p)\}.$$

Since $F(p; A)^*$ is a wedge the left side of (14.7) must be zero. This has two implications: first, that $f'(p; x_o) \leqslant -\delta < 0$, and second, that $x_o \in \overline{F(p; A)}$. This second fact follows from the bipolar theorem (**12C**) when we recognize that $F(p; A)^* \equiv -F(p; A)^\circ$. Now since $f'(p; \cdot)$ is continuous at θ (**14C**), it is everywhere continuous (**14A**). Hence $f'(p; x)$ is negative at all x in some x_o-neighborhood and in particular at some point $\bar{x} \in F(p; A)$. But this means that $f(p + t\bar{x}) - f(p)$ is negative for sufficiently small t; since $p + t\bar{x} \in A$ for such t we have arrived at a contradiction. □

Corollary. *Under the same hypotheses on A, f, and p, a necessary and sufficient condition for p to be a solution of the convex program (A, f) is that there exist $\phi \in \partial f(p)$ such that p is a solution of the program (A, ϕ).*

Again, the sufficiency of the condition does not depend on the continuity assumption. The effect of this corollary is to reduce the quest for solutions of the original convex program (A, f) to the quest for solutions to the linear program (A, ϕ). The practical application of this reduction depends of course on our knowledge of the subdifferential $\partial f(p)$. The most important special case is that where f is smooth, in the sense that $f'(p; \cdot) \equiv \nabla f(p)$ exists in X^*. Then, as we know (**14D**), $\partial f(p) = \{\nabla f(p)\}$ and so our program (A, f) reduces to the linear programs $(A, \nabla f(p))$, $p \in A$.

Example 1. Consider the special case where $A = x_o + M$ is an affine subspace (**1C**) of X. For any $p \in A$ the necessary and sufficient condition that p solve the program (A, f) is that $\partial f(p) \cap M^\circ \neq \varnothing$, where M° is the annihilator subspace of M in X^*. This is because $F(p; A) = M$ in this case. □

Example 2. Suppose that $g \in \text{Conv}(X)$ and that A is the set $\{x \in X : g(x) \leqslant 0\}$. An important special case occurs when we have a semi-norm ρ on X and we put $g(\cdot) = \rho(\cdot) - \lambda$, for some $\lambda > 0$; that is, $A = \lambda U_\rho$. We select $p \in A$ and try to determine $F(p, A)^*$. If $p \in \text{int}(A)$ then clearly $F(p, A) = X$ and $F(p, A)^* = \{\theta\}$, whence p is a solution of the program (A, f) if and only if $\theta \in \partial f(p)$. (Note that this conclusion does not depend on the special form of A.) Otherwise, and this is the more typical case, p is a boundary point of A. We assume that g is continuous so that $g(p) = 0$. We shall also assume that there is some point $q \in A$ such that $g(q) < 0$ ("Slater's regularity condition"). Then we assert that

(14.8)
$$\begin{aligned} F(p; A)^* &= (-\infty, 0]\partial g(p) \\ &\equiv \{\phi \in X^*: \phi = t\psi, t \leqslant 0, \psi \in \partial g(p)\}. \end{aligned}$$

The inclusion from right to left in (14.8) is clear, because if $x \in F(p; A)$ and $\psi \in \partial g(p)$, then

$$t\psi(x) \leqslant g(p + tx) \leqslant 0$$

for sufficiently small $t > 0$, whence $\psi(x) \leqslant 0$. To reverse the inclusion we note that since $q - p \in A$ we have

$$\psi(q - p) \leqslant g(q) < 0, \qquad \psi \in \partial g(p),$$

so that $\theta \notin \partial g(p)$. Let $\phi \in F(p; A)$ ($\phi \neq \theta$); we shall assume that $t\phi \notin \partial g(p)$ for any $t < 0$ and reason to a contradiction. We can separate the weak*-closed convex set $(-\infty, 0]\phi - \partial g(p)$ from θ and so obtain $x_o \in X$ such that

(14.9)
$$\sup\{\psi(x_o): \psi \in \partial g(p)\} < 0 \leqslant t\phi(x_o), \qquad t \leqslant 0.$$

By **14D** it follows that $g'(p; x_o) < 0$ and hence $x_0 \in \operatorname{cor}(F(p; A))$. But then $\phi(x_o) > 0$, in contradiction to (14.9).

We conclude that a point p with $g(p) \leqslant 0$ is a solution of the convex program

$$\min\{f(x): g(x) \leqslant 0\}$$

if and only if there are subgradients $\phi \in \partial f(p)$, $\psi \in \partial g(p)$ and a "multiplier" $\lambda \geqslant 0$ such that

(14.10)
$$\begin{aligned} \phi + \lambda\psi &= \theta \\ \lambda g(p) &= 0. \end{aligned}$$

□

This example could be further generalized by replacing g by a vector valued convex function on X, that is, a map from X into $\mathbb{R}^n$ each component of which is a convex function. We would then be dealing with the problem of minimizing f subject to n simultaneous convex constraints $g_i(x) \leqslant 0$, $i = 1, \ldots, n$. After a fair amount of work we would arrive at the natural generalization of (14.10), namely that the existence of multipliers $\lambda_i \geqslant 0$ such that

(14.11)
$$\begin{aligned} &\theta \in \partial f(p) + \sum_{i=1}^{n} \lambda_i \, \partial g_i(p), \\ &\lambda_i g_i(p) = 0, \qquad i = 1, \ldots, n \end{aligned}$$

is necessary and sufficient for p to be a solution (the necessity of (14.11) again requires a regularity assumption).

It would even be possible to go further, replacing $\mathbb{R}^n$ by a suitable ordered linear space Y, and g be a convex mapping of X into Y. But we do not feel the added generality justifies the effort involved, the above examples being adequate illustrations of the optimality principle. However, the concept of a Y-valued convex mapping is useful, and will be utilized in the following sub-sections; in particular, to develop some new principles of convex optimization.

F. We are now going to derive a very general principle of convex analysis whose usefulness will be amply illustrated by subsequent examples. Let A be a convex subset of a linear space X and let Y, Z be two linear topological spaces (all linear spaces are real). We assume that Y and Z are also ordered linear spaces (**5A**) with orderings induced by positive wedges $P \subset Y$, $Q \subset Z$. A mapping $S: A \to Y$ is a *convex mapping* if

$$S(\alpha u + (1 - \alpha)v) \leqslant \alpha S(u) + (1 - \alpha)S(v),$$

for every $u, v \in A$ and $0 \leqslant \alpha \leqslant 1$. The inequality here refers to the vector ordering on Y, and so we cannot determine the convexity of a mapping S until we have specified also the ordering on Y. Obviously all linear maps from X into Y are examples of convex mappings. We suppose given two convex mappings $S: X \to Y$ and $T: X \to Z$. The case $Y = \mathbb{R}^1$, $Z = \mathbb{R}^n$ is of special importance. In this case S is simply a convex function on A and T is an n-tuple $(g_1, \ldots, g_n)$ with each $g_i \in \text{Conv}(A)$.

A subset V_o of Y is said to be a *regularizing set* for the positive wedge P if $\theta \in \bar{V}_o$ and $P + V_o$ is a solid convex set. If we know that P is solid then this condition holds with $V_o \equiv \{\theta\}$ or P. In general, regularizing sets are introduced when P is known not to have interior. When V_o is a regularizing set for P we write

$$y_1 \leqslant y_2 + (V_o)$$
$$(\text{resp. } y_1 < y_2 + (V_o))$$

to indicate that $y_2 - y_1 \in P + V_o$ (resp. $y_2 - y_1 \in \text{int}(P + V_o)$).

Consider now an abstract inequality system

(14.12) $$S(x) < \theta, \qquad x \in A.$$

If V_o is a regularizing set for P we shall say that $x_0 \in A$ is a V_o-*solution* of the system (14.12) if $S(x_o) < \theta + (V_o)$. If such a solution exists, the system is V_o-*consistent.* Suppose that $W_o \subset Z$ is a regularizing set for the positive wedge Q. Then the system

(14.13) $$S(x) < \theta, \qquad T(x) < \theta, \qquad x \in A$$

is (V_o, W_o)-*consistent* if there exists a (V_o, W_o)-*solution* $x_o \in A$ in the sense that $S(x_o) < \theta + (V_o)$ and $T(x_o) < \theta + (W_o)$. The main result ("Tuy's inconsistency condition") is a necessary condition for the inconsistency of an abstract system of the form (14.13) under certain hypotheses.

Theorem. *Let S and T be convex mappings from the convex set A in the (real) linear space X into the ordered linear topological spaces Y and Z, respectively. Let V_o (resp. W_o) be a regularizing set for the positive wedge P (resp. Q) in Y (resp. Z) and, suppose that the system (14.12) is V_o-consistent. Then if the system (14.13) is (V_o, W_o)-inconsistent there exist continuous* *monotone linear functionals $\phi \in P^*$, $\psi \in Q^*$ such that $\psi \neq \theta$ and*

(14.14) $$\langle S(x), \phi \rangle + \langle T(x), \psi \rangle \geqslant 0, \qquad x \in A.$$

Proof. Let $\bar{x}$ be a V_o-solution of the system (14.12). We introduce the set

$$E \equiv \{(y, z) \in Y \times Z : y - S(x) \in P + V_o \text{ and } z - T(x) \in Q + W_o \text{ for some } x \in A\}.$$

We assert that E is a solid convex set and that (θ, θ) is not an interior point of E. The convexity of E follows from the convexity of S and T, and the second assertion follows from the hypothesis that the system (14.13) is (V_o, W_o)-inconsistent. To prove that E is solid we select an interior point w_o of $Q + W_o$ and define $\bar{z} = w_o + T(\bar{x})$. We then claim that $(\theta, \bar{z}) \in \operatorname{int}(E)$. To see this, we select θ-neighborhoods $V \subset Y$, $W \subset Z$ such that $V - S(\bar{x}) \subset P + V_o$ and $w_o + W \subset Q + W_o$. Then it easily follows that $V \times (\bar{z} + W) \subset E$. Thus E is a solid convex set. We therefore can separate E from (θ, θ) by a closed hyperplane (**11E**) and so find a non-zero linear functional $\Phi \in (Y \times Z)^*$ such that $\Phi(y, z) \geqslant 0$ for all $(y, z) \in E$. Now we define $\phi \in Y^*$, $\psi \in Z^*$ by $\phi(y) = \Phi(y, \theta)$ and $\psi(z) = \Phi(\theta, z)$. The remainder of the proof involves showing that ϕ and ψ have the desired properties.

By definition of regularizing sets there are nets $\{y_\delta\} \subset P + V_o$ and $\{z_\gamma\} \subset Q + W_o$ each convergent to the respective zero vectors. We have

$$(S(x) + y_\delta) - S(x) \in P + V_o, \qquad x \in A,$$
$$(T(x) + z_\gamma) - T(x) \in Q + W_o, \qquad x \in A,$$

so that

$$(S(x) + y_\delta, T(x) + z_\gamma) \in E, \qquad x \in A,$$

and hence

$$\langle S(x) + y_\delta, \phi \rangle + \langle T(x) + z_\gamma, \psi \rangle \geqslant 0, \qquad x \in A,$$

We thus obtain (14.14) by letting $y_\delta \to \theta$, $z_\gamma \to \theta$.

Next, select any $y \in P$, $z \in Q$ and $s, t > 0$. Then

$$(S(x) + sy, T(x) + tz) \in E, \qquad x \in A,$$

and so

$$\langle S(x) + sy, \phi \rangle + \langle T(x) + tz, \psi \rangle \geqslant 0.$$

Hence

(14.15) $$\langle S(x), \phi \rangle + \langle T(x), \psi \rangle + s\phi(y) + t\psi(z) \geqslant 0.$$

By letting first s, then t become large in (14.15) we see that $\phi(y) \geqslant 0$, $\psi(z) \geqslant 0$; that is, we have shown that $\phi \in P^*$, $\psi \in Q^*$.

Finally, we observe that if $\psi = \theta$, then from the fact that $(\theta, \bar{z}) \in \operatorname{int}(E)$ we would have ϕ non-negative on some θ-neighborhood in Y, whence $\phi = \theta$. But this contradicts $\Phi \neq \theta$. □

It is to be noted that the regularizing sets make a transient appearance in this argument; the conclusion (14.14) depends only on the data A, S, T, P, and Q.

Corollary 1. *If the positive wedges P and Q are both solid and if the system (14.12) is consistent, then the system (14.13) is inconsistent only if functionals $\phi \in P^*$, $\psi \in Q^*$ ($\psi \neq \theta$) exist and satisfy (14.14).*

We now proceed to several applications of the theorem and Corollary 1. The first is known as the "Farkas-Minkowski lemma".

Corollary 2. *Let A be a convex subset of the (real) linear topological space X and let $f \in \mathrm{Conv}(A)$. Let $S: A \to Y$ be a convex mapping with values in an ordered linear topological space Y. Assume that the associated system (14.12) is consistent and that $f(x) \geqslant 0$ whenever $x \in A$ satisfies $S(x) < \theta$. Then there exists a monotone linear functional $\phi \in Y^*$ such that*

$$f(x) \geqslant -\langle S(x), \phi \rangle, \qquad x \in A.$$

Proof. This is a direct consequence of Corollary 1 and the theorem if we take $Z = \mathbb{R}^1$ with the usual ordering and let $T = f$. □

This corollary contains as a special case the classical version of Farkas' lemma in matrix theory. Namely, let B be an $m \times n$ real matrix. Then a vector $b \in \mathbb{R}^n$ will satisfy $\langle b, x \rangle \geqslant 0$ for all x such that $Bx \geqslant \theta$ if and only if there is a non-negative vector $y \in \mathbb{R}^m$ such that $yB = b$.

G. For another application of Tuy's inconsistency condition we reconsider the general convex programming problem of **14E**. We shall assume that our program has the form

$$\min\{f(x): x \in A, S(x) \leqslant \theta\} \tag{14.16}$$

where A is a convex set in some linear space X, $S: A \to Y$ is a convex mapping, and Y is an ordered linear topological space with positive wedge P. As usual, the case $Y = \mathbb{R}^n$ is of special importance. The following optimality principle is the "Hurwicz saddle-point condition".

Theorem. *If $p \in A$ solves the program (14.16) and if the associated system (14.12) is consistent then there exists a linear functional $\phi \in P^*$ such that*

$$f(p) + \langle S(p), \psi \rangle \leqslant f(x) + \langle S(x), \phi \rangle \tag{14.17}$$

for all $x \in A$ and all $\psi \in P^*$. Conversely, if for some point $p \in A$ such a ϕ exists in P^*, and if P is closed in Y, Y being now a locally convex space, then p is a solution of (14.16).

Proof. Again the first assertion follows directly from Corollary 1 in **14F**, because if p is to be a solution of (14.16) then the system

$$S(x) < \theta, \qquad f(x) - f(p) < 0, \qquad x \in A \tag{14.18}$$

must be inconsistent.

To establish the second assertion we observe that (14.17) entails $\langle S(p), \psi \rangle \leqslant \langle S(p), \phi \rangle$ for all $\psi \in P^*$, whence $\langle S(p), \psi \rangle \leqslant 0$ for all $\psi \in P^*$. Then since P is closed it follows that $-S(p) \in P$, or, that $S(p) \leqslant \theta$. We now appeal to exercise 2.43 to conclude that the system

$$S(x) \leqslant \theta, \qquad f(x) - f(p) < 0, \qquad x \in A$$

is inconsistent and hence that p is a solution of (14.16). □

The consistency of the associated system (14.12), as a hypothesis for the necessity of (14.17) is again known as Slater's regularity condition (**14E**).

To see the reason for the saddle-point terminology employed just above we define a function L (the "Lagrangian function") on $A \times P^*$ by

(14.19) $$L(x, \phi) = f(x) + \langle S(x), \phi \rangle.$$

A point $(p, \phi) \in A \times P^*$ is called a *saddle-point* of L if for every $x \in A$ and $\psi \in P^*$ we have

$$L(p, \psi) \leqslant L(p, \phi) \leqslant L(x, \phi).$$

It thus appears that if we assume that Y is a locally convex ordered space with closed positive wedge P, and that the Slater condition holds, then $p \in A$ is a solution of the convex program (14.16) if and only if there exists $\phi \in P^*$ such that (p, ϕ) is a saddle point of the Lagrangian function (14.19).

H. We continue our study of the convex program (14.16) with the same assumptions as in the first paragraph of **14G**. Let v be the value (**8E**) of the program. We introduce the companion notion of weak value. We say that a net $\{x_\delta : \delta \in D\}$ is a *weak solution* of the system

(14.20) $$S(x) \leqslant \theta, \qquad x \in A$$

if $S(x_\delta) = y'_\delta + y''_\delta$ where $y'_\delta \leqslant \theta$ and $y''_\delta \to \theta$. The *weak value* of the program (14.16) is then

$$v' \equiv \inf \underline{\lim} \{ f(x_\delta) : \delta \in D \},$$

where the infimum is taken over all weak solutions (if there are none we set $v' \equiv +\infty$). In all cases we clearly have $v' \leqslant v$.

Suppose now that Y is locally convex and that its positive wedge P is closed. It follows that

$$g(x) \equiv \sup \{ f(x) + \langle S(x), \phi \rangle : \phi \in P^* \} = \begin{cases} f(x), \text{ if } S(x) \leqslant \theta \\ +\infty, \text{ if not} \end{cases}$$

and so

$$\inf\{ g(x) : x \in A \} = v.$$

In other words, in terms of the Lagrangian function (14.19)

$$v = \inf_{x \in A} \sup_{\phi \in P^*} L(x, \phi).$$

We are thus led to consider the variational pair (P^*, h) where $h(\phi) \equiv \inf\{ L(x, \phi) : x \in A \}$. The corresponding maximizing program is called the *dual* of (14.16). Let the dual value be denoted v^*:

$$v^* \equiv \sup_{\phi \in P^*} \inf_{x \in A} L(x, \phi).$$

We now have "Golštein's duality theorem". The proof constitutes another application of **14F** and illustrates the use of regularizing sets.

Theorem. *Assume that the system (14.20) is weakly consistent in that it possesses a weak solution. Then the weak value of the primal program (14.16) equals the value of the dual program:* $v' = v^*$.

Proof. Suppose first that $-\infty < v'$ and select some $\alpha < v'$. We claim that for some convex θ-neighborhood $V_0 \subset Y$ the system

(14.21) $$S(x) < \theta + (V_0), \qquad f(x) - \alpha < 0, \qquad x \in A$$

is inconsistent. For if not we let $\{V_\delta : \delta \in D\}$ be a θ-neighborhood base directed by inclusion and select a solution x_δ of (14.21) for each $\delta \in D$. The net $\{x_\delta : \delta \in D\}$ is then a weak solution of (14.20), but $\underline{\lim}\{f(x_\delta) : \delta \in D\} \leqslant \alpha < v'$, a contradiction of the definition of v'. (Note that any convex θ-neighborhood in Y is a regularizing set for the positive wedge P.) Now for the θ-neighborhood V_0 that makes (14.21) inconsistent the system

$$S(x) < \theta + V_0, \qquad x \in A$$

is consistent; this follows from the existence of a weak solution of (14.20). Hence we can apply Tuy's inconsistency condition and obtain $\phi \in P^*$ such that

$$f(x) + \langle S(x), \phi \rangle \geqslant \alpha, \qquad x \in A.$$

This proves that $v^* \geqslant \alpha$ for all $\alpha < v'$ and hence that $v^* \geqslant v'$.

For the converse let $\{x_\delta : \delta \in D\}$ be a weak solution of (14.20) so that $S(x_\delta) = y'_\delta + y''_\delta$, where $y'_\delta \leqslant \theta$ and $y''_\delta \to \theta$. For any $\phi \in P^*$ we have

$$\langle S(x_\delta) - y''_\delta, \phi \rangle + f(x_\delta) = \langle y'_\delta, \phi \rangle + f(x_\delta) \leqslant f(x_\delta),$$

which yields

$$\underline{\lim}\{\langle S(x_\delta) - y''_\delta, \phi \rangle + f(x_\delta) : \delta \in D\} \leqslant \underline{\lim}\{f(x_\delta) : \delta \in D\}.$$

Thus

$$\begin{aligned} h(\phi) &\equiv \inf\{L(x, \phi) : x \in A\} \\ &\leqslant \underline{\lim}\{L(x_\delta, \phi) : \delta \in D\} \leqslant \underline{\lim}\{f(x_\delta) : \delta \in D\}. \end{aligned}$$

This being true for all $\phi \in P^*$ we see that

$$v^* \equiv \sup\{h(\phi) : \phi \in P^*\} \leqslant v',$$

which completes the proof, even in the case where $v' = -\infty$. □

We say that the convex program (14.16) is *well-posed* if it has the same value as its dual program. Thus, being well-posed is equivalent to

(14.22) $$\inf_{x \in A} \sup_{\phi \in P^*} L(x, \phi) = \sup_{\phi \in P^*} \inf_{x \in A} L(x, \phi).$$

Let us also say that a sequence $\{(x_n, \phi_n)\}$ in $A \times P^*$ is a *weak saddle-point* for the Lagrangian L on $A \times P^*$ if there exists a numerical sequence $\varepsilon_n \geqslant 0$,

$\varepsilon_n \to 0$ such that for every n

$$\textbf{(14.23)} \qquad L(x_n, \psi) - \varepsilon_n \leqslant L(x, \phi_n) + \varepsilon_n, \qquad x \in A, \qquad \phi \in P^*.$$

We then have an alternative characterization of well-posed programs.

Corollary. *The convex program (14.16) is well-posed (with a finite value) if and only if there is a weak saddle-point for its Lagrangian.*

Proof. From (14.23) we see that the left side of (14.22) is not larger than the right side. Since the reverse inequality is always true for any function L we see that (14.22) is valid and hence that the program is well-posed. Conversely, assume the program to be well-posed and let $\{x_n\} \subset A$ be any minimizing sequence: $g(x_n) \to v > -\infty$. Similarly, let $\{\phi_n\} \subset P^*$ be a maximizing sequence for the dual program: $h(\phi_n) \to v$. Define

$$\varepsilon_n = \max\{|v - g(x_n)|, |v - h(\phi_n)|\}.$$

Then $0 \leqslant \varepsilon_n \to 0$ while $g(x_n) \leqslant v + \varepsilon_n$ and $v - \varepsilon_n \leqslant h(\phi_n)$; this leads to (14.23). □

I. In **6E** we summarized the equivalence of six versions of the basic separation theorem in linear spaces. In the present chapter we have obtained the topological forms of these principles along with several new versions. It remains true that all these versions are equivalent to one another. They (collectively) constitute the single most important general principle of geometric functional analysis. (We may also safely assert that the (extended) Krein-Milman theorem of **13B** is the second most important general principle of our subject.) For ease of reference we now list the ten topological formulations of our fundamental principle.

1) The separation theorem (**11E**);
2) the support theorem (**11E**, Cor. 2);
3) the Hahn-Banach theorem (**11G**);
4) the Krein-Rutman theorem (**6B** and exercise 2.46);
5) the subdifferentiability theorem (**11C**);
6) the Tuy inconsistency theorem (**14F**);
7) the Farkas-Minkowski lemma (**14F**, Cor. 2);
8) the Hurwicz saddle-point condition (**14G**);
9) the Golštein duality theorem (**14H**);
10) the Dubovitskii-Milyutin separation condition (exercise 2.47).

It is important to be convinced of the mutual equivalence of these theorems. Most of the techniques for establishing these equivalences have already been presented (particularly in §6), so we will be content with making a few additional suggestions.

We have shown in the preceding several sub-sections that 6) implies 7)–10). Now we can use either 7) or 8) to establish the linear space version of the Krein-Rutman theorem (**6B**), and from that as usual the topological version. To do this assume the data M, P, X and ϕ as in **6B**, and define $S: M \to X$ by $S(x) = -x$. If we assume 7) then the hypotheses that the system (14.12) is consistent and that $\phi(x) \geqslant 0$ at each of its solutions are both satisfied, and the conclusion of 7) immediately yields the conclusion of the Krein-Rutman theorem. On the other hand, if we assume 8) then we consider the convex program

$$\min\{f(x): x \in M, S(x) \leqslant \theta\}$$

where $f \in X'$ is any extension of ϕ. The value of this program is 0 and θ is a solution. By 8) we obtain a positive linear functional $\bar{\phi} \in X'$ such that

$$0 \leqslant f(x) + \langle S(x), \bar{\phi} \rangle \equiv f(x) - \bar{\phi}(x)$$

for all $x \in M$. Thus f is a positive extension of ϕ.

Finally, it is immediate that 10) implies 1), so that it only remains to see what we can do with 9). Let X be a real locally convex space. A real-valued affine function f on X (exercise 1.6) necessarily has the form $f(x) = \phi(x) + c$ for some $\phi \in X'$ and $c \in \mathbb{R}$. Let $\{f_j : j \in J\}$ be a family of continuous affine functions on X, put $f(x) = \sup\{f_j(x) : j \in J\}$, and $A = \{x : f(x) < +\infty\}$. Then if $A \neq \varnothing$ it is clear that $f \in \operatorname{Conv}(A)$ and is lower semicontinuous. We can use 9) to demonstrate the converse.

Lemma. *Let A be a convex set in X and assume that $f \in \operatorname{Conv}(A)$ is lower semicontinuous. Then there is a family $\{f_j : j \in J\}$ of continuous affine functions on X such that $f_j | A \leqslant f$, $j \in J$, and*

$$f(x) = \sup\{f_j(x) : j \in J\}, \qquad x \in A.$$

Proof. For each $\phi \in X^*$ define

(14.24) $$f^*(\phi) = \sup\{\phi(x) - f(x) : x \in A\}.$$

If there exists $\phi \in X^*$ such that $f^*(\phi)$ is finite then $f_\phi \equiv \phi | A - f^*(\phi)$ is a continuous affine minorant of f on A; we shall show in fact that

(14.25) $$f(x) = \sup\{f_\phi(x) : f^*(\phi) < \infty\}, \qquad x \in A.$$

In order to apply 9) we select $p \in A$ and set up a convex program of the type (14.16) with $S: A \to X$ defined by $S(x) = p - x$, and assume that X has the trivial positive wedge $P = \{\theta\}$ (so that $P^* = X^*$). Because of the lower semicontinuity of f on A, the hypotheses of **14H** are satisfied and so we may assert that

$$\begin{aligned} f(p) &= \sup_{\phi \in X^*} \inf_{x \in A} L(x, \phi) \\ &\equiv \sup_{\phi} \inf_{x} \{f(x) + \phi(p - x)\} \\ &= \sup_{\phi} \{\phi(p) - f^*(\phi)\}. \end{aligned}$$

Thus either $\{\phi \in X^* : f^*(\phi) < \infty\}$ is non-empty or else $A = \varnothing$; assuming the former to be the case we have then proved (14.25). □

This lemma in turn leads to a proof of 2) by the same method employed in **6D**. Thus let A be a solid convex set in X and $x_0 \notin \operatorname{int}(A)$. Assuming that

$\theta \in \text{int}(A)$ we can write (14.25) as

$$\rho_A(x) = \sup\{\phi(x) - \rho_A^*(\phi) : \rho_A^*(\phi) < \infty\}$$
$$= \sup\{\phi(x) : \phi \leqslant \rho_A\}, \qquad x \in X,$$

where the second equality follows from the positive homogeneity of the gauge ρ_A. Since the set $\{\phi \in X^* : \phi \leqslant \rho_A\} \equiv A^\circ$ is weak*-compact (**12D**) we can choose $\phi \leqslant \rho_A$ with $\phi(x_0) = \rho_A(x_0) \equiv 1$ and then $[\phi; 1]$ defines the desired supporting hyperplane to A at x_0.

A more direct proof of the lemma is indicated in exercise 2.47.

§15. Some More Applications

In this section we present a variety of applications illustrating the ideas and principles of this chapter.

A. In **7B** we studied a criterion (Fan's condition) for the consistency of a finite system of linear inequalities. We now give a generalization but, in keeping with the advice offered in **6E**, we formulate the problem in a suitable conjugate space, as in **7C**.

Let X be a real locally convex space, $\{x_j : j \in J\}$ a family of vectors in X, and $\{c_j : j \in J\}$ an accompanying family of real numbers. We inquire about the consistency of the system

(15.1) $$c_j \leqslant \phi(x_j), \qquad j \in J,$$

where ϕ is to belong to X^*. A condition for the consistency of (15.1) can be expressed in terms of the smallest closed wedge $P \subset X \times \mathbb{R}^1$ that contains each pair $y_j \equiv (x_j, c_j)$ for $j \in J$. That is, P is the closure of the set $[0, \infty)\text{co}(\{y_j : j \in J\})$. Our consistency criterion is based on the following simple consequence of the strong separation theorem (**11F**).

Lemma. *Let A be a subset of the real locally convex space Y. A point $q \in Y$ belongs to the smallest closed wedge containing A if and only if every $\phi \in Y^*$ satisfying $\phi(x) \geqslant 0$ for $x \in A$ also satisfies $\phi(q) \geqslant 0$.*

Corollary. *The inequality system (15.1) is consistent if and only if $(\theta, 1) \notin P$.*

Proof. Using the general form of linear functionals on the product space $Y \equiv X \times \mathbb{R}^1$ (**6D**) we see from the lemma that $(\theta, 1) \in P$ if and only if every real s satisfying

$$\phi(x_j) + sc_j \geqslant 0, \qquad j \in J$$

for some $\phi \in X^*$ satisfies $s \geqslant 0$. But it is clear that this last condition is equivalent to the inconsistency of (15.1). □

The consistency criterion described in the lemma will also be referred to as "Fan's condition". It is easily verified that this result contains the earlier result of **7B** as a special case. A further application to inequality systems is given in exercise 2.50.

B. In **13D** we derived a formula (13.5) for the distance $d(x, M)$ from a point x to a linear subspace of a normed linear space X. An interesting and important question is whether or not this distance is attained in the sense that there exists some $y \in M$ so that $\|x - y\| = d(x, M)$. Such a y is a *best approximation* to x from M. In other words, we ask: does the convex program

$$\textbf{(15.2)} \qquad \min\{\|x - y\| : y \in M\}$$

have a solution? The general theory of **14E** (cf. Ex. 1 there) implies that $p \in M$ is a solution of (15.2) if and only if $\partial f(p) \cap M^\circ \neq \varnothing$, where $f(y) \equiv \|x - y\|$, $y \in X$. Interpreting this condition we may assert that $p \in M$ is a solution if and only if there exists $\phi \in U(M^\circ)$ such that $\phi(x - p) = \|x - p\|$; this intuitively means that p must be chosen in M so that the error vector $x - p$ is in a certain sense "perpendicular" to M. But the optimality theory does not actually help us to decide whether such a p exists. This existence question is really quite difficult in general and particular cases must often be handled by *ad hoc* methods. The next result ("Godini's theorem") contains a pair of necessary and sufficient conditions for a subspace M to admit a best approximation to every $x \in X$. Such subspaces are said to be *proximinal* in X; they clearly must be closed in X.

Theorem. *Let M be a linear subspace of the real normed linear space X. The following conditions are equivalent*:

a) *M is proximinal in X*;

b) $Q_M(U(X)) = U(X/M)$;

c) *$Q_M(U(X))$ is closed in X/M.*

Proof. Assume that M is proximinal and select a coset $x + M$ with $1 = \|x + M\| \equiv d(x, M)$. Let $y \in M$ be a best approximation to x. Then $x - y \in U(X)$ and $Q_M(x - y) = x + M$. This proves that a) implies b) while it is trivial that b) implies both a) and c). It remains to show that c) implies b). If $Q_M(U(X))$ is closed but properly contained in $U(X/M)$ then there is a coset $x + M$ of norm one which can be strongly separated from $Q_M(U(X))$. Taking into account the duality formula of **1H** we can obtain $\phi \in M^\circ$ such that

$$\phi(x) > \sup\{\phi(u) : \|u\| \leqslant 1\} \equiv \|\phi\|.$$

But this results in a contradiction since $|\phi(x)| \leqslant \|\phi\| d(x, M) = \|\phi\|$. □

Let us see what this theorem says in the particular cases where M is either of finite dimension or finite codimension.

Corollary. a) *Every finite dimensional subspace of X is proximinal in X.*

b) *If M is closed and of codimension $n < \infty$ in X, then M is proximinal in X if and only if $S(U(M))$ is closed in $\mathbb{R}^n$, where $S: X \to \mathbb{R}^n$ is defined by*

$$S(x) = (\phi_1(x), \ldots, \phi_n(x))$$

for any given basis $\{\phi_1, \ldots, \phi_n\}$ of M°.

Proof. a) We show that $Q_M(U(X))$ is closed in X/M. Suppose that $x_n + M \to x + M$ where $\|x_n\| \leqslant 1$. Then $d(x_n - x, M) \to 0$ and so there exist vectors $y_n \in M$ and $\varepsilon_n \in X$, $\varepsilon_n \to \theta$, such that $x_n - x - y_n = \varepsilon_n$. The sequence $\{y_n\}$ is bounded and so contains a convergent subsequence (**9F**) with limit $y \in M$ (**9E**). Consequently, $\lim x_n = x + y \in U(X)$ and $x + M = Q_M(x + y)$.

b) Given a basis $\{\phi_1, \ldots, \phi_n\}$ for M° we can select vectors $v_1, \ldots, v_n \in X$ so that $\phi_i(v_j) = \delta_{ij}$ (exercise 1.41). The set $\{v_1 + M, \ldots, v_n + M\}$ is then a basis for X/M, and since

$$x - \sum_{j=1}^{n} \phi_j(x)v_j \in {}^\circ M^\circ = M$$

(**12C**), it follows that

$$x + M = \sum_{j=1}^{n} \phi_j(x)(v_j + M).$$

Now if $T: \mathbb{R}^n \to X/M$ is the isomorphism defined in **9E** in terms of the basis $\{v_1 + M, \ldots, v_n + M\}$, we have $Q_M = S \circ T$, and our assertion follows from Godini's theorem. □

Observe that the condition of b) may also be expressed as the condition that $U(X)$ be complete in the (non-Hausdorff) weak topology $\sigma(X, M^\circ)$. Also note the special case $n = 1$ of b): if $M = \ker(\phi)$ is a hyperplane in X then M is proximinal in X if and only if ϕ "attains its norm" in the sense that there is some non-zero $x \in U(X)$ such that $\phi(x) = \|\phi\|$. That this need not always happen is demonstrated by the example

$$\phi(x) = \int_0^1 tx(t)dt, \qquad x \in L^1([0, 1], \mathbb{R}).$$

C. In **13E** (Ex. 5) we introduced the notion of a strictly normed linear space X. Such spaces were noted to have the property that their unit ball $U(X)$ is rotund, that is, every boundary point (unit vector) is an extreme point of $U(X)$. It follows that if we try to minimize the distance from a point x in such a space to a linear subspace M or, indeed, to any convex set A, this distance can be attained by at most one point in M (resp. A). In other words, a convex subset of a strictly normed space contains at most one best approximation to any point. The reason for this is simply that if there were two best approximations to some particular point x then the line segment joining these two best approximations would consist entirely of best approximations and we would hence have a line segment lying on the boundary of a ball centered at x, in contradiction to its rotundity.

Lemma 1. *Let A be a closed locally compact convex subset of a strictly normed linear space. For each $x \in X$ there is a unique best approximation $P_A(x)$ to x in A and the map P_A is continuous on X.*

Proof. The sets $A \cap (x + \lambda U(X))$ are closed, convex, locally compact and non-empty for sufficiently large $\lambda > 0$. Since they have a trivial recession cone, they must also be compact (**13C**). Their intersection, taken over those

λ yielding non-empty sets, is therefore also non-empty, and consists exactly of the best approximations to x in A. Because X is strictly normed we see that the map $P_A: X \to A$ is well defined and single valued. It remains to verify the continuity of P_A. Suppose that $\lim_n x_n = x$. Then

$$\begin{aligned} \text{(15.3)} \qquad \big|\, \|x_n - P_A(x_n)\| - \|x - P_A(x)\| \,\big| &\equiv |d(x_n, A) - d(x, A)| \\ &\leqslant \|x_n - x\| \to 0, \qquad n \to \infty. \end{aligned}$$

Because of local compactness any subsequence of the sequence $\{P_A(x_N)\}$ has a cluster point $y \in A$ which, by (15.3), satisfies

$$\|x - y\| = \|x - P_A(x)\| \equiv d(x, A).$$

By uniqueness of approximation it follows that $y = P_A(x)$. □

The map P_A is called the *metric projection* of X on A.

Most normed spaces do not come equipped with strict norms. In order to be able to apply Lemma 1 to some interesting situations we show next that the normed spaces occuring in practice can be "renormed" with strict norms. This means that we can find a new norm on the space which defines the same topology as does the original norm and which is in addition a strict norm. In general, it is easy to see (**9B**) that two norms ρ and σ on a linear space define the same topology exactly when they are *equivalent* in the sense that positive constants a and b exist such that

$$\text{(15.4)} \qquad a\rho(x) \leqslant \sigma(x) \leqslant b\rho(x), \qquad x \in X.$$

The first inequality quantitatively expresses that the ρ-topology on X is weaker than the σ-topology, and the second that the σ-topology is weaker than the ρ-topology. We can alternatively state that the norms ρ and σ on X are equivalent if and only if the identity map $I:(X, \rho) \to (X, \sigma)$ is an isomorphism. Renorming a space in this sense thus changes the geometry but not the topology. We now have the "Clarkson-Rieffel renorming lemma".

Lemma 2. *Let X be a separable normed linear space.*

a) *There is an equivalent strict norm on X.*

b) *There is a strict norm ρ on X^*, weaker than the usual dual norm, such that the ρ-topology on any bounded subset of X^* coincides with the weak*-topology.*

Proof. a) By **12F** there is a weak*-dense sequence $\{\phi_n\}$ in $U(X^*)$. Define

$$\sigma(x) = \|x\| + \left(\sum_{n=1}^{\infty} 2^{-n}|\phi_n(x)|^2\right)^{1/2}, \qquad x \in X,$$

where $\|\cdot\|$ is the given norm on X. Then σ is a norm on X and since

$$\|x\| \leqslant \sigma(x) \leqslant 2\|x\|, \qquad x \in X,$$

the two norms are equivalent. Finally, we show that σ is a strict norm. Let us assume that $\sigma(x + y) = \sigma(x) + \sigma(y)$, and set $\xi_n = \phi_n(x)$, $\eta_n = \phi_n(y)$.

Then

$$\text{(15.5)}\quad \left(\sum_{n=1}^{\infty} 2^{-n}|\xi_n + \eta_n|^2\right)^{1/2} = \left(\sum_{n=1}^{\infty} 2^{-n}|\xi_n|^2\right)^{1/2} + \left(\sum_{n=1}^{\infty} 2^{-n}|\eta_n|^2\right)^{1/2}.$$

Because all $L^2(\Omega, \mu, \mathbb{F})$ spaces are strictly normed (in the present case, $\Omega = \{1, 2, \ldots\}$ and $\mu(\{n\}) = 2^{-n}$), equation (15.5) entails either $\eta \equiv (\eta_n) = \theta$ or else the existence of $t \geqslant 0$ such that $\xi = t\eta$. Thus, recalling that the sequence $\{\phi_n\}$ is total over X (**12C**), it follows that either $y = \theta$ or else $x = ty$, proving that σ is a strict norm.

b) Let $\{x_n\}$ be a dense sequence in $U(X)$ and define

$$\rho(\phi) = \left(\sum_{n=1}^{\infty} 2^{-n}|\phi(x_n)|^2\right)^{1/2}, \qquad \phi \in X^*.$$

Then ρ is a strict norm on X^* satisfying $\rho(\phi) \leqslant \|\phi\|$, $\phi \in X^*$. Consider now the identity map from $U(X^*)$ in its weak*-topology to $U(X^*)$ with its ρ-topology. If we show that this map is continuous it will follow that it is actually a homeomorphism because of the weak*-compactness. This will show that the two topologies agree on $U(X^*)$ and hence on any bounded subset of X^*. To prove the continuity of the identity map at $\phi_0 \in U(X^*)$, select $\varepsilon > 0$ and let $V = \{\phi \in U(X^*): \rho(\phi - \phi_0) < \varepsilon\}$. Then if an integer m is chosen so that

$$\sum_{n=m+1}^{\infty} 2^{-n} < \varepsilon^2/4,$$

and we define the weak*-ϕ_0-neighborhood

$$W \equiv \left\{\phi \in X^*: |\phi(x_n) - \phi_0(x_n)| < \frac{\sqrt{3}}{2}, 1 \leqslant n \leqslant m\right\},$$

we see that $U(X^*) \cap W$ is a (relative) weak*-ϕ_0-neighborhood contained in V. □

Note that the argument just given reproves in stronger form a special case of **12F**. We now outline some applications of these two lemmas.

Example 1. Let $f: \Omega \to \Omega$ be a map from a set Ω into itself. A point $p \in \Omega$ is a *fixed point* of f if $f(p) = p$. The solution of many non-linear equations can be obtained as fixed points of certain mappings. The existence of fixed points in Euclidean spaces is resolved by the classical "Brouwer fixed point theorem": every continuous map of a Euclidean unit ball into itself has a fixed point. Since every compact convex set in $\mathbb{R}^n$ is homeomorphic to some Euclidean unit ball, the fixed-point statement can be made about all such sets as well. We shall obtain a substantial generalization of this fact known as the "Schauder fixed point theorem".

Theorem. *Let A be a closed convex subset of a normed linear space X, and let f be a continuous map of A into a compact subset of A. Then f has a fixed point in A.*

Proof. We first reduce the problem to the case where A is bounded and X is a separable strictly normed space. This can be done noting that if $f(A) \subset K$, a compact set, then we can replace A by $B \equiv \overline{\text{co}}(K) \subset A$ and try to prove the theorem for B. Next we let $Y = \text{span}(B)$; this is a separable subspace of X and we can work entirely in Y. Finally, by Lemma 2, we can assume that Y is strictly normed.

Now $f(B)$ is totally bounded and hence contains a $\frac{1}{n}$-net $\{f(x_i): i = 1, \ldots, m = m(n)\}$ for each n (**10B**). Let Y_n be the linear hull of this $\frac{1}{n}$-net and put $B_n = B \cap Y_n$, a compact convex subset of the finite dimensional space Y_n. Let $P_n: Y \to B_n$ be the metric projection. Then the map

$$f_n \equiv (P_n \circ f)|Y$$

is a continuous map from B_n into itself (using Lemma 1) and so has a fixed point $u_n: f_n(u_n) = u_n$. By compactness we can assume that $v \equiv \lim_n f(u_n)$ exists in K. Now

$$\begin{aligned} \|u_n - v\| &= \|f_n(u_n) - v\| \\ &\leqslant \|f_n(u_n) - f(u_n)\| + \|f(u_n) - v\| \\ &\leqslant \frac{1}{n} + \|f(u_n) - v\|, \end{aligned} \tag{15.6}$$

because, for any $x \in B_n$,

$$\begin{aligned} \|f_n(x) - f(x)\| &\equiv \|P_n(f(x)) - f(x)\| \\ &= d(f(x), B_n) \leqslant \min\{\|f(x) - f(x_i)\|: 1 \leqslant i \leqslant m\} \leqslant \frac{1}{n}. \end{aligned}$$

The estimate (15.6) proves that $\lim_n u_n = v \equiv \lim_n f(u_n) = f(v)$; that is, v is a fixed point of f. □

Example 2. Let (Ω, d) be a metric space and let A be a closed subset of Ω. According to the "Tietze extension theorem" every continuous function: $A \to \mathbb{R}$ can be extended to a continuous function: $\Omega \to \mathbb{R}$. (This extension theorem actually characterizes normal Hausdorff spaces.) According to the "Borsuk-Dugundji extension theorem" this extension can be achieved for bounded functions via a "linear extension operator" from $C_b(A, \mathbb{R})$ into $C_b(\Omega, \mathbb{R})$. This means that there exists $T \in B(C_b(A, \mathbb{R}), C_b(\Omega, \mathbb{R}))$ such that $T(x)$ is an extension of x for every $x \in C_b(A, \mathbb{R})$. Further, it can be arranged that $\|T\| = 1$ and that the functions be permitted to take values in an arbitrary locally convex space instead of $\mathbb{R}$.

We give now a geometric proof of the latter extension theorem under the restriction that the metric space (Ω, d) be compact, and that the functions be real or complex-valued. The reason for the compactness restriction is that it is the only case where we can guarantee that $C(\Omega, \mathbb{F})$ is separable.

Lemma 3. *The Banach space $C_b(\Omega, \mathbb{F})$ is separable if and only if the metric space (Ω, d) is compact.*

Proof. From exercise 1.40 we see that $C_b(\Omega, \mathbb{C})$ is separable if and only if $C_b(\Omega, \mathbb{R})$ is, so we just work with the latter space. Suppose that Ω is compact. Then Ω is 2nd-countable and there is a countable base $\{V_n\}$ for its topology. We define $x_n(t) = d(t, \Omega \backslash V_n)$ and let $\mathscr{A}$ be the subalgebra of $C(\Omega, \mathbb{R})$ generated by the x_n's. Now $\mathscr{A}$ is by definition the linear hull of the "monomials" $x_1^{\alpha_1} \cdots x_m^{\alpha_m}$, where $\alpha_1, \ldots, \alpha_m$ are non-negative integers and m is arbitrary. This collection of monomials is countable and the linear combinations of them with rational coefficients also constitute a countable set which is, moreover, dense in $\mathscr{A}$. Thus it suffices to show that $\mathscr{A}$ is dense in $C(\Omega, \mathbb{R})$. But this is an immediate consequence of the Stone-Weierstrass theorem, since the functions $\{x_n\}$ evidently separate the points of Ω. This proves that $C(\Omega, \mathbb{R})$ is separable whenever Ω is a compact metric space.

Conversely, suppose that Ω is not compact. Then there is a sequence $\{t_n\}$ of distinct points in Ω with no cluster point in Ω. Centered at each t_n there is a ball B_n containing no other member of the sequence and such that $B_n \cap B_m = \varnothing$, $m \neq n$. As above, let $x_n(t) = d(t, \Omega \backslash B_n)$. Let $\mathscr{J}$ be the family of all non-empty subsets of the positive integers and, for each $J \in \mathscr{J}$, define

$$y_J(t) = \sum_{n \in J} x_n(t)/\|x_n\|_\infty, \qquad t \in \Omega. \tag{15.7}$$

The function y_J is a well-defined member of $C_b(\Omega, \mathbb{R})$ since at each $t \in \Omega$, at most one term on the right hand side of (15.7) is non-zero. But if J_1 and J_2 are distinct members of $\mathscr{J}$ then $\|y_{J_1} - y_{J_2}\|_\infty = 1$, showing that $\{y_J : J \in \mathscr{J}\}$ is an uncountable discrete subset of $C_b(\Omega, \mathbb{R})$, and hence that this space is not separable.

Now to achieve our geometric proof of the Borsuk-Dugundji theorem subject to the above restrictions, we fix a closed set $A \subset \Omega$ and consider the weak*-compact convex subset B of $C(\Omega, \mathbb{F})^*$ consisting of those positive linear functionals ϕ such that $\phi(e) = 1$ (where e is the constantly one function in $C(\Omega, \mathbb{F})$), and such that ϕ annuls every continuous function vanishing on A. Thus B contains in particular the set $\{\delta_t : t \in A\}$. Let Q be the restriction of the metric projection P_B (defined and weak*-continuous by Lemmas 1 and 2b)) to the set $\{\delta_t : t \in \Omega\}$. We can then define our linear extension operator $T: C(A, \mathbb{F}) \to C(\Omega, \mathbb{F})$ by the formula

$$T(x)(t) = \langle x, Q(\delta_t) \rangle, \qquad t \in \Omega.$$

It is easily checked that T has the desired properties and that, in addition, T maps the constantly one function on A into e. □

D. We give now the generalization of the strong separation theorem promised in **11F**. This result, known as "Dieudonne's separation theorem" is an immediate consequence of the general criterion for strong separation given in **11F** and the lemma below. Actually, this lemma has considerable independent interest. For example, we shall utilize it along with a result in Chapter III in discussing the existence of solutions to a certain type of convex program which serves as model for problems of optimal control and spline approximation.

Theorem. *Let X be a locally convex space and let A, B be disjoint closed convex subsets of X. We suppose that A is locally compact and that the recession cones have trivial intersection: $C_A \cap C_B = \{\theta\}$. Then A and B can be strongly separated by a closed hyperplane.*

It is clear that this theorem includes the strong separation theorem of **11F**, since the hypothesis there was that A is compact which implies $C_A = \{\theta\}$. On the other hand, the hypothesis that $C_A \cap C_B = \{\theta\}$ is crucial; omitting that, the conclusion can fail even in case B is finite dimensional. We now state and prove the key lemma.

Lemma. *Let X be a Hausdorff linear topological space and let A, B be closed convex subsets of X. If A is locally compact and $C_A \cap C_B = \{\theta\}$ then $B - A$ is closed in X.*

Proof. The conclusion is clear if A is compact (exercise 2.1) so we shall explicitly assume that A is non-compact. Let $c \in \overline{B - A}$; there are thus nets $\{b_\delta : \delta \in D\} \subset B$ and $\{a_\delta : \delta \in D\} \subset A$ such that $c = \lim_\delta (b_\delta - a_\delta)$. For any balanced θ-neighborhood V in X we define

$$M_V = \{x \in A : x + (c + V) \cap B \neq \varnothing\}.$$

The sets M_V are non-empty: they eventually contain the vectors a_δ. Suppose that some M_V is relatively compact. Then there is a cluster point $a \in A$ of the net $\{a_\delta : \delta \in D\}$ and so the net $\{b_\delta : \delta \in D\}$ also has a cluster point b, necessarily in B. Hence $c = b - a \in B - A$, and the proof is complete in this case. Thus we are reduced to the situation where none of the sets M_V is relatively compact.

Without loss of generality we may assume that $\theta \in A$. Let W be a closed balanced θ-neighborhood such that $A \cap W$ is compact. Now, for each positive integer n and each V as above, we define

$$P_{n,V} = M_V \cap A \cap X \backslash (nW).$$

Because $A \cap nW \subset n(A \cap W)$ which is compact, and because of our assumption that none of the sets M_V is relatively compact, it follows that none of the sets $P_{n,V}$ is empty. The proof is now concluded in two steps: a) the existence of a half-line $L \subset A$ such that $L \subset [0, \infty)P_{n,V}$, for every n and V; and b) use of this half-line to obtain a contradiction to the hypothesis $C_A \cap C_B = \{\theta\}$.

Proof of a). Let $K = A \cap W \cap (X \backslash \operatorname{int}(\frac{1}{2}W))$; this is a compact subset of A. We consider the directed family of sets $[0, \infty)P_{n,V} \cap K$. Because K is compact this family has a cluster point $x_0 \in K$; necessarily $x_0 \neq \theta$. Then the desired half-line L is $[0, \infty)x_0$. To see this, choose any $\lambda > 0$. Then, noting that the sets $P_{n,V}$ decrease as n increases and/or V shrinks, we see that λx_0 belongs to the closure of each set $[0, \infty)P_{n,V}$. Also, since $\lambda x_0 \in \lambda W$, λx_0 belongs in fact to the closure of $[0, \infty)P_{n,V} \cap nW$ for any $n > \lambda$. But this set lies in A, since $\theta \in A$ and A is convex. That is, $\lambda x_0 \in \bar{A} = A$.

Proof of b). We select an arbitrary $b \in B$ and show that $b + L \subset B$; this will yield the desired contradiction. Let $z \in L$ and choose an integer n_0 such that $z \in \operatorname{int}(n_0 W)$. Now for any integer n and any balanced θ-neighborhood V there exist $\lambda > 0$ and $v \in V$ such that $x \equiv \lambda(z + v) \in P_{n,V}$ and $z + v \in n_0 W$. This entails $\lambda \geqslant n/n_0$. Since $P_{n,V} \subset M_V$ we can write $x = y - c + v'$ with $y \in B$ and $v' \in V$. Then

$$
\begin{aligned}
b + z &= b + x/\lambda - v \\
&\equiv b + (y - c + v')/\lambda - v \qquad (15.8)\\
&= b + (y - b)/\lambda + (b - c)/\lambda + v'/\lambda - v.
\end{aligned}
$$

Now if $n > n_0$ then $\lambda > 1$ and so $b + (y - b)/\lambda \in B$; we can further take n so large that $(b - c)/\lambda$ belongs to V. Then by (15.8) we have $b + z \in B + 3V$, whence (**9A**) $b + z \in \bar{B} \equiv B$. □

We note one special case of this lemma as a corollary; it makes use of the local compactness of finite dimensional spaces (**9F**).

Corollary. *Let N be a finite dimensional subspace of X and P a closed wedge in X such that $N \cap P = \{\theta\}$. Then $N + P$ is closed in X.*

Note that if the wedge P is also a linear subspace then $N + P$ is always closed, whether or not $N \cap P = \{\theta\}$ (**9E**). Thus in this very special case we can obtain a stronger result than that provided by the lemma. In general, when N and P are closed subspaces of a normed space satisfying $N \cap P = \{\theta\}$, it is easy to see that $N + P$ is closed whenever $\inf\{\|x - y\| : x \in N, y \in P, \|x\| = \|y\| = 1\} > 0$. This condition certainly holds in particular when N has finite dimension.

E. As our final application of this chapter we utilize the Krein-Milman theorem to study the range of certain vector measures. The result to be presented below contains as a special case the famous Liapunov convexity theorem concerning finite dimensional vector measures. This theorem in turn has a wide variety of applications, notably to the optimal control of linear dynamical systems (where it is essentially equivalent to the "bang-bang principle"), and to statistical decision theory.

Let (Ω, Σ, ν) be a totally finite (positive) measure space. In **12C** we observed that the continuous dual of the Lebesgue space $L^1(\nu) \equiv L^1(\Omega, \nu, \mathbb{R})$ could be identified with the space $L^\infty(\nu)$ of ν-essentially bounded measurable real-valued functions on Ω. Since the positive wedge $\{\phi \in L^\infty(\nu) : \phi \geqslant \theta\}$ is weak*-closed (because, for example, it can be expressed as $\{\phi \in L^\infty(\nu) : \int_E \phi \, d\nu \equiv \int_\Omega \phi \chi_E \, d\nu \geqslant 0\}$, and the characteristic functions χ_E belong to

$L^1(\nu)$ for all $E \in \Sigma$), its intersection K with $U(L^\infty(\nu))$ is weak*-compact. According to exercise 2.56 the extreme points of K are just the characteristic functions:

(15.9) $$\text{ext}(K) = \{\chi_E : E \in \Sigma\}.$$

Now let M be a subset of $L^1(\nu)$ and E a measurable subset of Ω. We consider the subspaces $M^\perp(E)$ of $L^\infty(\nu)$ consisting of all those ϕ that vanish a.e. on the complement of E and that annul each member of M: $\int_\Omega f\phi \, d\nu = 0$, $f \in M$. We shall write $M^\perp(\Omega)$ as simply $M^\perp$. M is said to be *thin* provided that $M^\perp(E) \neq \{\theta\}$ whenever $\nu(E) > 0$. It is clear that M is thin if and only if span(M) is thin.

Lemma 1. *If M is thin then $K \subset$ ext$(K) + M^\perp$.*

Proof. Let $\phi \in K$. Then the set $A \equiv K \cap (\phi + M)$ is compact and so by **13A** there exists $\psi \in \text{ext}(A)$. We claim that ψ is a characteristic function and hence in ext(K). If not, there exist $\varepsilon > 0$ and $E \in \Sigma$ such that $\varepsilon \leqslant \psi/E \leqslant 1 - \varepsilon$. Now since M is thin there is a non-zero $\psi_1 \in M^\perp(E)$ and, since ψ_1 is essentially bounded, we can arrange that $|\psi_1(\cdot)| \leqslant \varepsilon$. But then $\psi \pm \psi_1 \in A$, contradicting $\psi \in \text{ext}(K)$. □

Let us recall that a set $E \in \Sigma$ is an *atom* of ν if $\nu(E) > 0$ and every measurable set $F \subset E$ satisfies $\nu(F) = 0$ or else $\nu(F) = \nu(E)$. Intuitively an atom is a point of positive mass. The measure ν is called *purely atomic* if the complement of all the atoms of ν is a null set. An example of such a measure occurs when Ω is a countable set and for all $E \subset \Omega$, $\nu(E) \equiv$ cardinality of E. Such a measure is naturally called a *counting measure.* At the other extreme, the measure ν is *non-atomic* if there are no atoms in Σ. Lebesgue measure on $\mathbb{R}^n$ is non-atomic and, more generally, given any Lebesgue integrable function f on $\mathbb{R}^n$, the map $E \to \int_E f(t)dt$ defines a non-atomic measure on the Lebesgue measurable subsets of $\mathbb{R}^n$. The next two lemmas indicate some of the relevance of these types of measures in our present framework.

Lemma 2. *If ν is non-atomic then any finite set $M \subset L^1(\nu)$ is thin.*

Proof. Suppose that $M = \{f_1, \ldots, f_n\}$ and that $E \in \Sigma$. Since ν is non-atomic we can partition E into disjoint sets $E_1, \ldots, E_{n+1}$ of positive measure. Let A be the $n \times (n+1)$ matrix with entries $\int_{E_j} f_i \, d\nu$. Then there is a non-zero solution x of the system $Ax = \theta$. Hence the function $\sum_1^{n+1} x_j \chi_{E_j}$ is a non-zero element of $M^\perp(E)$. □

Lemma 2 does not characterize thin sets. By partitioning Ω into countably many sets of positive measure we can easily construct infinite thin sets.

Lemma 3. a) ext(K) *is weak***-dense in* K *if (and only if)* ν *is non-atomic.*
b) ext(K) *is weak***-closed in* K *if (and only if)* ν *is purely atomic.*

Proof. a) Let $N \equiv \{\phi \in L^\infty(\nu) : |\int_\Omega f_i\phi \, d\nu| \leqslant 1, 1 \leqslant i \leqslant n\}$ be a weak*-θ-neighborhood in $L^\infty(\nu)$ and set $M = \{f_1, \ldots, f_n\}$. Then by Lemmas 1 and 2

$$K \subset \text{ext}(K) + M \subset \text{ext}(K) + N,$$

so that $K \subset \overline{\text{ext}(K)}^*$.

b) If ν is purely atomic the set Ω is a countable union of atoms $E_1, E_2, \ldots$. Since any measurable function is necessarily constant a.e. on an atom, it follows from (15.9) that

$$\begin{aligned}\text{ext}(K) &= \bigcap_{n=1}^{\infty} \{\phi \in L^\infty(\nu) : \phi(\cdot) = 0 \text{ or } 1 \text{ a.e. on } E_n\} \\ &= \bigcap_{n=1}^{\infty} \{\phi \in L^\infty(\nu) : \textstyle\int_\Omega \chi_{E_n}\phi \, d\nu = 0 \text{ or } \nu(E_n)\}.\end{aligned}$$

This exhibits ext(K) as an intersection of weak*-closed sets. □

The conclusion of the first part of Lemma 3 should be duly noted. It provides us with an example of a non-trivial compact convex set which is the closure of its extreme points. This is a surprising possibility and emphasizes once again the occasionally bizarre properties of weak*-topologies.

We now suppose given a family $\{\mu_j : j \in J\}$ of totally finite signed measures on (Ω, Σ), each member of which is ν-absolutely continuous. By the Radon-Nikodym theorem the densities $f_j \equiv d\mu_j/d\nu$ exist in $L^1(\nu)$ for each $j \in J$. Let $\bar{\mu} : \Sigma \to \mathbb{R}^J$ be defined by $\bar{\mu}(E) = (\mu_j(E) : j \in J)$ and, for any $E \in \Sigma$, let $R(E) = \{\bar{\mu}(F) : F \in \Sigma, F \subset E\}$. Thus $\bar{\mu}$ is a vector measure on (Ω, Σ) and $R(E)$ is the range of its restriction to E. We are interested in the nature of $R(E)$ as a subset of the product space $\mathbb{R}^J$ (which is assumed to have the product topology (**9D**)). Our answer is contained in the following theorem of Kingman and Robertson.

Theorem. *With the above notation let $M = \{f_j : j \in J\}$. Then $R(E)$ is a (compact) convex set in $\mathbb{R}^J$ for every set $E \in \Sigma$ if and only if M is thin.*

Proof. Suppose that M is thin. It is sufficient to prove the assertion for the case $E = \Omega$, since otherwise we can apply the following argument to the restriction of $\bar{\mu}$ to E. Now the map $T : L^\infty(\nu) \to \mathbb{R}^J$ defined by

(15.10) $$T(\phi) = (\textstyle\int_\Omega f_j \phi \, d\nu : j \in J)$$

is linear and weak*-continuous, and $\ker(T) = M^\perp$. Further, $R(\Omega) = T(\text{ext}(K))$. Since M is thin Lemma 1 implies that

$$T(K) \subset T(\text{ext}(K) + M^\perp) = T(\text{ext}(K)) \subset T(K).$$

That is, $R(\Omega) = T(\text{ext}(K)) = T(K)$ is the continuous linear image of a compact convex set and so is itself compact and convex.

Conversely, suppose that M is not thin; we shall find a set $E \in \Sigma$ for which $R(E)$ is not convex. Indeed, there is a set E with $\nu(E) > 0$ such that $M(E) = \{\theta\}$. Let $M_E = \{f_j|E : j \in J\}$. Then M_E is dense in $L^1(\nu|E)$. This implies that the map T_E defined on $L^\infty(\nu|E)$ as in (15.10) (with Ω there replaced by E) is injective. Hence if $R(E)$ were convex then $\text{ext}(K_E) = T_E^{-1}(R(E))$ would also be convex, where $K_E \equiv \{\phi \in L^\infty(v|E) : \theta \leqslant \phi \leqslant 1\}$; thus $\text{ext}(K_E) = \{\chi_F : F \in \Sigma, F \subset E\}$. But this set of characteristic functions is certainly not convex (consider, for example, the sets $F = E$ and $F = \varnothing$). □

Our approach in this theorem has been to study the vector measure $\bar{\mu} : \Sigma \to \mathbb{R}^J$ by postulating the existence of a density defined wrt some positive measure v. It is useful to note that this represents no restriction when the index set J is denumerable. For then such a measure always exists. We may take, for example,

$$v = \sum_{j \in J} |\mu_j| / 2^j |\mu_j|(\Omega), \tag{15.11}$$

where $|\mu_j|$ is the total variation (**10D**) of μ_j. Then each μ_j is certainly v-absolutely continuous. Also, if each μ_j is non-atomic the same is true for v.

The most important case of the theorem occurs when J is a finite set, say $J = \{1, \ldots, n\}$. Now $\bar{\mu} \equiv (\mu_1, \ldots, \mu_n)$ is an $\mathbb{R}^n$-valued measure on Σ. We then have the classical "Liapunov convexity theorem".

Corollary. *The* range $R(E)$ $(E \in \Sigma)$ *of a finite dimensional vector measure* $\bar{\mu} = (\mu_1, \ldots, \mu_n)$ *is a compact subset of* $\mathbb{R}^n$ *and is convex whenever each* μ_j *is non-atomic.*

Proof. The second assertion follows from Lemma 2 and the theorem. Now if some of the μ_j fail to be non-atomic the measure v defined by (15.11) will have atoms. However, the restriction of v to the complement of the set of its atoms is a non-atomic measure. Thus Ω can be partitioned as $\Omega_1 \cup \Omega_2$ such that v is non-atomic on Ω_1 and purely atomic on Ω_2. Then for any $E \in \Sigma$, $R(E) = R(E \cap \Omega_1) + R(E \cap \Omega_2)$ and we know that $R(E \cap \Omega_1)$ is compact (and convex). On the other hand, taking part b) of Lemma 3 into account, we see that $R(E \cap \Omega_2)$ is also compact, as the continuous image of a compact set of extreme points. □

To see some implications of Liapunov's theorem let us briefly consider a dynamical system governed by a set of linear differential equations

$$x(t) = A(t)x(t) + B(t)u(t), \tag{15.12}$$

where $x : [0, T] \to \mathbb{R}^n$, $u : [0, T] \to \mathbb{R}^m$, and A, B are appropriately shaped matrix functions of $t \in [0, T]$ with Lebesgue integrable entries. The vector $x(t)$ represents the "state" of our system at time t and $u(t)$ represents a "control" which we apply to the system in order to influence its state. If we know the transition matrix $\Phi(t)$ of the system (15.12), that is, the matrix solution of

$$\dot{\Phi}(t) = A(t)\Phi(t), \qquad \Phi(0) = I,$$

then we can express the effect of the control on the state by

$$x(t) = \Phi(t) \int_0^t \Phi^{-1}(s)B(s)u(s)ds, \tag{15.13}$$

where for simplicity we have assumed $x(0) = \theta$. In particular, if A is a constant matrix, then $\Phi(t)$ is just the matrix exponential $\exp(tA)$.

The controls u are chosen to be measurable and to take their values in some fixed compact convex subset $C \subset \mathbb{R}^m$. In order to know what effect we can hope to achieve on the state $x(\cdot)$ with such controls, we see from (15.13) that we must know about the set $\Xi(t; C)$ of functions $\{\int_0^t \Psi(s)u(s)ds\}$ where $\Psi \equiv \Phi^{-1}B$ and u runs through the measurable C-valued controls. The "bang-bang principle" asserts that

(15.14) $$\Xi(t; C) = \Xi(t; \text{ext}(C)), \qquad 0 \leqslant t \leqslant T.$$

This terminology arises from the special case where $C = \{z \in \mathbb{R}^m : |z_i| \leqslant 1, i = 1, \ldots, m\}$, the unit cube in $\mathbb{R}^m$. It means that any point in $\mathbb{R}^n$ to which we can drive the state by means of an admissible (that is, C-valued) control in time T can also be attained in the same time by a "bang-bang control", that is, intuitively one that uses full power at all times.

The proof of (15.14) in its full generality is difficult but the special case where C = unit cube is a fairly direct consequence of the Liapunov convexity theorem. The necessary argument is outlined in exercise 2.57.

Exercises

2.1. Let X be a linear topological space over the field $\mathbb{F}$ and let A, B be subsets of X. Prove the following assertions.
   a) $\overline{x + A} = x + \bar{A}$ for $x \in X$;
   b) if A is a linear subspace of X, so is $\bar{A}$;
   c) the interior of A is contained in cor(A);
   d) if B is open then so is $A + B$;
   e) if A is compact, B is open, and $A \subset B$, then there is a θ-neighborhood $U \subset X$ such that $A + U \subset B$;
   f) if A is compact and B is closed then $A + B$ is closed;
   g) if A and B are both compact, so is $A + B$.

2.2. Let the linear topological space X be the algebraic direct sum of its subspaces M and N, and let $P: X \to M$ be the associated projection.
   a) Show that the direct sum is topological exactly when P is continuous.
   b) If P is continuous show that M and N must be closed.
   c) Show that the conditions of a) are also equivalent to the following: the map $T: X/M \to N$ defined by $T(x + M) = (x + M) \cap N$ is an isomorphism (note that T is always continuous).
   d) If M is closed and of finite codimension in X show that the direct sum $X = M \oplus N$ must be topological.

2.3. Let A be a subset of the linear topological space X.
   a) if A is bounded or totally bounded then the same is true of $\bar{A}$ and of any continuous linear image of A;
   b) if $X = \prod_\alpha X_\alpha$ is a product of linear topological spaces X_α and X has the product topology then A is bounded if and only if $A \subset \prod_\alpha B_\alpha$, where each B_α is a bounded subset of X_α.
   c) If $X = \mathbb{F}^I$ for an arbitrary set I ($\neq \varnothing$) then any bounded set $A \subset X$ is actually totally bounded.

2.4. Let A be a convex absorbing subset of a linear topological space X, and let ρ_A be the gauge of A. Show that $\{x \in X : \rho_A(x) < 1\}$ is the interior of A and that $\{x \in X : \rho_A(x) \leqslant 1\} = \bar{A}$.

2.5. a) Give the details of the proof of the theorem in **10B**.

b) Establish the following metrizability criterion for a locally convex space $X : X$ is metrizable if and only if its given topology is Hausdorff and X contains a countable local base. (Assume that $\{U_1, U_2, \ldots\}$ is a local base of barrels. After replacing U_n by $U_1 \cap \cdots \cap U_n$ if necessary it may be assumed that $U_1 \supset U_2 \supset \cdots$. Let ρ_n be the gauge of U_n and define

$$d(x, y) = \sum_{n=1}^{\infty} 2^{-n} \min(\rho_n(x - y), 1)$$

for $x, y \in X$. d is the desired metric and is, in addition, translation invariant.)

c) Is the function $x \mapsto d(x, \theta)$ a semi-norm on X?

2.6. Show that every locally convex Hausdorff space is a completely regular topological space.

2.7. Prove the quotient norm formula (10.6).

2.8. Verify that the normed linear spaces in Examples 1 through 5 (**10D**) are actually Banach spaces. (For the L^p spaces of Example 2 note first that if $\theta \leqslant f_n \in L^p$ and $\sum_1^{\infty} \|f_n\|_p < \infty$, then $f \equiv \sum_1^{\infty} f_n$ belongs to L^p also and $\|f\|_p \leqslant \sum_1^{\infty} \|f_n\|_p$. This conclusion follows from Minkowski's inequality and the monotone convergence theorem. Now if $\{f_n\}$ is a Cauchy sequence in L^p it may be supposed that $\|f_{n+1} - f_n\|_p < 2^{-n}$ (by passing to a subsequence if necessary). If $g_n \equiv f_n - \sum_n^{\infty} |f_{i+1} - f_i|$ and $h_n \equiv f_n + \sum_n^{\infty} |f_{i+1} - f_i|$, then our remark implies $g_n, h_n \in L^p$ and $\|g_n - h_n\| < 2^{-n+2}$. Finally, $f \equiv \lim_n g_n$ exists in L^p and is the desired limit of $\{f_n\}$.)

2.9. Let $\mathbb{R}^n$ be normed by the p-norm (**3C**) with $p > 1$ and let $f : \mathbb{R}^n \to \mathbb{R}$ be continuously differentiable with gradient $\nabla f(x) \in \mathbb{R}^n$, $x \in \mathbb{R}^n$. Let Ω be a compact convex subset of $\mathbb{R}^n$. Prove that $f \in \operatorname{Lip}(\Omega, \|\cdot\|_p, \mathbb{R})$ by showing that the norm (10.4) is given by $\sup\{\|\nabla f(x)\|_q : x \in \Omega\}$, where $q = p/(p - 1)$. It follows that for smooth functions convergence in the metric defined by (10.5) is equivalent to uniform convergence of the functions and their first partial derivatives.

2.10. Every linear space X has a strongest locally convex topology, namely that generated by the family of all semi-norms on X. Establish the following properties of this unique topology, often called the *convex core topology* on account of g) below.

a) The topology is Hausdorff;
b) every linear (hence every affine) subspace is closed;
c) unless X is finite dimensional the topology is not metrizable;
d) any bounded set is necessarily finite dimensional;
e) X is complete (begin by showing that the projection of a given Cauchy net in X on any finite dimensional subspace has a limit in that subspace);
f) every linear functional in X' is necessarily continuous;
g) if A is a convex subset of X then $\text{int}(A) = \text{cor}(A)$.

2.11. Let X be a linear topological space and let $A \subset X$.
a) If A is open then so is $\text{co}(A)$.
b) If $\theta \in A$ and A is convex then $\bar{A} \supset \cap\{tA:t > 1\}$ and equality holds if A is a θ-neighborhood.
c) If A is solid and convex then $\partial(A)$ is nowhere dense in X.

2.12. Show that Corollary 1 in **11E** is false for all non-Hausdorff locally convex spaces (use **9A**).

2.13. Let A be a subset of a real locally convex space X. Show that $\text{co}(A)$ is the intersection of all the closed half-spaces in X that contain A. In case A is already closed and convex it follows that A can be determined by a family of linear constraints: $A = \{x \in X:\phi_\alpha(x) \leqslant c_\alpha\}$ for some family $\{\phi_\alpha\} \subset X^*$ and corresponding family $\{c_\alpha\} \subset \mathbb{R}$. This is the principle of "quasi-linearization of convex sets". When X is a separable normed linear space the family of determining linear functionals can always be taken to be denumerable. (Compare with exercise 1.35.)

2.14. Let X be an infinite dimensional normed linear space. Show that there exists a discontinuous linear functional on X (that is, $X' \backslash X^* \neq \varnothing$). Use this fact to construct discontinuous (unbounded) linear map from X into itself with a closed kernel (compare with **11D**).

2.15. Show that a linear functional ϕ defined on a real linear topological space X is discontinuous if and only if the set $\{x \in X:\phi(x) \neq 0\}$ is connected.

2.16. Let M be a linear subspace of the locally convex space X and let $\phi \in M^*$. Show that there exists an extension of ϕ in X^*.

2.17. A semi-norm ρ defined on a linear space X is *discrete* if $M \equiv \{x \in X: \rho(x) = 0\}$ has finite codimension in X.
a) Give examples of discrete semi-norms on the space $C_b(\Omega, \mathbb{R})$.
b) Show that $\phi \in X'$ is continuous in the ρ-topology on X exactly when $\phi \in M^\circ$.

2.18. Let A be a closed convex subset of the separable Banach space X. Show that either A is contained in some closed hyperplane or else A has a non-support point. (Assume the former not to be the case and that $\theta \in A$. Let $\{x_n\}$ be a dense sequence in A and set $y_n = x_n$ if $||x_n|| \leqslant 1$ and otherwise $y_n = x_n/||x_n||$. Define $p = \sum_1^\infty 2^{-n}y_n$. Then $p \in A$ but p is not a support point of A.)

2.19. Establish the following cancellation law for convex sets. Let A and B be closed convex sets of a linear topological space X. Suppose that $A + C = B + C$ for some (non-empty) bounded set $C \subset X$. Then $A = B$.

2.20. Let X be a locally convex space, A a compact convex subset of X, and M a closed linear subspace of codimension $\geqslant n$. If $A \cap M = \varnothing$ show that there exists a linearly independent set $\{\phi_1, \ldots, \phi_n\} \subset X^*$ such that each ϕ_j defines a hyperplane strongly separating A and M.

2.21. Let $A_1, \ldots, A_n$ be compact convex subsets of a linear topological space. Show that $\text{co}(A_1 \cup \cdots \cup A_n)$ is also compact (and convex).

2.22. Let X be a linear topological space.

a) If A is a countable subset of X show that the closed linear span of A is a separable subspace of X.

b) If X is locally convex show that X is separable if (and only if) X is weakly separable.

c) If X is separable and normed show that X^* is weak*-separable. Is the converse true?

2.23. Let X be an infinite dimensional normed linear space.

a) Endowed with its weak topology X is a set of first category in itself.

b) The weak closure of the set $\{x \in X : \|x\| = 1\}$ is the entire unit ball $U(X)$.

c) The analogues of a) and b) for X^* and its weak* topology are also valid.

2.24. Let (Ω, Σ, μ) be a positive σ-finite measure space.

a) The space $L^\infty(\mu)$ is either finite dimensional or not separable.

b) Define an equivalence relation on Σ by $E \sim F$ if and only if $\mu(E \Delta F) \equiv \mu(E \backslash F) + \mu(F \backslash E) = 0$ and let Σ_0 be the set of equivalence classes containing a set of finite measure. We can define a metric on Σ_0 by $d(E, F) = \mu(E \Delta F)$. Verify that d is a metric on Σ_0 and then prove that for $1 \leqslant p < \infty$ the space $L^p(\mu)$ is separable if and only if the metric space (Σ_0, d) is separable.

c) Show that $L^p(\mu)$ is separable whenever the σ-algebra Σ is countably generated, that is, whenever Σ is the smallest σ-algebra containing a given countable family of subsets of Ω.

d) Show that if $\Omega \subset \mathbb{R}^n$ and μ is Lebesgue measure then $L^p(\mu)$ is separable for $1 \leqslant p < \infty$.

e) Give an example of a finite measure space such that the corresponding L^p spaces are not separable.

2.25. Discuss the separability of the Banach spaces in Examples 3–5 of **10D**.

2.26. Prove that two disjoint closed convex subsets of a locally convex space, one of which is weakly compact, can be strongly separated by a closed hyperplane.

2.27. Let X be a locally convex space and J_X the canonical embedding (**12B**) of X into $X^{*\prime}$.

a) Show that a net $\{x_\delta : \delta \in D\}$ in X converges to $x \in X$ if and only if

$J_X(x_\delta)$ converges to $J_X(x)$ uniformly on each equicontinuous subset of X^*.

b) Show that each weakly compact subset of X is complete.

2.28. Establish the following working rules for polars in a locally convex space.

a) $B^\circ \subset A^\circ$ if $A \subset B$;

b) $(tA)^\circ = \frac{1}{t} A^\circ, t \neq 0$;

c) $(\cup\{A_j : j \in J\})^\circ = \cap\{A_j^\circ : j \in J\}$;

d) $(\cap\{A_j : j \in J\})^\circ = \overline{\text{co}}^*(\cup\{A_j^\circ : j \in J\})$.

2.29. Give an example of a compact convex set in $\mathbb{R}^3$ whose set of extreme points is not closed.

2.30. Let X be either the space $L^1([0, 1], \mu, \mathbb{F})$ (where μ is Lebesgue measure) or the space $C_0(\Omega, \mathbb{F})$ consisting of all continuous $\mathbb{F}$-valued functions on the non-compact locally compact Hausdorff space that vanish at infinity (that is, functions $f \in C_b(\Omega, \mathbb{F})$ such that $\{t \in \Omega : |f(t)| \geqslant \varepsilon\}$ is compact for all $\varepsilon > 0$), normed by the uniform norm (10.1). Show that in both cases the unit ball $U(X)$ has no extreme points. Thus "compact" in **13A** cannot be replaced by, for example, "closed and bounded".

2.31. In the space m of bounded sequences (**12F**) let e_n be the sequence with all terms 0 except the nth which is 1. Let $A = \text{co}(\{\theta, e_1, e_2/2, e_3/3, \ldots\})$. Show that A is compact (and convex) but that $A \neq \text{co}(\text{ext}(A))$. Thus in the Krein-Milman formula $A = \overline{\text{co}}(\text{ext}(A))$ of **13B** the closure operation cannot generally be omitted.

2.32. Let A be a compact convex subset of a locally convex space. Prove that A is the closed convex hull of its extreme support points (that is, support points belonging to ext(A)). Note that it is not claimed that every extreme point must be a support point.

2.33. Prove that, in contrast with exercise 1.36, there exists an unbounded closed convex set B in some Banach space whose recession cone C_B contains only the zero vector. (In the sequence space $\ell^2(\aleph_0)$ let $B = \{x = (\xi_1, \xi_2, \ldots) : |\xi_n| \leqslant n\}$.)

2.34. Let X be a locally convex space and P a wedge in X.

a) If P is solid then its dual wedge P^* is weak*-locally compact in X^*.

b) If P is a closed cone in X then the subspace $P^* - P^*$ is weak*-dense in X^*.

2.35. Let Ω be a completely regular Hausdorff space (a Tychonov space).

a) The map $q : t \to \delta_t$ is a homeomorphism of Ω into the ball $U(C_b(\Omega, \mathbb{F})^*)$ endowed with the weak* topology.

b) The image $q(\Omega)$ is weak*-closed exactly when Ω is compact.

c) Assume that Ω is not compact and set $\beta(\Omega) = \overline{q(\Omega)}^*$. Then $\beta(\Omega)$ is a compact space containing Ω as a dense subspace. Thus $\beta(\Omega)$ is, by definition, a compactification of Ω and is in fact the maximal or *Stone-Cech compactification* of Ω. (It is to be shown that if (Γ, h) is any compactification of Ω, so that h is a homeomorphism of Ω

onto a dense subspace of the compact Hausdorff space Γ, then Γ is a quotient space of $\beta(\Omega)$. In turn this may be achieved by constructing a continuous surjection $F:\beta(\Omega) \to \Gamma$ such that $h = F \circ q$. Finally, we can obtain F by considering the transpose (**1H**) of the map $g \mapsto g \circ h$ from $C(\Gamma, \mathbb{F})$ into $C_b(\Omega, \mathbb{F})$, restricting this transpose to $\beta(\Omega)$, and composing this restriction with the inverse of the canonical homeomorphism of Γ into $U(C(\Gamma, \mathbb{F})^*)$.)

d) Every $f \in C_b(\Omega, \mathbb{F})$ has a uniquely specified extension to $C(\beta(\Omega), \mathbb{F})$.

e) $\beta(\Omega)$ consists of the non-trivial homomorphisms defined on the algebra $C_b(\Omega, \mathbb{F})$. It then follows that $\beta(\Omega)$ may be identified with the set of extreme points of $\{\phi \in U(C_b(\Omega, \mathbb{F})^*): \phi(e) = 1\}$, where e is, as usual, the constantly one function on Ω.

f) If f is any continuous map from Ω into a compact space Γ then there exists a uniquely specified continuous extension of f defined on $\beta(\Omega)$. (Same argument as in c).) Hence this extension property characterizes $\beta(\Omega)$ up to homeomorphism.

g) Let Ω be the space of positive integers with the discrete topology. Then $m \equiv C_b(\Omega, \mathbb{F})$ can be identified with $C(\beta(\Omega), \mathbb{F})$ (use d)).

h) If $\Omega = (0, 1]$ with the usual topology then $\beta(\Omega)$ is not (homeomorphic to) $[0, 1]$. (use d)).

All of which goes to show, among other things, that when Ω is not compact the extreme point structure of $U(C_b(\Omega, \mathbb{F})^*)$ is much more complicated than in the compact case, when formula (13.9) provides a complete description. We may also note, in reference to Lemma 3 of **15C**, that from **12F** and a) above it follows that in the case where Ω is completely regular, the metrizability of Ω is a necessary condition for the separability of $C_b(\Omega, \mathbb{F})$.

2.36. Show that the unit ball of a normed linear space X is rotund if and only if X is strictly normed.

2.37. Let X be a normed linear space, let $\{\phi_1, \ldots, \phi_n\} \subset X^*$ and $\{c_1, \ldots, c_n\} \subset \mathbb{R}$. Let V be the affine subspace $\{x \in X: \phi_j(x) = c_j, j = 1, \ldots, n\}$. For any $x_0 \in X$ show that

$$d(x_0, V) = \sup \frac{|\sum t_j(c_j - \phi_j(x_0))|}{\|\sum t_j \phi_j\|}$$

where the supremum is taken over all sets $\{t_1, \ldots, t_n\} \subset \mathbb{R}$ such that $\sum t_j \phi_j \neq \theta$. Conclude that a sequence $\{x_n\} \subset X$ converges weakly to x_0 if and only if its distance from any closed finite codimensional flat through x_0 tends to 0. (By contrast, a sequence converges to x_0 in the norm topology if and only if its distance to any closed flat through x_0 tends to 0.)

2.38. Let X be either a space $L^1(\Omega, \mu, \mathbb{F})$ for μ σ-finite, or a space $C(\Omega, \mathbb{F})$ for Ω compact. Show that each extreme point of $U(X^*)$ is actually a vertex of $U(X)$ (in the sense of **13E**, Ex. 6). More precisely, for any $\phi \in X^*$ define $B_\phi = \{x \in X: \phi(x) = \|x\|\}$. Then show that $\phi \in \operatorname{ext}(U(X^*))$ if and only if $X = B_\phi - B_\phi$.

2.39. Let A and B be compact convex subsets of the locally convex space X. Prove that, in the product space $X \times X$, $\text{ext}(A \times B) = \text{ext}(A) \times \text{ext}(B)$.

2.40. Let X be a locally convex space with the property that every nonempty, closed, bounded, convex subset of X has an extreme point. Show that every such subset is then the closed convex hull of its extreme points. (Proceed by contradiction and use the strong separation theorem. This result was originally formulated for Banach spaces, and provided the basis for a proof that every closed bounded convex subset of $\ell^1(\aleph_0)$ is the closed convex hull of its extreme points. Compare with exercise 3.10).

2.41. Let f be a convex function defined on a neighborhood of a point x_0 in some normed linear space and continuous at x_0. Show that there exist an x_0-neighborhood V and a positive constant λ such that whenever x and y belong to V we have the Lipschitz inequality $|f(x) - f(y)| \leqslant \lambda \|x - y\|$. (It may be assumed that $x_0 = \theta$; choose $\delta > 0$ so that $|f(x) - f(\theta)| < 1$ if $\|x\| < \delta$. Then we may take $V = \dfrac{\delta}{2} U(X)$ and $\lambda = 8/\delta$.) It follows that the restriction of a continuous convex function to a compact convex set in X satisfies a Lipschitz condition uniformly on that set.

2.42. Prove the formula (14.5).

2.43. The solvability of any optimization problem is always a topological matter. Thus, let (A, f) be a variational pair. We define a topology $\tau = \tau(f)$ on the set A by taking as a subbase all sets of the form $\{x \in A: f(x) > \lambda\}$ as λ runs through $\mathbb{R}$.

a) τ is the weakest topology on A in which f is lower semicontinuous.

b) There is a solution to the program (A, f) if and only if A is τ-compact.

2.44. With the terminology and notation of Tuy's inconsistency theorem (**14F**) suppose that the wedge Q is solid and that there exist linear functionals $\phi \in P^*$, $\psi \in Q^*$ ($\psi \neq \theta$) satisfying (14.14). Show that the system

$$S(x) \leqslant \theta, \qquad T(x) < \theta, \qquad x \in A$$

is inconsistent.

2.45. With the same notation and the assumption that Q is solid, suppose also that the system

$$T(x) < \theta, \qquad x \in A$$

is weakly inconsistent (in other words, has no weak solution in the sense of (**14H**)). Prove that there exists a non-zero $\psi \in Q^*$ such that $\langle T(x), \psi \rangle \geqslant 0$, $x \in A$.

2.46. Apply the theory of **14F** to establish the topological form of the Krein-Rutman theorem: let M be a linear subspace of the linear topological space X, ordered by the solid positive wedge P; if $\text{int}(P) \cap M \neq \varnothing$ then any positive linear functional in M^* has an extension belonging to P^*.

2.47. Establish the "Dubovitskii-Milyutin separation condition": let A_0, $A_1, \ldots, A_n$ be convex sets in a real linear topological space X with $\theta \in \bar{A}_j$ for all j and $A_1, \ldots, A_n$ open; then $\cap\{A_j : j = 0, 1, \ldots, n\} = \varnothing$ if and only if the sets are separated in the sense that there exist linear functionals $\phi_0, \phi_1, \ldots, \phi_n \in X^*$, not all θ, such that $\phi_j(x) \geqslant 0$ for all $x \in A_j$ and $\phi_0 + \phi_1 + \cdots + \phi_n = \theta$. (For the necessity, apply a separation theorem in the product space $X \times \cdots \times X$ (n times) to the sets $A_1 \times \cdots \times A_n$ and $\{(x_1, \ldots, x_n) : x_1 = \cdots = x_n \in A_0\}$.)

2.48. Give a direct proof the lemma in **14I**. (Apply exercise 2.13 to the epigraph of f.)

2.49. Consider a convex program (A, f) where A is a convex subset of a locally convex space X and f is lower semicontinuous. Unless the value of this program is $-\infty$, the *conjugate function* f^* defined by (14.24) will be finite at $\phi = \theta$ in X^*. Assuming this, prove that the program is solvable if and only if f^* is subdifferentiable at θ, and then $\partial f^*(\theta)$ is exactly the set of solutions of the program.

2.50. Let $f \in \mathrm{Conv}(\mathbb{R}^n)$ be differentiable. Show that the conjugate function f^* is finite at $y \in \mathbb{R}^n$ exactly when $y = \nabla f(x)$, for some $x \in \mathbb{R}^n$. Thus f^* is everywhere finite exactly when $\nabla f : \mathbb{R}^n \to \mathbb{R}^n$ is surjective. (**3A** is helpful for the sufficiency; for the necessity, assume that $f^*(y) < \infty$ and show that the supremum in (14.24) is attained at some $x \in \mathbb{R}^n$.)

2.51. a) Verify that the consistency condition of **7B** is a special case of that in **15A**.

b) Let the system (15.1) be consistent and let $S \subset X^*$ be the set of all its solutions. Let $(y, b) \in X \times \mathbb{R}^1$. Then the inequality $b \leqslant \phi(y)$ is a consequence of (15.1) (in the sense that this inequality is valid for every $\phi \in S$) if and only if there exists $a \geqslant b$ such that (y, a) belongs to the wedge P of **15A**. (If no such a exists then the wedge P and the segment $[(\theta, 1), (y, b)]$ in $X \times \mathbb{R}^1$ can be strongly separated.)

2.52. Let X be a normed linear space and A a (non-empty) weak*-closed subset of X^*. Prove that A is proximinal in X^*. (Use **12D**. This is a very versatile and powerful optimization principle, which illustrates again the value of the recommendation in **6E**. Note that A need not be a linear subspace nor even convex.)

2.53. Let T be a continuous map of a normed linear space X into itself. Suppose that T maps bounded sets into compact sets and that

$$\lim_{\|x\| \to \infty} \frac{\|T(x)\|}{\|x\|} = 0.$$

Show that for any $\lambda > 0$ and $y \in X$ the equation $x = \lambda T(x) + y$ has a solution $x \in X$. (Apply the Schauder fixed point theorem to the map $f(x) \equiv \lambda T(x) + y$ after observing that f must map some multiple of the unit ball into itself.)

2.54. Let $(s_0, t_0) \in \mathbb{R}^2$ and let $g : \mathbb{R}^2 \to \mathbb{R}$ be continuous on a neighborhood of (s_0, t_0). According to "Peano's theorem" the differential equation

$y'(s) = g(s, y)$ has a solution h defined on a neighborhood of s_0 and satisfying $h(s_0) = t_0$. To prove this, choose $\varepsilon > 0$ so that $|g(s, t)| \leqslant 1$ on the square $[s_0 - \varepsilon, s_0 + \varepsilon] \times [t_0 - \varepsilon, t_0 + \varepsilon]$. Let $X = C([s_0 - \varepsilon, s_0 + \varepsilon], \mathbb{R})$ and let $A = \{x \in X : |t_0 - x(s)| \leqslant \varepsilon, |s - s_0| \leqslant \varepsilon\}$. Then define $f : A \to A$ by

$$f(x)(s) = t_0 + \int_{s_0}^{s} g(s, x(s))ds$$

and show that a solution may be obtained by applying the Schauder fixed point theorem to f.

2.55. Let ρ and σ be two norms on a finite dimensional linear space. Show that ρ and σ must be equivalent, in that constants $a, b > 0$ exist so that (15.4) holds.

2.56. Prove formula (15.9).

2.57. Establish the special case of the bang-bang principle (15.14) where C is the unit cube in $\mathbb{R}^m$. (Let u be a C-valued measurable function on $[0, T]$ and, for definiteness, take $t = T$ in (15.14). We are to find a C-valued measurable function v on $[0, T]$ such that each component v_j of v satisfies $|v_j(\cdot)| = 1$ a.e. on $[0, T]$ and

$$\int_0^T \Psi(s)u(s)ds = \int_0^T \Psi(s)v(s)ds.$$

Let us concentrate on a particular v_j. We must choose a measurable set $B_j \subset [0, T]$ (the set on which v_j equals 1) such that

$$\int_0^T \Psi_{1j}(s)u_j(s)ds = \int_{B_j} \Psi_{1j}(s)ds - \int_{B_j'} \Psi_{1j}(s)ds,^{(4)}$$

$$\vdots$$

$$\int_0^T \Psi_{nj}(s)u_j(s)ds = \int_{B_j} \Psi_{nj}(s)ds - \int_{B_j'} \Psi_{nj}(s)ds,$$

where Ψ_{ij} is the $(i, j)^{\text{th}}$-entry of the matrix Ψ. Thus the requirement on B_j is that

$$\int_{B_j} \Psi_{ij}(s)ds = \tfrac{1}{2}\int_0^T (1 + u_j(s))\Psi_{ij}(s)ds, \; i = 1, \ldots, n.$$

In order to apply Liapunov's theorem we define a vector measure $\bar{\mu}$ on $[0, T]$ by

$$\bar{\mu}(E) = \left(\int_E \Psi_{1j}(s)ds, \ldots, \int_E \Psi_{nj}(s)ds\right);$$

then we know $R([0, T])$ is a compact convex set in $\mathbb{R}^n$. We also define a weak*-continuous affine map $\Phi : U(L^\infty([0, T], \mathbb{R})) \to \mathbb{R}^n$ by

$$\Phi(w) = \tfrac{1}{2}\int_0^T (1 + w)d\bar{\mu}.$$

We can finish by proving that the range of Φ lies in $R([0, T])$, and for this it suffices to prove that Φ maps every extreme point of $U(L^\infty([0, T], \mathbb{R}))$ into $R([0, T])$. This is not hard as these extreme points can be characterized in a way analogous to that of (15.9).)

---

(4) B_j' is the complement of B_j.

2.58. Let (Ω, Σ, μ) be a positive non-atomic measure space and let M be a finite dimensional linear subspace of $L^1(\mu)$. Establish the existence of arbitrarily small "determining sets" for M. Precisely, given $\varepsilon > 0$, there exists $E \in \Sigma$ such that $0 < \mu(E) < \varepsilon$ with the property that if x, $y \in M$ and $x|E = y|E$ then $x = y$. (Proceed by induction on $\dim(M)$.)

2.59. Let (Ω, Σ, μ) be a positive measure space and consider the unit ball $U \equiv U(L^1(\Omega, \mu, \mathbb{F}))$. In exercise 2.30 it was shown that this set may have no extreme points.

a) Let $\mathbb{F} = \mathbb{C}$ and choose $f \in U$, $\|f\|_1 = 1$. Then $f \in \text{ext}(U)$ if and only if $|f(\cdot)| \in \text{ext}(U(L^1(\Omega, \mu, \mathbb{R})))$.

b) Let $\mathbb{F} = \mathbb{R}$ and choose $f \in U$, $\|f\|_1 = 1$. Then $f \in \text{ext}(U)$ if and only if $f = \pm\chi_A/\mu(A)$, where A is an atom in Σ.

2.60. Show that the conclusion of Dieudonne's separation theorem (**15D**) remains valid if the hypothesis concerning the recession cones is weakened to: $M \equiv C_A \cap C_B$ is a linear subspace of X. (Work in the quotient space X/M.)

Chapter III

Principles of Banach Spaces

The theme of our presentation up to this point may be described as a study of the interplay between the algebraic-geometric notions of convex sets and mappings, extreme points, etc. and the topological notions of openness, compactness, continuity, etc. For such a study the correct setting is, as we have seen, the linear topological space (frequently required also to be locally convex). The resulting theory is broad and powerful, as we hope has been demonstrated by Chapter II. Further development now requires some additional specialization of our setting. The crucial new hypothesis which we now bring in is that of completeness. We shall also generally limit our considerations to normed linear spaces, unless the results under consideration can be clearly and cleanly extended to locally convex spaces. More typically, locally convex topologies will play a vital supporting role in our theory of Banach spaces, particularly the weak and weak* topologies.

Of the numerous important results presented in this chapter the most important for applications (as we hope to illustrate in the examples and problems) are the various category theorems of §17. For further theoretical developments in the study of Banach spaces, the profound characterizations of weak compactness due to Eberlein and James and the theorems of Bishop-Phelps are important.

§16. Completion, Congruence, and Reflexivity

In this section we present a miscellany of general facts about Banach spaces and linear maps defined thereon. The most important notion is that of reflexivity of a Banach space. A number of characterizations of reflexive spaces are collected together in **16F**, and others occur in later sections. Because of their many useful properties and generally tidy theory, it is both a shame and a challenge that many of the important Banach spaces are not reflexive.

A. Let X and Y be normed linear spaces over the same scalar field and let $B(X, Y)$ be the normed space of bounded linear maps from X into Y (**11D**). Our first problem is to decide when $B(X, Y)$ is a Banach space (that is, complete). The answer depends only on the range space Y.

Theorem. *$B(X, Y)$ is a Banach space if and only if Y is a Banach space.*

Proof. Suppose that Y is complete and let $\{T_n\}$ be a Cauchy sequence in $B(X, Y)$. For any $x \in X$ we have

$$\begin{aligned} \|T_n(x) - T_m(x)\| &= \|(T_n - T_m)(x)\| \\ &\leqslant \|T_n - T_m\| \, \|x\|, \end{aligned}$$

from which we conclude that $\{T_n(x)\}$ is a Cauchy sequence in Y. Hence $T(x) \equiv \lim_n T_n(x)$ exists in Y. The correspondence $x \mapsto T(x)$ clearly defines a linear map from X into Y. Now a Cauchy sequence in any metric space is bounded; in our case we have therefore that $\beta \equiv \sup_n \|T_n\| < \infty$. Thus $\|T_n(x)\| \leqslant \beta\|x\|$ and so

$$\|T(x)\| \equiv \|\lim_n T_n(x)\| = \lim_n \|T_n(x)\| \leqslant \beta\|x\|,$$

for all $x \in X$; that is, $\|T\| \leqslant \beta$ and so T belongs to $B(X, Y)$.

Conversely, suppose that $B(X, Y)$ is complete and let $\{y_n\}$ be a Cauchy sequence in Y. Select a vector $x_0 \in X$ with $\|x_0\| = 1$ and then, by **11E**, a functional $\phi \in X^*$ such that $\phi(x_0) = 1$. Then if we define $T_n \in B(X, Y)$ by

$$T_n(x) = \phi(x)y_n, \qquad x \in X,$$

we see that $\|T_n - T_m\| \leqslant \|\phi\| \, \|y_n - y_m\|$. Consequently, $\{T_n\}$ is a Cauchy sequence in $B(X, Y)$ and therefore has a limit $T \in B(X, Y)$. Finally,

$$\|y_n - T(x_0)\| = \|T_n(x_0) - T(x_0)\| \to 0, \qquad n \to \infty,$$

whence $\lim_n y_n = T(x_0) \in Y$. □

Corollary. *Let X be a normed linear space. Then X^* is a Banach space.*

B. Again, let X and Y be normed spaces over the same field and $T \in B(X, Y)$. As in **9E** T is an isomorphism if it identifies X and Y as linear topological spaces, that is, if T is both an algebraic isomorphism and a homeomorphism. Otherwise put, T is an isomorphism exactly when T^{-1} exists and belongs to $B(Y, X)$. More generally we say that T is an isomorphism *into* Y if T is injective and T^{-1} is bounded on its domain (which is, by definition, the image $T(X)$ of X in Y).

Lemma. *$T \in B(X, Y)$ is an isomorphism into Y if and only if $\gamma \equiv \inf\{\|T(x)\| : \|x\| = 1\} > 0$.*

Proof. By its definition γ satisfies $\|T(x)\| \geqslant \gamma\|x\|$ for all $x \in X$. Hence if $\gamma > 0$ then T is certainly injective and $\|T^{-1}(T(x))\| \equiv \|x\| \leqslant \gamma^{-1}\|T(x)\|$, so that T^{-1} is bounded. Similarly, if T^{-1} exists and is bounded then $\|x\| \equiv \|T^{-1}(T(x))\| \leqslant \|T^{-1}\| \, \|T(x)\|$ so that $\gamma \geqslant 1/\|T^{-1}\| > 0$. □

We can thus write for comparison

(16.1)
$$\frac{1}{\|T^{-1}\|} = \inf\{\|T(x)\| : \|x\| = 1\} \leqslant \sup\{\|T(x)\| : \|x\| = 1\} \equiv \|T\|$$

when T is an isomorphism into, and in particular

(16.2)
$$1 \leqslant \|T\| \, \|T^{-1}\|.$$

We see also that T is an isomorphism into exactly when T has a bounded inverse.

Dealing as we are with normed spaces rather than more general linear topological spaces, it is important to know when two such spaces can be identified with all their structure. Such identifications are made by means of a *congruence*, that is, a norm-preserving isomorphism. Thus $T \in B(X, Y)$ is a congruence when T is surjective and

(16.3) $$\|T(x)\| = \|x\|, \qquad x \in X.$$

This condition is clearly equivalent to

(16.4) $$\|T\| = \|T^{-1}\| = 1.$$

Again, when T is not surjective but satisfies (16.3) or (16.4) we shall say that T is a congruence into. In this case T identifies X with a subspace of Y.

Note that, because of its linearity, a congruence is actually an isometry on the underlying metric space of X. Thus it preserves all the geometric features of X and in particular it maps the unit ball $U(X)$ onto the unit ball of its range. This of course need not happen when T is merely an isomorphism; in that case we can only be sure that topological properties are preserved. In the remainder of this section we are going to see several important examples of congruences. (We have previously encountered a congruence between the spaces L^{1*} and L^∞ in **12C**.)

C. Let X, Y, T be as in **16B**. According to **1H** there is a linear map $T' : Y' \to X'$ (the transpose of T) defined by the equation

$$\langle x, T'(\psi)\rangle = \langle T(x), \psi\rangle, \qquad x \in X, \qquad \psi \in Y'.$$

We set

$$T^* = T' | Y^*,$$

and call T^* the *conjugate* (*adjoint*, *dual*) of T. Directly from the definitions it follows that T^* is *weak*-continuous*, that is, T^* is continuous when both X^* and Y^* are given their weak* topologies.

Lemma. *The mapping $T \mapsto T^*$ is a congruence of $B(X, Y)$ with the subspace of $B(Y^*, X^*)$ consisting of weak*-continuous linear maps.*

Proof. The mapping is clearly linear and its range consists of weak*-continuous maps on Y^* as was just noted. Next, we have

$$\begin{aligned} \|T^*\| &\equiv \sup\{\|T^*(\psi)\| : \|\psi\| \leqslant 1\} \\ &= \sup\{|\langle x, T^*(\psi)\rangle| : \|x\| \leqslant 1, \|\psi\| \leqslant 1\} \\ &\equiv \sup\{|\langle T(x), \psi\rangle| : \|\psi\| \leqslant 1, \|x\| \leqslant 1\} \\ &= \sup\{\|T(x)\| : \|x\| \leqslant 1\} \equiv \|T\|. \end{aligned}$$

(Note that the penultimate equality here requires the Hahn-Banach theorem **11G** to guarantee the existence of $\psi \in U(Y^*)$ such that $\langle T(x), \psi \rangle = \|T(x)\|$.) Finally, we must show that any weak*-continuous $S \in B(Y^*, X^*)$ is the conjugate of some $T \in B(X, Y)$. Such T must satisfy

(16.5) $$\langle T(x), \psi\rangle = \langle x, S(\psi)\rangle$$

for all $x \in X$ and $\psi \in Y^*$. Now the right-hand side of (16.5) is by assumption a weak*-continuous linear functional on Y^* for any fixed $x \in X$. By **12B** it follows that there is a vector in Y, which we call $T(x)$, such that (16.5) holds. The map $T: X \to Y$ so defined is single-valued (**11E**), linear, and continuous since

$$\begin{aligned}\|T(x)\| &= \sup\{|\langle T(x), \psi\rangle| : \|\psi\| \leqslant 1\} \\ &\equiv \sup\{|\langle x, S(\psi)\rangle| : \|\psi\| \leqslant 1\} \leqslant \|S\|\,\|x\|,\end{aligned}$$

so that $\|T\| \leqslant \|S\|$. □

This lemma implies that $B(Y^*, X^*)$ is generally a more complicated space than $B(X, Y)$. However, it is a direct corollary to **16F** below that whenever Y is reflexive every map in $B(Y^*, X^*)$ is weak*-continuous, so that this space is congruent to $B(X, Y)$ under the above mapping. The converse is also valid.

The following result belongs in this sub-section as it connects the ideas of conjugate maps and isomorphisms. However, part of its proof makes use of a later result, which we must temporarily take on faith.

Theorem. *$T \in B(X, Y)$ has a bounded inverse if and only if T^* is surjective.*

Proof. Suppose that T has a bounded inverse T^{-1}, defined on a subspace R ($\equiv$ range (T)) of Y. For any $\phi \in X^*$ the functional $\phi \circ T^{-1}$ belongs to R^* and may therefore be extended to a functional $\psi \in Y^*$ (exercise 2.16). We claim that $T^*(\psi) = \phi$. Indeed, for any $x \in X$,

$$\langle x, T^*(\psi)\rangle \equiv \langle T(x), \psi\rangle \equiv \phi \circ T^{-1}(T(x)) = \phi(x).$$

Thus T^* is seen to be surjective.

Conversely, suppose that T^* is surjective. If T fails to have a bounded inverse then by exercise 3.3 there is a sequence $\{x_n\}$ in X with

$$\textbf{(16.6)} \qquad \lim_n \|x_n\| = \infty, \qquad \lim_n T(x_n) = \theta.$$

Now for any $\phi \in X^*$ we have by assumption that $\phi = T^*(\psi)$ for for some $\psi \in Y^*$. Hence

$$\lim_n \phi(x_n) = \lim_n \langle x_n, T^*(\psi)\rangle = \lim_n \langle T(x_n), \psi\rangle = 0.$$

According to the uniform boundedness principle (**17C**) the sequence $\{x_n\}$ is bounded since we have just shown that it converges weakly to θ. We thus have a contradiction to (16.6). □

D. We now give a second important example of congruence in normed linear space theory. Let X be a normed linear space and consider once again the canonical embedding $J_X: X \to X^{*\prime}$ defined by

$$\langle \phi, J_X(x)\rangle = \phi(x), \qquad x \in X, \qquad \phi \in X^*.$$

For fixed x we see that

$$|\langle \phi, J_X(x)\rangle| = |\phi(x)| \leqslant \|\phi\|\,\|x\|$$

which proves that $J_X(x)$ is a continuous function on X^* with norm $\leqslant \|x\|$. Thus range$(J_X) \subset X^{**} \equiv (X^*)^*$. Further, as was noted in **16C**, it is an elementary consequence of the Hahn-Banach theorem that given $x \in X$ there is $\phi \in U(X^*)$ with $\phi(x) = \|x\|$. This entails

$$\text{(16.7)} \qquad \sup\{|\langle \phi, J_X(x)\rangle| : \|\phi\| \leqslant 1\} = \|x\|,$$

and proves that J_X is a congruence of X into X^{**}.

It is convenient to write $\hat{x} = J_X(x)$ for $x \in X$ and to denote the closure of range(J_X) by $\hat{X}$. Since X^{**} is a Banach space (**16A**) and $\hat{X}$ is closed (by definition) in X^{**}, it follows that $\hat{X}$ is complete and hence a Banach space. Consequently, $\hat{X}$ satisfies all the requirements for a *completion* of X: it is a Banach space containing a dense subspace congruent to X.

A routine argument shows that, for any normed space Y, any $T \in B(X, Y)$ has exactly one extension $\bar{T}$ belonging to $B(\hat{X}, Y)$ and satisfying $\|\bar{T}\| = \|T\|$. This fact in turn shows that, up to congruence, $\hat{X}$ is the only completion of X. It also shows that $(\hat{X})^*$ is congruent to X^*. But, let it be noted carefully that if $X \neq \hat{X}$ (that is, if X is not complete) then the respective weak* topologies on X^* and on $(\hat{X})^*$ are not the same, since the former is the topology of pointwise convergence on X while the latter is the topology of pointwise convergence on the properly larger space $\hat{X}$. However, the two topologies do agree on any ball B in X^*, since B is weak*-compact by **12D** and so the identity map of B, being $\sigma(X^*, \hat{X}) - \sigma(X^*, X)$-continuous, is actually a homeomorphism.

E. We are going to exhibit a few other congruences important in the general theory of normed spaces. First, however, let us agree on some terminology. A bounded linear map between two normed spaces will henceforth be called an *operator*. Given an operator $T \in B(X, Y)$ we define an operator $\hat{T}$ called the 1–1 *operator induced by* T by

$$\hat{T} : X/\ker(T) \to Y,$$
$$\hat{T}(x + \ker(T)) \equiv T(x), \qquad x \in X.$$

This new operator $\hat{T}$ is well-defined, linear, injective, and has the same range as T. Further, it is easy to see that $\|\hat{T}\| = \|T\|$. Do not confuse the notation $\hat{T}$ with the completion notation $\hat{X}$ just introduced.

The notation $X \cong Y$ was used in **1H** to signify canonical isomorphism of linear spaces. In our present context of normed linear spaces we shall use this notation to mean that there is a canonical isomorphism between X and Y which is also a congruence. Thus we have, for example, $X \cong J_X(X)$, because of (16.7).

Theorem. *Let M be a linear subspace of the normed linear space X. Then*
a) $(X/M)^* \cong M^\circ$;
b) $M^* \cong X^*/M^\circ$;
c) $M^{**} \cong \overline{J_X(M)}^*$.

Proof. a) The proof runs parallel to that of the linear space analogue given in **1H**, but also makes use of **16C**. Let $Q_M: X \to X/M$ be the quotient map. Then $Q_M^*: (X/M)^* \to X^*$ is an injective operator with range M° and it remains to see that it is a congruence. This requires proving that for each $\phi \in M^\circ$

(16.8) $$||\phi|| = \sup\{|\phi(x)|/d(x, M): x \in X \backslash M\}.$$

Now the left-hand side of (16.8) is $\leqslant$ the right hand side since $||Q_M^*|| = ||Q_M|| = 1$. On the other hand, if $m \in M$, $|\phi(x)| = |\phi(x - m)| \leqslant ||\phi|| \, ||x - m||$, so that $|\phi(x)| \leqslant ||\phi|| d(x, M)$.

b) Again we follow **1H** and consider the identity injection $I_M: M \to X$. Then $I_M^*: X^* \to M^*$ is just the restriction operator (that is, $I_M^*(\phi) = \phi|M$) and the 1–1 operator $\hat{I}_M^*: X^*/M^\circ \to M^*$ is an (algebraic) isomorphism. To prove that it is a congruence we must verify that for each $\phi \in X^*$

(16.9) $$d(\phi, M^\circ) = \sup\{|\phi(y)|: y \in U(M)\} \equiv ||\phi|M||.$$

Let us just check that the left-hand side of (16.9) cannot exceed the right-hand side. Let τ be a norm-preserving extension (**11G**) of $\phi|M$ to all of X and set $\psi = \phi - \tau$. Then $\psi \in M^\circ$ and hence $d(\phi, M^\circ) \leqslant ||\phi - \psi|| = ||\tau|| \equiv ||\phi|M||$.

c) Let $M^{\circ\circ} \equiv (M^\circ)^\circ$ be the annihilator of M° in $X^{**}: M^{\circ\circ} = \{\Phi \in X^{**}: \Phi(\phi) = 0, \phi \in M^\circ\}$. Then by combining a) and b) we see that

(16.10) $$M^{**} \cong M^{\circ\circ},$$

and the proof can be completed by showing that

(16.11) $$M^{\circ\circ} = \overline{J_X(M)}^*.$$

Now the left-hand side of (16.11) certainly contains the right-hand side since $M^{\circ\circ}$ is weak*-closed and contains $J_X(M)$. If the inclusion were proper there would exist a $\Phi \in M^{\circ\circ}$ that could be strongly separated from $\overline{J_X(M)}^*$ by a weak*-closed hyperplane in X^{**}. Using **12B** we could then find a $\phi \in X^*$ such that $\phi(y) = \langle \phi, \hat{y} \rangle = 0$, $y \in M$, and $\langle \phi, \Phi \rangle \neq 0$. In other words, $\phi \in M^\circ$ yet $\langle \phi, \Phi \rangle \neq 0$, which contradicts $\Phi \in M^{\circ\circ}$. □

A few remarks should be made about this theorem. First, the inverses of the congruences in a) and b) are easily described and could be used as the basis of alternate proofs. Thus in a) we have simply $\phi \mapsto \hat{\phi} \circ \tau$ as the congruence from M° to $(X/M)^*$, where $\tau: X/M \mapsto X/\ker(\phi)$ is the norm-decreasing surjection defined by $\tau(x + M) = x + \ker(\phi)$. In b) the congruence from M^* to X^*/M° sends each element ψ of M^* into the affine subspace of X^* consisting of all possible extensions of ψ to X.

Secondly, the congruence in c) can also be explicitly described. For example, in the direction $M^{\circ\circ} \mapsto M^{**}$ it sends $\Phi \in M^{\circ\circ}$ into the functional on M^* whose value at $f \in M^*$ is the value assumed by Φ on any extension of f in X^*. This value is of course independent of the particular extension since any two extensions of f must differ by an element of M°. (The inverse congruence is used in the proof of the theorem in **16F** below; see also exercise 3.16c.)

Finally, note that the theorem identifies the second conjugate space M^{**} with a subspace of X^{**}, while the first conjugate M^* can only be identified with a quotient space of X^*. It is unusual that this quotient can in turn be identified with a subspace of X^*.

A generalization of formula (16.11) pertaining to convex sets containing θ is given in exercise 3.7.

F. A normed linear space X is *reflexive* if the canonical embedding $J_X: X \to X^{**}$ is surjective. This requirement should be compared with the result in **1D** for the purely algebraic situation. There it was noted that when X is infinite dimensional there will always be linear functionals on X' which are not evaluation functionals. But our reflexivity definition does not ask so much; the requirement is merely that every continuous linear functional on the subspace X^* of X' should be an evaluation functional. This is a situation which does in fact occur in a number of common normed spaces (**16G** below). Such spaces have several pleasant and useful properties as we shall see.

It is clear that any reflexive normed space must be complete, by virtue of being congruent to a complete space (namely, X^{**}). It should also be clear that the condition of reflexivity is equivalent to the identity of the weak and weak* topologies on X^*. Indeed, if these topologies are the same then any functional in X^{**} is $\sigma(X^*, X^{**})$-continuous (since any operator between normed spaces is weakly continuous, that is, continuous when both spaces are given their weak topologies), and so $\sigma(X^*, X)$-continuous (by assumption). Hence the functional must be an evaluation functional by **12B**.

It is clear that every finite dimensional normed space is reflexive (**1D**), and we shall see other examples later in this section. On the other hand, the space $L^1 \equiv L^1([0, 1], \mathbb{R})$ is not reflexive. To see this we recall (**12C**) that L^{1*} can be identified with (is congruent to) $L^\infty \equiv L^\infty([0, 1], \mathbb{R})$. We shall prove the existence of functionals $\Phi \in L^{\infty*}$ which are not of the form

$$\textbf{(16.12)} \qquad \Phi(g) = \int_0^1 f(t)g(t)dt, \qquad g \in L^\infty,$$

for some $f \in L^1$. Now the subspace $C \equiv C([0, 1], \mathbb{R})$ is a proper closed subspace of L^∞ and so there is a non-zero $\Phi \in C^\circ$ (**11F**). If Φ were an evaluation functional defined by some $f \in L^1$ as in (16.12) then we would have in particular

$$\int_0^1 f(t)g(t)dt = 0, \qquad g \in C.$$

But, as was observed in **12C**, this last condition forces $f = \theta$; that is, a representation of Φ as in (16.12) does not exist, and so L^1 is not reflexive.

We are going to present a few necessary and sufficient conditions for the reflexivity of a given Banach space. It is convenient to first establish a useful general fact, known as the "Goldstine-Weston density lemma". For X a given normed space, we let V be a subspace of X^* and define the canonical embedding $J_{X,V}: X \to V^*$ by the usual formula:

$$\langle \phi, J_{X,V}(x) \rangle = \phi(x), \qquad x \in X, \qquad \phi \in V.$$

Clearly $\|J_{X,V}\| \leqslant 1$ and $J_{X,V}$ is injective exactly when V is total.

Lemma. *For every $\Psi \in V^*$ there is a net $\{x_\delta : \delta \in D\}$ in X such that*

$$\sup\{\|x_\delta\| : \delta \in D\} \leqslant \|\Psi\|$$

and

$$\lim\{\phi(x_\delta) : \delta \in D\} = \Psi(\phi), \qquad \phi \in V.$$

In particular, $J_{X,V}(U(X))$ is weak-dense in $U(V^*)$.*

Proof. Let $\Psi_0 = \Psi/\|\Psi\|$ (the result is trivial if $\Psi = \theta$). Let N be an arbitrary weak*-neighborhood of $\Psi_0 : N = \{\Phi \in V^* : |\Phi(\phi_i) - \Psi_0(\phi_i)| < \varepsilon_0\}$ for some finite set $\{\phi_1, \ldots, \phi_n\} \subset V$. We shall exhibit $x_0 \in U(X)$ such that $J_{X,V}(x_0) \in N$, thereby proving that Ψ_0 belongs to the weak*-closure of $J_{X,V}(U(X))$. Let $r = \max\{\|\phi_i\| : i = 1, \ldots, n\}$. Because $\|\Psi_0\| = 1$ we can apply Helly's condition (**7A**) to conclude that for every $\varepsilon > 0$ there exists x_ε with $\|x_\varepsilon\| \leqslant 1 + \varepsilon/r$ and $\phi_i(x_\varepsilon) = \Psi_0(\phi_i)$, $i = 1, \ldots, n$. We now set $x_0 = (r/(r + \varepsilon_0))x_{\varepsilon_0}$. Then $\|x_0\| \leqslant 1$ and

$$\begin{aligned} |\phi_i(x_0) - \Psi_0(\phi_i)| &= |\phi_i(x_0) - \phi_i(x_{\varepsilon_0})| \\ &= |\phi_i(x_0) - (1 + \varepsilon_0/r)\phi_i(x_0)| \leqslant r\left|\frac{-\varepsilon_0}{r}\right| = \varepsilon_0. \end{aligned}$$

Thus $\Psi_0 \in \overline{J_{X,V}(U(X))}^*$ and so there is a net $\{y_\delta : \delta \in D\}$ in $U(X)$ that converges weak* to Ψ_0. Hence we may set $x_\delta = \|\Psi\| y_\delta$ to conclude. ☐

The most important application of this lemma is of course to the case where $V = X^*$. Also, we note that whenever V is separable then the net $\{x_\delta : \delta \in D\}$ can be taken to be a sequence, since, in this case, $U(V^*)$ is compact and weak*-metrizable (**12F**). An alternate proof of the lemma is suggested in exercise 3.7.

Theorem. *The following properties of a Banach space X are equivalent.*
a) *X is reflexive;*
b) *X^* is reflexive;*
c) *M and X/M are reflexive for every closed linear subspace M of X;*
d) *M and X/M are reflexive for some closed linear subspace M of X;*
e) *$U(X)$ is weakly compact;*
f) *X is weakly quasi-complete.*

Proof. Assume that X is reflexive. To show that X^* is reflexive, choose any $G \in X^{***} \equiv (X^{**})^*$ and define $\phi = G \circ J_X \in X^*$. Then if $\Phi \in X^{**}$, $\Phi = \hat{x}$ for some $x \in X$, and

$$\begin{aligned} \langle \Phi, G \rangle &= \langle \hat{x}, G \rangle = \langle x, G \circ J_X \rangle \\ &\equiv \phi(x) = \langle \phi, \hat{x} \rangle = \langle \phi, \Phi \rangle; \end{aligned}$$

that is, $G = J_{X^*}(\phi)$, and thus J_{X^*} is surjective. Next, let M be a closed subspace of X. If $\Phi \in M^{\circ\circ} \subset X^{**}$ then $\Phi = \hat{x}_0$ for some $x_0 \in X$. Hence if $\phi \in M^\circ$, $0 = \langle \phi, \hat{x}_0 \rangle \equiv \phi(x_0)$, whence $x_0 \in {}^\circ M^\circ = M$ by **12C**. Now, given any $F \in M^{**}$, we define $\Phi \in M^{\circ\circ}$ by $\Phi(\phi) \equiv F(\phi|M)$, $\phi \in X^*$. Then $\Phi = \hat{x}_0$ for some $x_0 \in M$, as we have just seen. Therefore, $F = J_M(x_0)$ and J_M is surjective.

Suppose that X^* is reflexive. Then X^{**} is reflexive and so is its closed subspace $J_X(X)$. Since this subspace is congruent to X, X must also be reflexive. Also, if M is again a closed subspace of X, then $(X/M)^* \cong M^\circ$ is a closed subspace of the reflexive space X^*, hence is reflexive. Since $(X/M)^*$ is reflexive so is X/M.

To complete the proof of equivalence of a) through d) we show that d) implies a). Select any $\Phi \in X^{**}$. Then $\Phi|M^\circ \in M^{\circ *} \cong (X/M)^{**} = J_{X/M}(X/M)$, and so there exists $x_0 \in X$ such that $\langle \phi, \Phi \rangle = \phi(x_0)$, for all $\phi \in M^\circ$. Consequently, $\Phi - \hat{x}_0 \in M^{\circ\circ} \cong M^{**} = J_M(M)$, and it follows that there is $y_0 \in M$ such that $\Phi - \hat{x}_0 = \hat{y}_0$. Therefore, $\Phi = \hat{x}_0 + \hat{y}_0 \in J_X(X)$, and J_X is surjective.

It remains to establish the equivalence of a), e), and f). For e) the keys are the Goldstine-Weston density lemma (with $V = X^*$) and the observation (**12B**) that J_X is a homeomorphism between X and $J_X(X)$ when these spaces are given the weak and weak* topologies respectively. Thus $U(X)$ is weakly compact if and only if $J_X(U(X))$ is weak* compact. Since $J_X(U(X))$ is weak*-dense in $U(X^{**})$ which is weak*-compact (**12D**), it follows that $U(X)$ is weakly compact exactly when $J_X(U(X)) = U(X^{**})$. This last condition is of course equivalent to the surjectivity of J_X. Next, assume that X is weakly quasi-complete. Then in particular $U(X)$ is weakly complete. The same argument then shows that $J(U(X)) = U(X^{**})$ and hence that X is reflexive.

Finally, suppose that X is reflexive, and let A be any weakly bounded and weakly closed subset of X. We want to see that A is weakly complete. Appealing once more to the forthcoming uniform boundedness principle (**17C**), the set A is norm-bounded. Thus A is a weakly closed subset of some ball which, by e), is weakly compact and, in particular, weakly complete. Hence A must also be weakly complete. □

In exercise 3.37 there is given a useful necessary and sufficient condition for an arbitrary subspace of a Banach space to be reflexive.

G. In **12C** we outlined an argument showing that the spaces $L^1(\Omega, \mu, \mathbb{F})^*$ and $L^\infty(\Omega, \mu, \mathbb{F})$ are congruent provided that the measure μ is σ-finite. That some restriction on μ is necessary can be seen from the following example.

Example. Let $\Omega = [0, 1]$. Let the σ-algebra Σ of subsets of Ω consist of all countable subsets of Ω and their complements. Then any Σ-measurable function defined on Ω has a countable range. Now let μ be the counting measure on Ω so that $\mu(E) \equiv$ cardinality of E if $E \in \Sigma$ is finite, and otherwise $\mu(E) \equiv \infty$. Then the functional

$$f \mapsto \int_0^1 tf(t)d\mu(t), \qquad f \in L^1(\Omega, \mu, \mathbb{F}),$$

does not arise from any Σ-measurable function, and so the usual congruence from $L^\infty(\Omega, \mu, \mathbb{F})$ into $L^1(\Omega, \mu, \mathbb{F})^*$ is not surjective. □

In spite of examples such as this, any L^1 space that fails to have the corresponding L^∞ space as its dual (via the usual congruence) does so for an essentially trivial reason. In other words, the difficulty is more apparent than real. Precisely, the following result is true: for any measure space (Ω, Σ, μ) there is another such space $(\Gamma, \mathcal{T}, \nu)$ such that $L^1(\Gamma, \nu, \mathbb{F}) \cong L^1(\Omega, \mu, \mathbb{F})$ and $L^1(\Gamma, \nu, \mathbb{F})^* = L^\infty(\Gamma, \nu, \mathbb{F})$ (via the usual congruence). We shall not prove this result but in exercise 3.11 we ask that such a measure space be determined for the preceding example.

Let us now consider the problem of determining $L^p(\Omega, \mu, \mathbb{F})^*$ for a given measure space (Ω, Σ, μ) and $1 < p < \infty$. In this case the answer requires no restrictions whatever on the nature of the underlying measure space, and can be used to show that all such spaces are reflexive. This is the single most important class of infinite dimensional reflexive Banach spaces. Given p with $1 < p < \infty$ we define $q = p/(p-1)$, and for $g \in L^q \equiv L^q(\Omega, \mu, \mathbb{F})$ we define a linear functional Φ_g on $L^p \equiv L^p(\Omega, \mu, \mathbb{F})$ by the rule

(16.13) $$\Phi_g(f) = \int_\Omega fg\, d\mu, \qquad f \in L^p.$$

Theorem. *The map $g \mapsto \Phi_g$ is a congruence of L^q with $(L^p)^*$.*

Proof. Hölder's inequality shows that $\Phi_g \in (L^p)^*$ and that $\|\Phi_g\| \leqslant \|g\|_q$. Let us assume first that $\mu(\Omega) < \infty$. Given any $\Psi \in (L^p)^*$ we proceed as in **12C** by considering the measure ν on Σ defined by $\nu(E) = \Psi(\chi_E)$, $E \in \Sigma$. ν is μ-absolutely continuous and the Radon-Nikodym theorem yields $g \equiv d\nu/d\mu \in L^1$ such that $\nu(E) = \int_E g\, d\mu$. Then $\Psi(f) = \int_\Omega f\, d\nu = \int_\Omega fg\, d\mu$ for all simple functions f, and hence for all $f \in L^1$ (which includes L^p). Let us next see that $g \in L^q$. Select any $h \in L^\infty$ with $0 \leqslant h \leqslant |g|$. Then

$$\begin{aligned}\|h\|_q^q \leqslant \int_\Omega h^{q-1}|g|\, d\mu &= \int_\Omega h^{q-1}\frac{d|\nu|}{d\nu}\, d\nu \\ &\equiv \Psi\left(h^{q-1}\frac{d|\nu|}{d\nu}\right) \leqslant \|\Psi\| \left\|h^{q-1}\frac{d|\nu|}{d\nu}\right\|_p \\ &= \|\Psi\|\, \|h\|_q^{q/p},\end{aligned}$$

whence $\|h\|_q \leqslant \|\Psi\|$. Taking the supremum over all such h we see that $\|g\|_q \leqslant \|\Psi\|$. This completes the proof for the case $\mu(\Omega) < \infty$.

In the general case we consider an arbitrary set $E \in \Sigma$ with $0 < \mu(E) < \infty$. Such sets are called *chunks*. Restricting the given $\psi \in (L^p)^*$ to the subspace $\{f\chi_E : f \in L^p\}$ we obtain by the preceding method an element $g_E \in L^q$ with $g_E = g_E\chi_E$ and $\Psi(f\chi_E) = \int_\Omega fg_E\, d\mu$. If E_1 is any other chunk with corresponding g_{E_1} then on $E \cap E_1$ the elements g_E and g_{E_1} must be equal. Now let $E_1, E_2, \ldots,$ be a sequence of chunks chosen so that

$$\lim_n \|g_{E_n}\|_q = \sup\{\|g_E\|_q : 0 < \mu(E) < \infty\} \equiv \gamma.$$

Then $\gamma \leqslant \|\Psi\|$ and the sequence $\{g_{E_n}\}$ is a Cauchy sequence in L^q since, assuming $m \leqslant n$,

$$\|g_{E_m} - g_{E_n}\|_q^q \leqslant \int_{E_n \setminus E_m} |g_{E_n}|^q \, d\mu = \int_{E_n} |g_{E_n}|^q \, d\mu - \int_{E_m} |g_{E_m}|^q \, d\mu.$$

The limit $g \in L^q$ satisfies $\|g\|_q = \gamma$ and g vanishes outside the σ-finite set $A \equiv \bigcup_n E_n$. This g is the desired L^q representer of Ψ. Indeed, by the maximal property of g ($\|g\|_q = \gamma$) it follows that any chunk E disjoint from A must have its corresponding $g_E = \theta$. Thus, if we select any $f \in L^p$ such that f vanishes outside some chunk E, then

$$\begin{aligned} \Psi(f) &= \Psi(f\chi_E) = \int_\Omega f g_E \, d\mu \\ &= \int_{A \cap E} f g_E \, d\mu + \int_{E \setminus A} f g_E \, d\mu \\ &= \int_{A \cap E} f g_{A \cap E} \, d\mu + 0 = \int_{A \cap E} f g \, d\mu \\ &= \int_\Omega f g \, d\mu, \end{aligned} \tag{16.14}$$

where the penultimate equality follows from the relation $g_{A \cap E} = g_{E_n}$ in the set $E \cap E_n$ (which implies that, on $A \cap E$, $g_{A \cap E} = \lim_n g_{E_n} = g$). Equation (16.14) shows that $\Psi(f) = \Phi_g(f)$ whenever f vanishes off a chunk. The set of such f contains in particular the space of the characteristic functions of chunks and so is dense in L^p. Hence by continuity $\Psi = \Phi_g$ and so the map $g \mapsto \Phi_g$ is surjective in the general case as well. □

Corollary. *For $1 < p < \infty$ the space $L^p(\Omega, \mu, \mathbb{F})$ is reflexive.*

The proof is a direct consequence of the theorem; the details are left to exercise 3.12.

H. We finally consider some function spaces defined on an arbitrary set Ω and determine congruent representations of their dual spaces. It will follow from these representations that none of the spaces under consideration is reflexive.

We define $c_0 \equiv c_0(\Omega, \mathbb{F})$ to be the set of all $\mathbb{F}$-valued functions x on Ω that vanish at infinity in the sense that $[t \in \Omega : |x(t)| > \varepsilon\}$ is finite for each $\varepsilon > 0$, and $m \equiv m(\Omega, \mathbb{F})$ to be the set of all bounded $\mathbb{F}$-valued functions on Ω. These are special cases of earlier examples of spaces of continuous functions if Ω is assumed to have the discrete topology. In both cases, of course, the spaces are normed by the uniform norm $\|x\|_\infty \equiv \sup\{|x(t)| : t \in \Omega\}$, and c_0 is a closed linear subspace of the Banach space m. We also define $\ell^1 \equiv L^1(\Omega, \mu, \mathbb{F})$ where μ is the counting measure with domain 2^Ω, the family of all subsets of Ω. Note that $m = L^\infty(\Omega, \mu, \mathbb{F})$ but that in general (unless Ω is countable) μ is not σ-finite.

If X and Y are two normed spaces we define the *ℓ^p-product* $(X \times Y)_p$ to be the product space $X \times Y$ with the norm $\|(x, y)\|_p = \|(\|x\|, \|y\|)\|_p \equiv (\|x\|^p + \|y\|^p)^{1/p}$ for $1 \leqslant p < \infty$ (a special case of formula (10.8)). We can now represent the dual spaces of c_0, ℓ^1, and m.

Theorem. a) $c_0^* \cong \ell^1$;

b) $^{1*} \cong m$;

c) $m^* \cong (\ell^1 \times (c_0)^\circ)_1$.

Proof. a) The congruence of this part (and of b)) is defined in the expected way; namely, given $y \in \ell^1$ we define $\Phi_y \in c_0^*$ by

$$\Phi_y(x) = \int xy\, d\mu \equiv \sum_{t\in\Omega} x(t)y(t), \qquad x \in c_0. \tag{16.15}$$

Clearly, $\|\Phi_y\| \leqslant \sum_{t\in\Omega} |y(t)| \equiv \|y\|_1$. Consider now any $\Psi \in c_0^*$. We define a function y on Ω by $y(t) \equiv \Psi(e_t)$, where $e_t \equiv \chi_{\{t\}}$, $t \in \Omega$. For each finite set $E \subset \Omega$ define an operator T_E on c_0 by $T_E(x) = x\chi_E$. Then T_E^* is an operator on c_0^* (**16C**) and

$$\begin{aligned}\langle x, T_E^*(\Psi)\rangle &= \langle T_E(x), \Psi\rangle \\ &= \langle \sum_{t\in E} x(t)e_t, \Psi\rangle = \sum_{t\in E} x(t)\Psi(e_t) \\ &\equiv \sum_{t\in E} x(t)y(t), \qquad x \in c_0.\end{aligned} \tag{16.16}$$

Setting $x(t) = |y(t)|/y(t)$ whenever $y(t) \neq 0$ and $x(t) = 0$ otherwise, it follows from (16.16) and (**16C**) that

$$\|\Psi\| \geqslant \|T_E^*\Psi\| \geqslant \sum_{t\in E} |y(t)| \equiv \int_E |y|\, d\mu.$$

Thus $y \in \ell^1$ and $\|y\|_1 \leqslant \|\Psi\|$. This proves that the map $y \mapsto \Phi_y$ is a surjective congruence and completes the proof of a).

b) The proof of this assertion is quite similar to that of a). The congruence from m onto ℓ^{1*} is defined as in (16.15) except that now $y \in m$ and $x \in \ell^1$.

c) The notation c_0° refers to the annihilator of c_0 in m^* when c_0 is considered as a subspace of m. The congruence is defined as follows: let $y \in \ell^1$ and $\phi \in c_0^\circ$, then to this pair we associate the functional $\Phi_{y,\phi}$ in m^* defined by

$$\Phi_{y,\phi}(x) = \langle y, x\rangle + \langle x, \phi\rangle \equiv \int_\Omega xy\, d\mu + \phi(x), \qquad x \in m.$$

Clearly $\|\Phi_{y,\phi}\| \leqslant \|y\|_1 + \|\phi\|$. Consider now any $\Psi \in m^*$. The restriction $\Psi|c_0$ defines an element Φ_y of c_0^* (notation as in part a)) for some $y \in \ell^1$. We can then let $\phi = \Psi - J_m(y)$ and ϕ will belong to c_0°. Thus $\Psi = \Phi_{y,\phi}$ and so the mapping $(y, \phi) \mapsto \Phi_{y,\phi}$ is linear and surjective. It remains to see that $\|\Phi_{y,\phi}\| \geqslant \|y\|_1 + \|\phi\|$.

Let $\varepsilon > 0$. Since every integrable function on a measure space can be approximated in the mean (of order 1) by an integrable simple function, there is a finite set $\Gamma \subset \Omega$ such that $\sum_{t\notin\Gamma} |y(t)| < \varepsilon$. We define a function $x \in c_0$ *by*

$$x(t) = \begin{cases} \operatorname{sgn}(y(t)), & t \in \Gamma \\ 0, & \text{otherwise} \end{cases}$$

so that $\|x\|_\infty = 1$. Next, we choose any $\tilde{z} \in U(m)$ for which $\phi(\tilde{z}) > \|\phi\| - \varepsilon$, and then put $z \equiv \chi_{\Omega\setminus\Gamma} \cdot \tilde{z}$. Then z has the properties that $z \in U(m)$, $x \cdot z = \theta$, and $\phi(z) = \phi(\tilde{z})$, since $z - \tilde{z} \in c_0$. Now the function $x + z$ has unit norm in m and

$$\Phi_{y,\phi}(x + z) = \langle x, y\rangle + \langle y, z\rangle + \phi(z)$$

since $\phi(x) = 0$. Thus

$$||\Phi_{y,\phi}|| \geqslant \Phi_{y,\phi}(x + z) > ||y||_1 - \varepsilon - \varepsilon + ||\phi|| - \varepsilon$$
$$= ||y||_1 + ||\phi|| - 3\varepsilon. \qquad \square$$

The preceding representation of m^* is of somewhat limited usefulness when Ω is of infinite cardinality due to the mysterious nature of the linear functionals belonging to c_0°. Indeed, it is only by means of non-constructive arguments such as the Hahn-Banach or separation theorems that we can even establish the existence of such functionals. An alternative perspective on c_0° can be gained by employing exercise 2.35 to identify m congruently with $C(\beta(\Omega), \mathbb{F})$, where $\beta(\Omega)$ is the Stone-Cech compactification of Ω (Ω being considered to have the discrete topology). Since the boundary of the unit ball $U(c_0^\circ)$ is an extremal subset of $U(m^*)$, the extreme points of $U(c_0^\circ)$ correspond via formula (13.9) to the functionals $\alpha\delta_t \in C(\beta(\Omega), \mathbb{F})^*$ where $|\alpha| = 1$ and $t \in \beta(\Omega)\backslash\Omega$.

I. We make two final remarks about reflexive spaces. First, we have defined a Banach space X to be reflexive if the canonical embedding J_X: $X \to X^{**}$ is a congruence between X and (all of) X^{**}. It is important to stress that this definition requires that the congruence between X and X^{**} be implemented by this particular mapping. For example, there is a non-reflexive Banach space Y (the "James space") which is congruent to Y^{**}. However, the canonical embedding $J_Y(Y)$ of Y in Y^{**} is a subspace of co-dimension one in Y^{**}. In general, a Banach space whose canonical image in its second conjugate space has finite codimension is called *quasi-reflexive*. In analogy with the theorem of **16F** it can be shown that X is quasi-reflexive if and only if X^* is quasi-reflexive if and only if M and X/M are quasi-reflexive for some (hence every) closed linear subspace $M \subset X$ (see exercise 3.16).

Second, reflexive spaces are an ideal setting for optimization problems involving norms. This is because every (non-empty) weakly closed set in a reflexive space is proximinal and, in particular, contains an element of minimal norm. This assertion follows from exercise 2.51. From **12A** we then see that every closed convex set in a reflexive space is proximinal. Moreover, as we shall establish later in this chapter (**19C**), this last property is actually characteristic of reflexive spaces.

§17. The Category Theorems

In this section we present a collection of results of decisive importance in the theory of Banach spaces. Most are derived from a single basic fact (Lifshits' lemma) which in turn depends on the Baire category theorem. The section also contains a variety of applications of these and other results based on the Baire theorem (recall, in fact, that we have already had occasion in **16C** and **16F** to look ahead to the uniform boundedness principle of **17C**).

A. We begin by reviewing the notions of category and the theorem of Baire. Let Ω be a topological space. A subset A of Ω is *nowhere dense* in Ω if

$$\operatorname{int}(\bar{A}) = \phi.$$

Thus a closed subset A of Ω is nowhere dense if and only if $\Omega \backslash A$ is dense (and open) in Ω. From this we conclude that the closed nowhere dense sets in Ω are the boundaries of the open sets in Ω. Specific examples of such sets are (i) a finite set, (ii) the Cantor set in $[0, 1]$, (iii) a rectifiable curve in $\mathbb{R}^n$, and (iv) a closed proper affine subspace of a linear topological space.

In general, any finite union of nowhere dense sets is nowhere dense. We define a set of *first category* in Ω to be a countable union of nowhere dense subsets of Ω. All other subsets of Ω are said to be of *second category*. Roughly speaking, sets of first category play a role in topology analogous to that of null sets in measure theory, although there is no direct overlap even in the case $\Omega = [0, 1]$. For although any countable set (in any topological space) is of first category, there exists a first category subset of $[0, 1]$ which has (Lebesgue) measure 1. (Namely, the union of a sequence of Cantor sets, the measures of which form a sequence increasing to one.)

The topological space Ω is a *Baire space* if the complement of every first category subset of Ω is dense in Ω. Such subsets are called *residual sets* in Ω. Since a set of first category in a Baire space Ω can have no interior, it follows that any residual subset of Ω (in particular Ω itself) is of second category (as well as dense in Ω). Also, any open subset of a Baire space is clearly a Baire space. Finally we can apply DeMorgan's rules of set theory to deduce that Ω is a Baire space if and only if the intersection of any countable family of dense open subsets of Ω is itself dense in Ω.

In the applications we typically use the following property of Baire spaces: if a Baire space Ω is the countable union of closed subsets A_n then some one of the A_n has non-empty interior. In fact, $\bigcup_n \operatorname{int}(A_n)$ is dense in Ω. To see this we let B equal the union of the boundaries of the A_n, so that B is a set of first category in Ω. It follows that $\bigcup_n \operatorname{int}(A_n)$ contains the residual set $\Omega \backslash B$.

For the purposes of functional analysis the most important Baire spaces are those topological spaces Ω whose topology is defined by a complete metric (or pseudo-metric). This is the content of the "Baire category theorem".

Theorem. *Let Ω be a complete metric (or pseudo-metric) space. Then Ω is a Baire space.*

Proof. Let $\{\mathcal{O}_1, \mathcal{O}_2, \ldots\}$ be a sequence of dense open subsets of Ω and let $\mathcal{O}$ be an open ball in Ω of radius $r > 0$. We must show that $(\bigcap_n \mathcal{O}_n) \cap \mathcal{O}$ is not empty. Now there exists $t_1 \in \mathcal{O} \cap \mathcal{O}_1$ such that $\mathcal{O} \cap \mathcal{O}_1$ contains the ball B_1 at t_1 of positive radius $r_1 < r/2$. Next, there exists $t_2 \in B_1 \cap \mathcal{O}_2$ such that $B_1 \cap \mathcal{O}_2 \cap \mathcal{O}_1 \cap \mathcal{O}$ contains the ball B_2 at t_2 of positive radius $r_2 < r/4$. Proceeding inductively we find a point $t_n \in B_{n-1} \cap \mathcal{O}_n$ and a positive number $r_n < r/2^n$ such that the ball B_n at t_n of radius r_n is contained in $B_{n-1} \cap \mathcal{O}_n \cap \cdots \cap \mathcal{O}_1 \cap \mathcal{O}$. The sequence $B_1, B_2, \ldots$ is a nested sequence of sets whose diameters tend to 0. By virtue of the completeness of Ω we can conclude that there is a point common to all the B_n and hence to all the sets $\mathcal{O}_n$ and $\mathcal{O}$. $\square$

In exercise 3.17 it is to be shown that any locally compact Hausdorff space is also a Baire space.

B. The Baire category theorem can be used to provide non-constructive existence proofs in somewhat the same way as the various versions of the Hahn-Banach and separation theorems. However, there is a difference in that the Baire theorem is applied after showing that the desired object belongs to a residual set in some Baire space; it is then concluded that the objects being sought comprise a large subset (precisely, dense subset of second category) of the particular Baire space. In other words the Baire theorem does not produce examples one by one but, so to speak, in bunches.

An interesting and classical illustration of the technique is provided by the problem of the existence of continuous functions on $[0, 1]$ that are nowhere differentiable. Examples of such functions have been known for a long time; the original example

$$f(t) = \sum_{n=0}^{\infty} b^n \cos(a^n \pi t)$$

where a is an odd integer and b satisfies $0 < b < 1$, $ab > 1 + 3\pi/2$ was given by Weierstrass. The nature of this and similar examples as contrasted with the familiar smooth functions encountered in calculus tends to suggest that such examples are the exception rather than the rule. But in fact the actual situation is quite the reverse. We are going to see that the class of nowhere differentiable functions in $C \equiv C([0, 1], \mathbb{R})$ is a residual set there, so that we should be surprised in some sense whenever we encounter a function that is differentiable at a single point in $[0, 1]$, let alone one that is, say, of class C^1.

The precise statement that we prove is the following: let E_n be the subset of C consisting of those functions f such that for some $t \in \left[0, 1 - \frac{1}{n}\right)$ and every $h \in \left(0, \frac{1}{n}\right)$, we have

$$\left|\frac{f(t + h) - f(t)}{h}\right| \leqslant n. \tag{17.1}$$

We claim that for $n = 1, 2, \ldots$, E_n is a nowhere dense subset of C. Granting this is follows that $E \equiv \bigcup_n E_n$ is a set of first category. But any function with even a finite right-hand derivative at some point in $[0, 1]$ must belong to E.

The proof that each of the sets E_n is nowhere dense is achieved by showing that each E_n is closed in C and has a dense complement. The closure of E_n follows from the observation that the set $E_{n,h}$ of f's satisfying inequality

(17.1) for some $t \in \left[0, 1 - \frac{1}{n}\right)$ and for a fixed $h \in \left(0, \frac{1}{n}\right)$ is closed under uniform convergence, and so $E_n = \cap\left\{E_{n,h} : 0 < h < \frac{1}{n}\right\}$ is also closed. Next, select any $f \in C$ and any $\varepsilon > 0$. We shall find $g \in C \backslash E_n$ such that $\|f - g\|_\infty < 2\varepsilon$ and thereby show the density of $C \backslash E_n$. To construct such a g we first find a polynomial p such that $\|f - p\|_\infty < \varepsilon$ (classical Weierstrass theorem). We then let q be a continuous function satisfying $\|q\|_\infty < \varepsilon$ and $|q'(t)| > n + 2\|p'\|_\infty$ for all but finitely many $t \in [0, 1]$. For example, p could be a piecewise-linear (saw-tooth) function, the straight line segments of whose graph have sufficiently large slope (in absolute value). Now we set $g = p + q$; clearly $\|f - g\|_\infty \leqslant \|f - p\|_\infty + \|q\|_\infty < 2\varepsilon$. On the other hand, if we select any $t \in \left[0, 1 - \frac{1}{n}\right)$ and choose $h > 0$ sufficiently small we see that

$$\left|\frac{p(t+h) - p(t)}{h}\right| \leqslant \|p'\|_\infty,$$

$$\left|\frac{q(t+h) - q(t)}{h}\right| > n + 2\|p'\|_\infty,$$

whence

$$\left|\frac{g(t+h) - g(t)}{h}\right| \geqslant \left|\frac{q(t+h) - q(t)}{h}\right| - \left|\frac{p(t+h) - p(t)}{h}\right| > n,$$

which proves that $g \in C \backslash E_n$.

C. Let X be a Banach space and $f \in \operatorname{Conv}(X)$. It was remarked in **14A** that if it is known that f is lower semi-continuous on X then actually f is continuous. We can see this by noting that

$$X \times \mathbb{R}^1 = \bigcup_{n=1}^{\infty} (\operatorname{epi}(f) - (\theta, n)).$$

Since $\operatorname{epi}(f)$ is closed on account of the lower semicontinuity of f (**14A**), this formula and the Baire category theorem imply that some translate of $\operatorname{epi}(f)$, and hence $\operatorname{epi}(f)$ itself, is solid. Consequently, by **14A** again, f is continuous. (See exercise 3.50 for a further generalization.)

We are now ready for the first of the important category theorems known as the "principle of uniform boundedness."

Theorem. *Let X be a Banach space, Y a normed linear space, and let $\mathscr{G} \subset B(X, Y)$. Then the following assertions are equivalent:*

a) $\sup\{\|T\| : T \in \mathscr{G}\} < \infty$;

b) $\sup\{\|T(x)\| : T \in \mathscr{G}\} < \infty, \quad x \in X$;

c) $\sup\{|\langle T(x), \psi\rangle| : T \in \mathscr{G}\} < \infty, \quad x \in X, \quad \psi \in Y^*$.

Proof. Trivially a) implies b) and c). Assume that b) holds and define $f_T(x) = \|T(x)\|$, $T \in \mathscr{G}$. The functions f_T are continuous and sublinear on X

and our hypothesis guarantees that

$$f(x) \equiv \sup\{f_T(x): T \in \mathscr{G}\} < \infty, \qquad x \in X.$$

Now f, as the supremum of a family of continuous convex functions on X is a lower semicontinuous convex function on X, and hence is continuous. In particular, there exists $\delta > 0$ such that

$$\sup\{\|T(x)\|: T \in \mathscr{G}\} \leqslant 1, \qquad \|x\| < \delta.$$

Hence

$$\sup\{\|T(x)\|: T \in \mathscr{G}\} \leqslant \frac{1}{\delta}\|x\|, \qquad x \in X,$$

and so $\sup\{\|T\|: T \in \mathscr{G}\} \leqslant 1/\delta$. Thus b) implies a). Finally, assume that c) holds and fix $x \in X$. Then the function

$$\psi \mapsto \sup\{|\langle T(x), \psi\rangle|: T \in \mathscr{G}\}$$

is continuous on Y^*. As before, there exists $\delta > 0$ such that

$$\sup\{|\langle T(x), \psi\rangle|: T \in \mathscr{G}\} \leqslant 1, \qquad \|\psi\| < \delta.$$

Therefore,

$$\|T(x)\| = \sup\{|\langle T(x), \psi\rangle|: \|\psi\| = 1\} \leqslant 1/\delta, \qquad T \in \mathscr{G}. \qquad \square$$

Corollary. a) *Let A be a subset of a normed linear space X. If $\sup\{|\phi(x)|: x \in A\} < \infty$ for all $\phi \in X^*$, then A is a bounded set.*

b) *Let B be a subset of X^* where X is a Banach space. If $\sup\{|\phi(x)|: \phi \in B\} < \infty$ for all $x \in X$, then B is a bounded set.*

The corollary may be paraphrased by stating that any weakly bounded subset of a normed linear space is bounded, and any weak*-bounded set in the conjugate space of a Banach space is bounded. The assumption of completeness of X in part b) here is vital: recall that in **12E** an example was given of an unbounded weak*-convergent sequence in the conjugate space of an incomplete normed space.

Example. Let X be a Banach space. Then X^* is weak*-quasi-complete. Indeed, if B is any weak*-closed and bounded subset of X^* then by the preceding corollary B is (norm)-bounded. Hence B is weak*-compact (**12D**) and *a fortiori* B is weak*-complete. This result was promised in **12G** and complements the negative result given there concerning the failure of (infinite) dimensional) conjugate spaces to be weak*-complete. $\square$

D. Here we take a slightly deeper look at the equivalence of parts a) and b) of the principle of uniform boundedness. With the notation of **17C** suppose that $\mathscr{G}$ is an unbounded set in $B(X, Y)$, that is, we suppose that part a) is not valid. Then it follows that the set of $x \in X$ for which $\sup\{\|T(x)\|: T \in \mathscr{G}\} = \infty$ is a residual set in X. Indeed, its complement M is a linear subspace of X which can be expressed as the union of the closed sets $M_k \equiv \{x \in X: \|T(x)\| \leqslant k, T \in \mathscr{G}\}$; since $\mathscr{G}$ is by assumption unbounded, none of the sets M_k can be solid, and therefore M is of first category.

This type of refinement of the uniform boundedness principle leads us to the classical result known as the "principle of condensation of singularities".

Theorem. *Let $X, Y_1, Y_2, \ldots$ be normed linear spaces with X complete and let $\mathcal{G}_n \subset B(X, Y_n)$ for $n = 1, 2, \ldots$. If $\mathcal{G}_n$ is unbounded for each n then the set*

$$\{x \in X : \sup_{T \in \mathcal{G}_n} ||T(x)|| = \infty, n = 1, 2, \ldots\} \tag{17.2}$$

is a residual set in X.

Proof. Put

$$A_n = \{x \in X : \sup_{T \in \mathcal{G}_n} ||T(x)|| < \infty\}$$

By hypothesis each A_n is a proper subset of X and the argument just given above shows that A_n is a set of first category in X. Consequently $\cup\{A_n : n = 1, 2, \ldots\}$ is of first category and therefore its complement, which is exactly the set in (17.2), is a residual set in X. □

In practice the set $\mathcal{G}_n$ usually consist of a sequence $\{T_{m,n} : m = 1, 2, \ldots\}$ of operators in $B(X, Y_n)$, and we are able to find a point x_n in X such that

$$\limsup_{m \to \infty} ||T_{m,n}(x_n)|| = \infty$$

for each n. We may then conclude that

$$\{x \in X : \limsup_{m \to \infty} ||T_{m,n}(x)|| = \infty \text{ for all } n\}$$

is a residual set in X.

Example. A classical application of the principle of condensation of singularities is to the problem of the pointwise convergence of Fourier series. We consider elements x of the Banach space $L^1 = L^1([-\pi, \pi], \mu, \mathbb{R})$ where μ is Lebesgue measure. For each such x the *Fourier coefficients* of x are

$$a_k = a_k(x) = \frac{1}{\pi}\int_{-\pi}^{\pi} x(t)\cos kt\, dt$$

$$b_k = b_k(x) = \frac{1}{\pi}\int_{-\pi}^{\pi} x(t)\sin kt\, dt$$

and we let

$$S_m(t; x) = \sum_{k=0}^{m} a_k \cos kt + b_k \sin kt, \qquad -\pi \leqslant t \leqslant \pi,$$

be the mth-partial sum of the Fourier series of x.

We recall that if $x \in L^2$ then the Fourier series converges to x in the mean of order 2:

$$\lim_{m \to \infty} ||S_m(\cdot; x) - x||_2 = 0.$$

Thus for large m the trigonometric sums $S_m(\cdot; x)$ represent x satisfactorily in a mean square sense. In particular, a subsequence of $\{S_m(\cdot; x)\}$ will converge pointwise a.e. to x. However, this approach yields no information as to whether

(17.3) $$\lim_{m\to\infty} S_m(t; x) = x(t)$$

for a particular $t \in [-\pi, \pi]$. Of course this is not really a well-defined question as x is itself only defined to within a.e. equivalence. Therefore, let us specialize to the case $x \in C_{2\pi}$, the (closed) subspace of $C([-\pi, \pi], \mathbb{R})$ consisting of functions with equal values at $t = -\pi$ and $t = \pi$. Thus $C_{2\pi}$ can be thought of as the linear space of real 2π-periodic continuous functions on $\mathbb{R}$; equipped with the uniform norm $C_{2\pi}$ is a Banach space.

In this setting it is known that the pointwise convergence (17.3) takes place under a variety of further hypotheses on x; for example, it is valid if x is of bounded variation on a neighborhood of t ("Jordan's test"). But continuity alone is definitely inadequate as we shall now see. Let $\{t_1, t_2, \ldots\}$ be an arbitrary (possibly dense) sequence in $[-\pi, \pi]$. We shall prove that there exists $x \in C_{2\pi}$ whose Fourier series diverges on an uncountable subset D of $[-\pi, \pi]$, and that D can be chosen to contain the sequence $\{t_1, t_2, \ldots\}$.

We recall that

$$S_m(t; x) = \frac{1}{2\pi}\int_{-\pi}^{\pi} x(s)D_m(s - t)ds$$

where

$$D_m(u) \equiv 1 + 2\sum_{k=0}^{m} \cos ku$$
$$= \frac{\sin\left(m + \frac{1}{2}\right)u}{\sin\left(\frac{u}{2}\right)}$$

is the Dirichlet kernel. According to exercise 3.19 $S_m(t; \cdot)$ is a bounded linear functional on $C_{2\pi}$ with

(17.4) $$\|S_m(t; \cdot)\| = \frac{1}{2\pi}\int_{-\pi}^{\pi} |D_m(s - t)|\, ds$$

Next we show that

$$\lim_{m\to\infty} \textstyle\int_{-\pi}^{\pi} |D_m(s - t)|\, ds = \infty.$$

Indeed,

$$\int_{-\pi}^{\pi} |D_m(s - t)|\, ds = 2\int_{\frac{-\pi-t}{2}}^{\frac{\pi-t}{2}} \left|\frac{\sin(2m + 1)u}{\sin u}\right| du \geqslant 2\int_{\frac{-\pi-t}{2}}^{\frac{\pi-t}{2}} \frac{|\sin(2m + 1)u|}{|u|}\, du$$

$$= 2 \sum_{k=0}^{2m} \int_{\frac{-\pi-t}{2}+\frac{k\pi}{2m+1}}^{\frac{\pi-t}{2}+\frac{(k+1)\pi}{2m+1}} \frac{|\sin(2m+1)u|}{|u|}\,du$$

$$\geqslant 2 \sum_{k=0}^{2m} \left[\frac{\pi - t}{2} + \frac{(k+1)\pi}{2m+1}\right]^{-1} \int_{\text{same}} |\sin(2m+1)u|\,du$$

$$\geqslant 2c \sum_{k=0}^{2m} \frac{1}{k+1},$$

where c is an appropriate positive constant. It follows that $\{S_m(t;\cdot):m = 1, 2, \ldots\}$ is not a bounded subset of $C_{2\pi}^*$ for any given $t \in [-\pi, \pi]$ and hence, by the uniform boundedness principle, there exists some $x \in C_{2\pi}$ whose Fourier series diverges at t.

Now consider the sequence $\{S_m(t_n;\cdot)\}$ in $C_{2\pi}^*$. The principle of condensation of singularities implies that

$$\left\{x \in C_{2\pi}: \limsup_{m\to\infty} |S_m(t_n; x)| = \infty \text{ for all } n\right\}$$

is a residual set in $C_{2\pi}$, and evidently the Fourier series of every x in this set diverges at each t_n. Select any such x and put

$$D = \left\{t: -\pi \leqslant t \leqslant \pi, \limsup_{m\to\infty} |S_m(t; x)| = \infty\right\};$$

we wish to show that D is uncountable. Let

$$D_{m,k} = \{t: -\pi \leqslant t \leqslant \pi, |S_m(t; x)| \leqslant k\},$$
$$D_k = \cap\{D_{m,k}: m = 1, 2, \ldots\}.$$

By continuity each $D_{m,k}$ is closed, hence so is D_k. We claim that each D_k is a set of first category. Granting this it follows that $\cup\{D_k: k = 1, 2, \ldots\}$ is also of first category; but this set contains the set of all t at which the Fourier series of x is convergent. Thus the set of t at which the Fourier series diverges contains a set of second category in $[-\pi, \pi]$ and such a set is necessarily uncountable.

To prove that each of the sets D_k is of first category we assume that the sequence $\{t_n\}$ is augmented (if necessary) by an additional sequence so as to be dense in $[-\pi, \pi]$. Then if D_k were not of first category it would not, in particular, be nowhere dense, and so, being closed, it would contain a non-trivial interval. Hence $|S_m(t; x)| \leqslant k$ for all $m = 1, 2, \ldots$ and all t in this interval. But this is a contradiction since any interval must contain one of the points t_n and each $t_n \in D$. □

E. The remaining category theorems will be derived from the following basic geometric principle due to Lifshits. If X is a linear topological space a subset A of X is called *ideally convex* if for any bounded sequence $\{x_n\} \subset A$ and sequence $\{\lambda_n\}$ of non-negative numbers with $\sum_1^\infty \lambda_n = 1$, the series $\sum_1^\infty \lambda_n x_n$ either converges to an element of A, or else does not converge at all. One general reason for the usefulness of such sets is that they satisfy certain preservation rules analogous to those for general convex sets; such rules are assembled in exercise 3.70. Further, we can note that any convex set which is either open or closed is ideally convex, and that if X is finite dimensional, any convex set is ideally convex.

We come now to the fundamental lemma of Lifshits.

Lemma. *Let A be an ideally convex subset of a Banach space X. Then*

$$\text{(17.5)} \qquad \operatorname{int}(A) = \operatorname{cor}(A) = \operatorname{cor}(\bar{A}) = \operatorname{int}(\bar{A}).$$

Proof. As remarked in **14A** it is a consequence of the Baire category theorem that the core of any closed convex set (such as $\bar{A}$) is equal to its interior. Indeed if $p \in \operatorname{cor}(\bar{A})$ then $\bar{A} - p$ is absorbing and so $X = \bigcup_1^\infty n(\bar{A} - p)$. Hence $\bar{A} - p$ is solid by Baire's theorem. Therefore, by **11A**, $\theta \in \operatorname{cor}(\bar{A} - p) = \operatorname{int}(\bar{A} - p)$ whence $p \in \operatorname{int}(\bar{A})$. Now to complete the proof of (17.5) it will suffice to show $\operatorname{int}(\bar{A}) \subset \operatorname{int}(A)$.

Suppose that $p \in \operatorname{int}(\bar{A})$; without loss of generality we may assume that $p = \theta$. Then for some $\varepsilon > 0$ we have $B \equiv \varepsilon U(X) \subset \bar{A}$. Hence $B \subset \bar{A} \cap B \subset \overline{A \cap B} \subset (A \cap B) + \frac{1}{2}B$ (**9A**). Consequently, for any $\delta > 0$ we have

$$\text{(17.6)} \qquad \delta B \subset \delta(A \cap B) + \frac{\delta}{2} B.$$

Now consider any $x \in \frac{1}{2}B$, that is, $\|x\| \leqslant \frac{\varepsilon}{2}$. By (17.6) there exist $x_1 \in A \cap B$ and $y_1 \in \frac{1}{4}B$ such that $x = \frac{1}{2}x_1 + y_1$. Proceeding inductively using (17.6) with $\delta = \frac{1}{4}, \frac{1}{8}, \ldots$, we obtain the sequences $\{x_n\} \subset A \cap B$ and $\{y_n\} \subset 2^{-(n+1)}B$ for which

$$y_n = 2^{-(n+1)}x_n + y_{n+1}.$$

But then

$$\left\| x - \sum_{n=1}^{m} 2^{-n}x_n \right\| = \|y_m\| \leqslant 2^{-(m+1)}\varepsilon,$$

whence $x = \sum_1^\infty 2^{-n}x_n$ and therefore, by the ideal convexity of A, $x \in A$. This proves that the θ-neighborhood $\frac{1}{2}B$ lies in A and so completes the proof. □

F. Let X and Y be normed linear spaces, and let T be a linear map defined on a subspace $D(T) \subset X$ and taking values in Y. The *graph* of T is the linear subspace of $X \times Y$ defined by

$$\operatorname{gr}(T) = \{(x, T(x)) : x \in D(T)\}.$$

We say that T is *closed on X* if $\operatorname{gr}(T)$ is a closed subspace of $X \times Y$ (in the product topology). We leave the proof of the following simple lemma as exercise 3.20.

Lemma. a) *T is closed on X if and only if $\{x_n\} \subset D(T)$, $\lim_n x_n = x \in X$, and $\lim_n T(x_n) = y$ imply $x \in D(T)$ and $T(x) = y$.*

b) *If T is closed on X and injective then T^{-1} is closed on Y.*

c) *If T is closed on X then* $\ker(T)$ *is a closed subspace of X.*

d) *If $D(T)$ is a closed subspace of X and T is continuous then T is closed on X.*

The notion of closed linear maps is used extensively in the theory of ordinary and partial differential equations, where it often serves as an effective substitute for the more familiar notion of continuous linear maps.

Example. The simplest kind of differential operator is the map $T = \frac{d}{dt}$ defined on $D(T) \equiv C^1([0, 1], \mathbb{R}) \subset X \equiv C([0, 1], \mathbb{R}) \equiv Y$. Thus for each continuously differentiable $x \in X$ we have

$$T(x)(t) \equiv x'(t).$$

T is certainly not continuous on $D(T)$ since if $p_n(t) \equiv t^n$, $n = 1, 2, \ldots$, we have $\|p_n\|_\infty = 1$ and $\|T(p_n)\|_\infty = \sup\{nt^{n-1}: 0 \leqslant t \leqslant 1\} = n$. But we claim that T is closed on X. To see this, suppose that $x_n \in D(T)$ and that $\lim_n x_n = x$, $\lim_n T(x_n) = y$. By definition this means that $\{x_n\}$ converges uniformly to x and that $\{x_n'\}$ converges uniformly to y on $[0, 1]$. It follows that

$$\begin{aligned} x(t) - x(0) &= \lim(x_n(t) - x_n(0)) \\ &= \lim \textstyle\int_0^t x_n'(s)ds = \int_0^t y(s)ds, \qquad 0 \leqslant t \leqslant 1, \end{aligned}$$

whence $x \in D(T)$ and $y = x' \equiv T(x)$. □

We shall not pursue this area of application of closed linear maps any further. Rather we shall proceed directly to the famous result which allows us to prove that certain closed linear maps are necessarily continuous. This fact, known as the "closed graph theorem" has a number of useful and surprising consequences, as we shall soon see.

Theorem. *Let X and Y be Banach spaces and suppose that the linear map $T: X \to Y$ is closed on X. Then $T \in B(X, Y)$; that is, T is continuous.*

Proof. To show that T is continuous at θ (and hence at every point in X) it will suffice to prove that θ is interior to $T^{-1}(U(Y))$. Now this set is certainly absorbing so we can complete the proof via Lifshits' lemma (**17E**) by showing that $T^{-1}(U(Y))$ is ideally convex. To do so we select a bounded sequence $\{x_n\}$ for which $\|T(x_n)\| \leqslant 1$ and a sequence $\{\lambda_n\}$ of non-negative numbers with $\sum_1^\infty \lambda_n = 1$. Let $x = \sum_1^\infty \lambda_n x_n$; we must show that $\|T(x)\| \leqslant 1$. Now the series $\sum_1^\infty \lambda_n T(x_n)$ is absolutely convergent, hence convergent to some element $y \in Y$ (exercise 3.1b). Because T is closed it follows that

$$T(x) = y \equiv \sum_{n=1}^{\infty} \lambda_n T(x_n),$$

from which it is clear that $\|T(x)\| \leqslant \sum_1^\infty \lambda_n = 1$. □

A most important consequence of this theorem is the following "inverse mapping theorem".

Corollary 1. *Let T be a continuous algebraic isomorphism between Banach spaces X and Y. Then T is a topological isomorphism.*

Proof. It must be shown that the inverse mapping T^{-1} is continuous. By hypothesis T^{-1} is defined on all of Y and is closed on Y by part b) of the preceding lemma. Therefore, T^{-1} is continuous by the closed graph theorem. □

G. We give two more general operator-theoretic consequences of the method of ideally convex sets (an application of a different nature occurs in exercise 3.71). The first of these is known as the "open mapping theorem".

Theorem. *Let X and Y be Banach spaces and suppose that $T \in B(X, Y)$ is surjective. Then*

a) *T is an open mapping; and*

b) *there exists a constant $\gamma > 0$ such that to every $y \in Y$ there corresponds an $x \in X$ with $T(x) = y$ and $\|x\| \leqslant \gamma\|y\|$.*

Proof. a) Because T is surjective the set $T(U(X))$ is absorbing. Since this set is easily seen to be ideally convex it follows from Lifshits' lemma (**17E**) that $\theta \in \operatorname{int}(T(U(X)))$. Hence, from the linearity of T, it follows that the T-image of any open ball in X is open in Y, and consequently that T is an open mapping.

b) By what we have just shown there exists a $\delta > 0$ such that $\partial U(Y) \subset T(U(X))$. Hence for any non-zero $y \in Y$ there is some $\bar{x} \in U(X)$ such that $T(\bar{x}) = \delta y/\|y\|$. If we put $x = \|y\|\bar{x}/\delta$ we see that $T(x) = y$ and $\|x\| \leqslant \|y\|/\delta \equiv \gamma\|y\|$. □

Corollary. *Let X and Y be Banach spaces and suppose that $T \in B(X, Y)$ has a dense but proper range in Y. Then there exists $y \in Y$ with the property that whenever $\lim_n T(x_n) = y$ it follows that $\lim_n \|x_n\| = +\infty$.*

Proof. If no such y exists then the set $\overline{T(U(X))}$ is absorbing. By **17E** then, the set $T(U(X))$ is a θ-neighborhood and consequently T is surjective. This is a contradiction. □

H. A mistake that occasionally occurs in applications of the open mapping theorem is the assumption that, because a linear mapping transforms open sets into open sets, it necessarily transforms closed sets into closed sets. Of course this would be true if the closed set were compact. Otherwise, the simplest examples in $\mathbb{R}^2$ show that it is possible to have a closed convex set map into a non-closed set. Specifically we may let T be the orthogonal projection of $\mathbb{R}^2$ onto $\mathbb{R}^1$ ($T(x, y) \equiv (x, 0)$) and let $A = \{(x, y) \in \mathbb{R}^2 : x > 0, xy \geqslant 1\}$. Then A is the closed region in the first quadrant bounded by a branch of the hyperbola $xy = 1$, and $T(A)$ is the positive x-axis $\{(x, 0) : x > 0\}$, which is not closed.

If we specialize our closed set further by assuming that it is an affine subspace then its T-image will also be an affine subspace. Thus we can assert that the linear image of any finite dimensional affine subspace is closed (using **9E**). But without the finite dimensionality restriction this assertion too can fail. This will prove to be a particular consequence of the following general necessary and sufficient condition.

Lemma. *Let X and Y be Banach spaces and let $T \in B(X, Y)$ be surjective with kernel N. Then a subset A of X has the property that $T(A)$ is closed in Y if and only if $A + N$ is closed in X.*

Proof. The key to the proof is once again the factorization $T = \hat{T} \circ Q_N$. If $T(A)$ is closed in Y then certainly $T^{-1}(T(A))$ is closed in X, since T is continuous. But $T^{-1}(T(A)) = Q_N^{-1}(Q_N(A)) = A + N$. Conversely, assume that $A + N$ is closed in X. We must show that $Q_N(A)$ is closed in X/N, as then $T(A) = \hat{T}(Q_N(A))$ will be closed in Y because $\hat{T}$ is an isomorphism (**17F**). Suppose that we have $\{x_n\} \subset A$ such that $\lim_n Q_N(x_n) = Q_N(x)$ for some $x \in X$. Then $\lim_n d(x_n - x, N) = 0$ and we must see that $Q_N(x) \in Q_N(A)$, or that $x \in A + N$. But we have sequences $\{z_n\} \subset N$ and $\{\varepsilon_n\} \subset X$, with $\lim_n \varepsilon_n = \theta$, such that $x_n - x - z_n = \varepsilon_n$ or $x_n - z_n = x - \varepsilon_n$. Since $\{x_n - z_n\} \subset A + N$ which is closed, we have $x = \lim_n(x - \varepsilon_n) \in A + N$. □

Example. Let $X = \ell^2(\aleph_0)$, let Y be the subspace of X consisting of sequences with zero components in the odd numbered places, and let $T \in B(X, Y)$ be defined as that operator which multiplies each $x \in X$ by the characteristic function of the even integers:

$$T((\xi_1, \xi_2, \xi_3, \xi_4, \ldots)) = (0, \xi_2, 0, \xi_4, 0, \ldots).$$

Then $N \equiv \ker(T) = \{x = (\xi_1, \xi_2, \ldots) \in X : \xi_{\text{even}} = 0\}$. Let

$$A = \overline{\text{span}} \left\{ \left(\cos \frac{1}{n} \right) \chi_{\{2n-1\}} + \left(\sin \frac{1}{n} \right) \chi_{\{2n\}} : n = 1, 2, \ldots \right\}.$$

Then $A + N$ is not closed in X (exercise 3.25) and so $T(A)$ cannot be closed in Y. □

The preceding lemma leads to another interesting question which we discuss briefly. Namely, given a Banach space X, which subsets A of X have the property that $T(A)$ is always closed whenever $T: X \to Y$ is a surjective operator and Y is a Banach space? Certainly the compact subsets of X have this property. A satisfactory answer is known only under certain restrictions: X is finite dimensional and A is considered *a priori* to be both convex and closed in X. Under these restrictions we say that a half-line $L \subset X$ is a *boundary ray* (resp. *asymptote*) of A if $L \subset \partial(A)$ (resp. if $L \subset X \backslash A$ but $\text{dist}(A, L) = 0$). Suppose that A has no boundary ray or asymptote. Then it can be proved that $A + N$ is closed in X for any closed convex set $N \subset X$ (exercise 3.26). In particular, when $N = \ker(T)$, we see that $T(A)$ must be closed in Y.

Finite dimensional convex sets A in $\mathbb{R}^n$ lacking both boundary rays and asymptotes are called *continuous* convex sets, because they can be alternatively characterized as sets whose *support function* σ_A is continuous on the unit sphere:

$$\sigma_A(u) \equiv \sup\{(x, u): x \in A\}, \qquad \|u\|_2 = 1.$$

The simplest example of a continuous but non-compact convex set is (the closure of) the region in $\mathbb{R}^2$ interior to a parabola.

We shall return to this general problem below in **21B** when we discuss spline minimization problems. In particular, we shall see some examples of infinite dimensional non-compact sets whose continuous linear image in a Banach space is always closed.

I. We now give a simple but useful application of the category theorems. Many other applications occur in later sections and in the exercises.

Let the normed linear space X be the direct sum of two subspaces M and N, $X = M \oplus N$, and let $P: X \to M$ be the associated projection:

$$P(m + n) \equiv m, \qquad m \in M, \qquad n \in N.$$

In exercise 2.2 it was shown that if P is continuous then M and N are necessarily closed in X (indeed, $N = P^{-1}(\theta)$, etc.). Of greater interest is the converse, which is valid under a completeness hypothesis.

Theorem. *Let X be a Banach space. If $X = M \oplus N$ where M and N are closed subspaces of X, then the associated projection is continuous.*

Proof. By the closed graph theorem (**17F**) it is sufficient to show that P is closed on X. Thus suppose that $\lim_n x_n = x \in X$ and $\lim_n P(x_n) = y \in M$. Then $\lim_n(x_n - P(x_n)) = x - y \in N$. Hence $x = y + (x - y)$ yields $P(x) = P(y) = y$, which proves that P is closed on X. □

An alternative proof can be based on part c) of exercise 2.2 and the open mapping theorem (**17G**). Also we note that if we define the *angle* between the (disjoint) subspaces M and N to be

$$\gamma(M, N) \equiv \inf\{\|m - n\|: m \in M, n \in N, \|m\| = \|n\| = 1\}$$

then

$$\textbf{(17.7)} \qquad \gamma(M, N) \geqslant \|P\|^{-1}.$$

Indeed, for any unit vectors $m \in M$, $n \in N$, we have

$$\|P\| \, \|m - n\| \geqslant \|P(m - n)\| = \|m\| = 1.$$

Inequality (17.7) establishes (for Banach spaces) the converse of the assertion made at the end of **15D**. Thus:

Corollary. *Let M and N be disjoint closed subspaces of a Banach space X. Then $M \oplus N$ is closed in X exactly when $\gamma(M, N)$ is positive.*

J. Our final topic of this section is a striking application of the Baire category theorem, due to Lindenstrauss and Phelps. We consider a solid closed convex subset A of a Banach space X and again raise the question: does A have any extreme points? We shall assume that A is bounded to rule out examples such as a half-space which has no extreme points. Even so, $\text{ext}(A)$ may still be empty as we know (exercise 2.30). However, if we further assume that X is reflexive, then $\text{ext}(A) \neq \varnothing$ by virtue of the Krein-Milman theorem. As we now see, it turns out that quite a stronger assertion is true, if X has infinite dimension.

Theorem. *Let X be a real reflexive Banach space of infinite dimension. If A is a solid, closed, bounded and convex subset of X then* $\text{ext}(A)$ *is uncountable.*

Proof. Suppose that the theorem has been proved whenever A is the unit ball in a reflexive space. Then given A and X as in the theorem, let Y be the reflexive space $X \times \mathbb{R}^1$ with norm $\|(x, t)\| \equiv \max\{\|x\|, |t|\}$. Let $A_1 = \{(x, 1): x \in A\}$ and define $B = \text{co}(A_1 \cup (-A_1))$. In Y, B is a solid absolutely convex bounded set and hence its gauge is an equivalent norm on Y. So normed, Y is still reflexive and now $U(Y) = B$. Therefore, $\text{ext}(B)$ is uncountable. Since $\text{ext}(B) = \text{ext}(A_1) \cup \text{ext}(-A_1) \equiv \text{ext}(A_1) \cup -\text{ext}(A_1)$ (formulas (2.1) and (13.1)), it follows that $\text{ext}(A_1)$ is uncountable. But $\text{ext}(A_1) = \{(x, 1): x \in \text{ext}(A)\}$, so that $\text{ext}(A)$ is also uncountable.

Thus the proof is reduced to the case $A = U(X)$. Suppose that $\text{ext}(U(X))$ is countable: $\text{ext}(U(X)) = \{x_1, x_2, \ldots\}$. Let $G_n = \{\phi \in U(X^*): |\phi(x_n)| = \|\phi\|\}$; these sets are weak*-closed subsets of $U(X^*)$ and hence are weak*-compact (**12D**). Since $U(X)$ is weakly compact (**16F**), each $\phi \in X^*$ attains its supremum on $U(X)$ at an extreme point (**13A**), and so $U(X^*) = \bigcup_n G_n$. By the Baire category theorem some G_n, say G_1, has non-empty interior relative to the weak topology on $U(X^*)$. Let ϕ_0 belong to this interior. Since scalar multiplication is weakly continuous, we can assume that $\|\phi_0\| < 1$. Now there exist $z_1, \ldots, z_m \in X$ such that $\phi \in G_1$ whenever $\|\phi\| \leqslant 1$ and $|\langle z_j, \phi - \phi_0\rangle| < 1, j = 1, \ldots, m$. Let V be the flat $\{\phi \in X^*: \phi(z_j) = \phi_0(z_j), j = 1, \ldots, m, \phi(x_1) = \phi_0(x_1)\}$. Because V has finite codimension in the infinite dimensional space X^*, there exists $\phi_1 \in V$, $\phi_1 \neq \phi_0$, and $\phi_t \equiv \phi_0 + t(\phi_1 - \phi_0) \in V$ for all t. Since $\|\phi_0\| < 1$ there exists t such that $\|\phi_t\| = 1$. But now $\phi_t \in G_1$ and so $1 = \|\phi_t\| = |\phi_t(x_1)| = |\phi_0(x_1)| \leqslant \|\phi_0\| \, \|x_1\| = \|\phi_0\|$. This is a contradiction to $\|\phi_0\| < 1$. □

This theorem does not initially appear to be of interest for complex spaces since, for example, if $x \in \text{ext}(U(X))$ then $\alpha x \in \text{ext}(U(X))$, for all α, $|\alpha| = 1$. Thus if there are any extreme points at all, there are uncountably many. However, if we say that two extreme points x and y are equivalent provided that $y = \alpha x$ for some $|\alpha| = 1$, then the theorem directly implies that if X is reflexive and infinite dimensional there must be uncountably many such equivalence classes of extreme points.

The theorem has various implications, some of which we mention now (see also exercise 3.59). First, suppose that A and X are as in the theorem, and that X is also separable. Then $\text{ext}(A)$ cannot consist entirely of isolated points. This is because any set of isolated points in a separable metric space is necessarily countable. Surprisingly enough, this consequence fails in general: there are inseparable reflexive spaces for which every pair x, y of distinct extreme points of the unit ball satisfies $\|x - y\| \geqslant \delta$ for some constant $\delta > 0$.

There is a class $\{X\}$ of real Banach spaces with the property that any three balls in X will intersect if each two of them intersect. This class contains all $C(\Omega, \mathbb{R})$ and $L^1(\Omega, \mu, \mathbb{R})$ spaces but does not contain, for example, any Euclidean space $(\mathbb{R}^n, \|\cdot\|_2)$. It is known that every space in this class contains a separable subspace again in the class. It is also known that if X belongs to this class and if $x \in \text{ext}(U(X))$, $\varphi \in \text{ext}(U(X^*)$, then $|\varphi(x)| = 1$. Now, given any pair $x, y \in \text{ext}(U(X))$, there is some $\varphi \in \text{ext}(U(X^*))$ for which $\varphi(x) \neq \varphi(y)$ (**13B**), whence $\|x - y\| \geqslant |\varphi(x - y)| = 2$. It follows that $\text{ext}(U(X))$ consists of isolated points and so no infinite dimensional space in this class can be reflexive.

§18. The Šmulian Theorems

In this section we establish two of the most profound and useful theorems in normed linear space theory. These are the Eberlein-Šmulian theorem characterizing weakly compact subsets of normed spaces and the Krein-Šmulian theorem characterizing weak*-closed convex subsets of the conjugate space of a Banach space.

A. Let Ω be a topological space. A subset A of Ω is called (*relatively*) *countably compact* if every sequence in A has a cluster point in A (in Ω). If Ω is metrizable this property of the subset A is equivalent to the (relative) compactness of A and to the (relative) sequential compactness of A. Further, in the metrizable case, A is then *sequentially dense*, meaning that every point in $\bar{A}$ is the limit of a sequence of points in A.

Supposed now that X is a normed linear space and that X_w is the topological space obtained by endowing X with the weak topology $\sigma(X, X^*)$ (**12A**). According to exercise 3.29 this topology is not metrizable unless X is finite dimensional. Nevertheless it turns out that the three types of compactness indicated above are equivalent properties of any subset of X_w, and further any subset having these properties is weakly sequentially dense. We now set out to prove this assertion; along the way we shall establish a few other equivalent properties of subsets of X_w.

Lemma *Let X be a normed linear space such that X^* contains a countable total subset. If A is a weakly compact set in X then the* (*relative*) *weak topology on A is metrizable.*

Proof. Let $\{\phi_1, \phi_2, \ldots\}$ be the countable total subset of X^*. Without loss of generality we can assume that $\|\phi_n\| = 1$ for all n. We define a norm ρ on X by

$$\rho(x) = \sum_{n=1}^{\infty} 2^{-n}|\phi_n(x)|;$$

clearly $\rho \leqslant \|\cdot\|$. Now the set A, being weakly compact, is in particular weakly bounded and hence norm bounded (**17D**). Hence the identity map from A in its (relative) weak topology to A in its (relative) ρ-topology is continuous and therefore is a homeomorphism (because A is weakly compact). □

The use of the norm ρ and the subsequent argument are quite analogous to the corresponding features of the Clarkson-Rieffel lemma of **15C**. Observe that the lemma applies to any space X which is either separable or conjugate to a separable space (for example, $X = m$).

Corollary. *Let A be a weakly compact subset of the normed linear space X. Then A is weakly sequentially compact.*

Proof. Let $\{x_n\}$ be a sequence in A. We must show that some subsequence converges weakly to a point in A. Let $M = \overline{\text{span}}(\{x_n\})$. Since M is weakly closed (**12A**) the set $A \cap M$ is a weakly compact subset of the separable space M. By the lemma, then, $A \cap M$ is weakly metrizable in M. Hence $\{x_n\}$ contains a subsequence that converges weakly to a point in $A \cap M$. Obviously this subsequence is also convergent in the weak topology on X. □

We next show that weak sequential compactness is equivalent to several other properties of subsets of normed spaces.

Theorem. *Let X be a real locally convex space and $A \subset X$. Then each of the following properties implies its successor. If X is normed all the properties are equivalent.*

a) *A is weakly sequentially compact;*

b) *A is weakly countably compact;*

c) *for any sequence $\{x_n\} \subset A$ there exists $\bar{x} \in A$ such that*

$$\varliminf_{n\to\infty} \phi(x_n) \leqslant \phi(\bar{x}) \leqslant \varlimsup_{n\to\infty} \phi(x_n), \qquad \phi \in X^*; \tag{18.1}$$

d) *if $\{C_n\}$ is a decreasing sequence of closed convex sets in X such that $A \cap C_n \neq \varnothing$ for all n then $A \cap (\bigcap_n C_n) \neq \varnothing$;*

e) *if M is a separable closed linear subspace of X and $\{H_n\}$ is a sequence of closed half-spaces such that $A \cap M \cap H_1 \cap \cdots \cap H_n \neq \varnothing$ for every n then $A \cap M \cap (\bigcap_n H_n) \neq \varnothing$.*

Proof. Directly from the definitions involved a) implies b). Now given a sequence $\{x_n\} \subset A$, any cluster point $\bar{x}$ of this sequence will meet the requirements of c). To see this select any $\phi \in X^*$ and let $r = \varlimsup_n \phi(x_n)$. If, for some $\varepsilon > 0$, $\phi(\bar{x}) > r + \varepsilon$, then for infinitely many n we would have $\phi(x_n) > r + \varepsilon$, contradicting the definition of r. This proves the right-hand side of (18.1) and the proof of the left-hand side is analogous. Next, assume that c) holds and for each n choose $x_n \in A \cap C_n$. Then we claim that the $\bar{x}$ of (18.1) corresponding to the sequence $\{x_n\}$ belongs to $A \cap (\bigcap_n C_n)$. Indeed, part of the conclusion of c) is that $\bar{x} \in A$. Further, if $\bar{x} \notin C_n$ for some n then by the strong separation theorem (**11F**) there exists $\phi \in X^*$ such that

$$\phi(\bar{x}) > \sup\{\phi(x): x \in C_n\} \geqslant \sup\{\phi(x): x \in C_{n+1}\} \geqslant \cdots,$$

whence $\overline{\lim}_n \phi(x_n) < \phi(\bar{x})$, contradicting (18.1) and proving d). It is clear that d) implies e) if we take $C_n \equiv M \cap H_1 \cap \cdots \cap H_n$.

It remains to show that e) implies a) if X is normed. Let $\{x_n\}$ be a sequence in A and put $M = \overline{\text{span}}(\{x_n\})$. Then M is separable and hence there exists a total sequence $\{\phi_m\}$ in M^*. Now we note that A is weakly bounded (if not, then for some $\phi \in X^*$, $\sup\{\phi(x): x \in A\} = \infty$; setting $H_n = \{x \in X: \phi(x) \geqslant n\}$, the sequence $\{H_n\}$ would then satisfy e) yet $\bigcap_n H_n = \varnothing$). In particular, for each m the sequence $\{\phi_m(x_n)\}$ is bounded. The Cantor diagonal process now yields a subsequence $\{y_n\}$ of $\{x_n\}$ such that $c_m \equiv \lim_n \phi_m(y_n)$ exists for every m.

We next claim that there exists $y \in A$ such that $c_m = \phi_m(y)$ for every m. This follows from e) by arranging the half-spaces $\left\{x \in X: \pm(\phi_m(x) - c_m) \leqslant \frac{1}{k}\right\}$, $(k, m = 1, 2, \ldots)$ in a sequence $\{H_n\}$ and using the vectors y_n; it follows that $A \cap M \cap (\bigcap_n H_n) \neq \varnothing$ and we can take y to be any point in this intersection.

We complete the proof by showing that weak-$\lim_n y_n = y$. For this it will suffice to prove that for every $\phi \in X^*$ $\overline{\lim}_n \phi(y_n - y) \leqslant 0$ (since then it will follow that $\underline{\lim}_n \phi(y_n - y) \geqslant 0$, whence $\lim_n \phi(y_n - y) = 0$). If not, there exists some $\phi \in X^*$, some $\varepsilon > 0$, and a subsequence $\{z_n\}$ of $\{y_n\}$ such that $\phi(z_n) \geqslant \phi(y) + \varepsilon$ for all n. Let H be the half-space $\{x \in X: \phi(x) \geqslant \phi(y) + \varepsilon\}$ and adjoin H to the sequence $\{H_n\}$ above. Then for each n, $A \cap H \cap H_1 \cap \cdots \cap H_n$ contains terms of $\{z_n\}$, so that by e) again there exists a point $z \in A \cap M \cap H \cap (\bigcap_n H_n)$. For this z we have

$$\textbf{(18.2)} \qquad \phi_m(z) = \lim_n \phi_m(z_n) = \lim_n \phi_m(y_n) = c_m = \phi_m(y).$$

But the sequence $\{\phi_m\}$ is total over M so that (18.2) entails $y = z$, which contradicts $z \in H$. □

For a comment on this theorem see exercise 3.31.

B. At this stage we know that weakly compact subsets of normed linear spaces are weakly sequentially compact and that this property in turn is equivalent to several others. We now complete this chain of implications by proving that weakly countably compact sets are weakly compact ("Eberlein's theorem"). The resulting collection of characterizations of weakly compact sets is the "Eberlein-Šmulian theorem". The essentials of the remaining proof are contained in the following technical lemma to be called "Whitley's construction".

Before the lemma we make one simple observation. Suppose that M is a finite dimensional subspace of a conjugate space X^*. Since $\partial U(M)$ is compact (**9E**) it contains a finite $\frac{1}{4}$-net$\{\phi_1, \ldots, \phi_n\}$; thus for each $\phi \in M$ with $\|\phi\| = 1$ there is some ϕ_k such that $\|\phi - \phi_k\| \leqslant \frac{1}{4}$. If we now select $x_1, \ldots, x_n \in \partial U(X)$ such that $\phi_k(x_k) > \frac{3}{4}$, we shall have for any $\phi \in M$

$$\textbf{(18.3)} \qquad \max\{|\phi(x_k)|: 1 \leqslant k \leqslant n\} \geqslant \tfrac{1}{2}\|\phi\|.$$

Lemma. *Let A be a relatively countably compact subset of the normed linear space X, and let $\Phi \in \overline{J_X(A)}^*$. Then there is a sequence $\{x_n\} \subset A$ with a unique weak cluster point $x \in X$ such that $\hat{x} = \Phi$.*

Proof. Choose any $\phi_1 \in \partial U(X^*)$. There is then $x_1 \in A$ such that $|\langle \phi_1, \hat{x}_1 - \Phi \rangle| < 1$ (since $\Phi \in \overline{J_X(A)}^*$). Applying now our preliminary observation above, there exist $\phi_2, \ldots, \phi_{n(2)} \in \partial U(X^*)$ such that for every $\Psi \in \text{span}(\{\Phi, \hat{x}_1 - \Phi\})$ we have (by (18.3))

$$\max\{|\Psi(\phi_k)| : 2 \leqslant k \leqslant n(2)\} \geqslant \tfrac{1}{2}\|\Psi\|.$$

Next we can find $x_2 \in A$ such that

$$\max\{|\langle \phi_k, \hat{x}_2 - \Phi \rangle| : 1 \leqslant k \leqslant n(2)\} < \tfrac{1}{2}.$$

Then by the observation again there exist $\phi_{n(2)+1}, \ldots, \phi_{n(3)} \in \partial U(X^*)$ such that (by (18.3) again)

$$\max\{|\Psi(\phi_k)| : n(2) < k \leqslant n(3)\} \geqslant \tfrac{1}{2}\|\Psi\|,$$

for every $\Psi \in \text{span}(\{\Phi, \hat{x}_1 - \Phi, \hat{x}_2 - \Phi\})$. Then we choose $x_3 \in A$ so that

$$\max\{|\langle \phi_k, \hat{x}_3 - \Phi \rangle| : 1 \leqslant k \leqslant n(3)\} < \tfrac{1}{3},$$

and continue. In this way we inductively construct a sequence $\{x_n\} \subset A$.

By hypothesis there is at least one weak cluster point x of the sequence $\{x_n\}$. Since $\overline{\text{span}}(\{x_n\})$ is weakly closed we have x in this subspace and consequently

$$\hat{x} - \Phi \in \overline{\text{span}}(\{\Phi, \hat{x}_1 - \Phi, \hat{x}_2 - \Phi, \ldots\}). \tag{18.4}$$

Now by construction any Ψ in this span satisfies

$$\sup\{|\Psi(\phi_k)| : k = 1, 2, \ldots\} \geqslant \tfrac{1}{2}\|\Psi\|, \tag{18.5}$$

so the same inequality remains valid for Ψ in the closed span of (18.4). In particular, (18.5) is true for $\Psi = \hat{x} - \Phi$.

We next prove that $\langle \phi_k, \hat{x} - \Phi \rangle = 0$ for every k; granting this it will follow from (18.5) that $\hat{x} = \Phi$. Since $p \leqslant n(p)$ for $p = 1, 2, \ldots$, we have for $k \leqslant n(p) < n$ that $|\langle \phi_k, \hat{x}_n - \Phi \rangle| < \dfrac{1}{p}$ and hence

$$\begin{aligned} |\langle \phi_k, \hat{x} - \Phi \rangle| &\leqslant |\langle \phi_k, \hat{x}_n - \Phi \rangle| + |\phi_k(x_n - x)| \\ &\leqslant \frac{1}{p} + |\phi_k(x_n - x)|. \end{aligned} \tag{18.6}$$

Since x is a weak cluster point of $\{x_n\}$, given ϕ_k and $p > k$ there exists x_n such that $|\phi_k(x_n - x)| < \dfrac{1}{p}$ and $k \leqslant n(k) \leqslant n(p) < n$ (because the sequence $\{x_n\}$ is frequently in the weak x-neighborhood defined by the functional ϕ_k and the number $\dfrac{1}{p}$, and so there will be some x_n in this neighborhood for arbitrarily large n). For this x_n we then have by (18.6) that $|\langle \phi_k, \hat{x} - \Phi \rangle| \leqslant \dfrac{2}{p}$. This proves that $\langle \phi_k, \hat{x} - \Phi \rangle = 0$ and hence that $\hat{x} = \Phi$.

Thus we have shown that any cluster point x of $\{x_n\}$ satisfies $\hat{x} \equiv J_X(x) = \Phi$; since J_X is injective the cluster point x must be unique. □

Notice that a particular consequence of Whitley's construction is

$$\textbf{(18.7)} \qquad \overline{J_X(A)}^* \subset J_X(X).$$

Theorem *Let A be a relatively weakly countably compact subset of a normed space X. Then the weak closure of A is weakly compact.*

Proof. For any $\phi \in X^*$ the image $\phi(A)$ is relatively countably compact, hence in particular bounded. Thus A is weakly bounded and therefore bounded by **17D**. Thus $\overline{J_X(A)}^*$ is a weak*-closed and norm-bounded subset of X^{**} and hence is weak*-compact. Because J_X is a homeomorphism from the weak topology on X to the (relative) weak* topology on $J_X(X)$ (**12B**), we see from (18.7) that the weak closure of A is weakly compact. □

It is now also clear that any such set A must be weakly sequentially dense. Indeed, if x belongs to the weak closure of A then $\hat{x} \in \overline{J_X(A)}^*$, and so Whitley's construction yields a sequence in A that converges weakly to x. It is also possible to use this result to give another proof that a weakly compact set is weakly sequentially compact (**18A**); however, we leave this to exercise 3.32.

We note an important consequence of the Eberlein-Šmulian theorem: one of the strongest characterizations of reflexive spaces (the others appear in **19C**).

Corollary. *A normed linear space X is reflexive if and only if $U(X)$ is weakly sequentially compact.*

Of course any reflexive space is necessarily complete (**16A**), but it is easy to see that if $U(X)$ is weakly sequentially compact then X must already be complete. The condition of the corollary is frequently stated in the form that every bounded sequence in X has a weakly convergent subsequence.

C. The Eberlein-Šmulian theorem can be used to give a negative answer to the question of whether every infinite dimensional Banach space must contain a reflexive subspace (of infinite dimension).

Example. Let $\ell^1 = \ell^1(\aleph_0)$. We prove that every reflexive subspace of ℓ^1 must be finite dimensional. Indeed, the following lemma ("Schur's lemma") shows that ℓ^1 has the peculiar property that every weakly convergent sequence is actually norm-convergent. Granting this for a moment, let M be any reflexive subspace of ℓ^1. Then $U(M)$ is weakly compact (**16F**), hence weakly sequentially compact (**18B**), and hence compact. Thus M must be finite dimensional (**9F**).

Lemma. *Let $\{x_n\} \equiv \{(\xi_1^{(n)}, \xi_2^{(n)}, \ldots)\}$ be a sequence of vectors in ℓ^1. Then $\{x_n\}$ is weakly convergent if and only if it converges in the norm topology.*

Proof. Suppose that $\{x_n\}$ is weakly convergent; we may assume that its weak limit is θ. We shall prove that

$$\textbf{(18.8)} \qquad \lim_{n\to\infty} \|x_n\|_1 \equiv \lim_{n\to\infty} \sum_{i=1}^{\infty} |\xi_i^{(n)}| = 0.$$

Given $\varepsilon > 0$ define

$$F_k = \left\{\phi \in U(m) : |\phi(x_n)| \leqslant \frac{\varepsilon}{3}, n \geqslant k\right\}.$$

The sets F_k, $k = 1, 2, \ldots$, are weak*-closed in m and their union exhausts $U(m)$. Since $U(m)$ is a compact metric space in its relative weak* topology, we can apply the Baire category theorem to conclude that some F_k has non-empty weak*-interior. For this k there then exists $\phi \in U(m)$, an integer N, and $\delta > 0$ such that

$$\textbf{(18.9)} \qquad \{\psi \in U(m) : |\phi_i - \psi_i| < \delta, 1 \leqslant i \leqslant N\} \subset F_k.$$

Because $w\text{-}\lim_n x_n = \theta$ we can arrange that $\sum_1^N |\xi_i^{(n)}| < \frac{\varepsilon}{3}$ for all $n \geqslant p$, say. Now fix any $n \geqslant \max(k, p)$ and define $\psi \in U(m)$ by

$$\psi_i = \begin{cases} \phi_i, & 1 \leqslant i \leqslant N \\ \operatorname{sgn}(\xi_i^{(n)}), & N < i. \end{cases}$$

This ψ belongs to the left-hand side of (18.9) and hence to F_k. Therefore,

$$\left| \sum_{i=1}^{N} \phi_i \xi_i^{(n)} + \sum_{i=N+1}^{\infty} |\xi_i^{(n)}| \right| \leqslant \frac{\varepsilon}{3}.$$

whence

$$\sum_{i=N+1}^{\infty} |\xi_i^{(n)}| \leqslant \frac{\varepsilon}{3} + \sum_{i=1}^{N} |\xi_i^{(n)}| < \frac{2\varepsilon}{3}.$$

It follows that

$$\|x_n\|_1 \equiv \sum_{i=1}^{\infty} |\xi_i^{(n)}| < \varepsilon, \qquad n \geqslant \max(k, p). \qquad \square$$

D. We now begin our preparations for the other major result of this section which concerns the conjugate space of a Banach space. We shall need a new locally convex topology on conjugate spaces. This topology, the *bounded weak*-topology*, is defined on the conjugate of any normed linear space X by declaring that a subset of X^* is bw*-closed if and only if its intersection with every weak*-compact set is again weak*-compact. Thus, formally, the bw*-topology is the inductive topology on X^* defined by the family of weak*-compact subsets of X^* together with the injection maps defined on these subsets; hence it is the strongest topology on X^* for which all these injection maps are continuous (equivalently, it is the strongest topology on X^* agreeing with the weak*-topology on every weak*-compact set).

By elementary manipulations of complements we see that a set in X^* is bw*-open if and only if its intersection with every ball centered at $\theta \in X^*$ is a relatively weak*-open subset of that ball. It is further easy to see that a set $G \subset X^*$ is bw*-closed exactly when every bounded weak*-convergent net in G has its limit in G. Of greater significance is the following lemma which shows in particular that the bw*-topology is locally convex and is stronger than the weak*-topology.

In order to encompass complex linear spaces in the following discussion let us define the *absolute polar* A^a of a set $A \subset X$ by

$$A^a = \{\phi \in X^* : |\phi(x)| \leqslant 1, x \in A\}.$$

Thus when X is real we have $A^a = (A \cup -A)^\circ$. Rules and results for absolute polars parallel those for polars as developed in **12C**; see exercise 3.34. However, they will not be important in what is to transpire. We are rather interested in the convenience of the notation.

Lemma. *Let X be a normed linear space. Then $\{A^a : A \subset X, A \text{ compact}\}$ is a local base for the bw*-topology on X^*.*

Proof. Let A be a compact subset of X. First we check that A^a is a bw*-θ-neighborhood. Given $t > 0$ we can select a finite $\left(\frac{1}{t}\right)$-net $\{x_1, \ldots, x_{n(t)}\}$ for the compact set $2A$. Now consider any $\phi \in X^*$ with $\|\phi\| \leqslant t$ and $|\phi(x_j)| < 1, j = 1, \ldots, n(t)$. If $x \in A$ we can find some x_i within a distance of $\frac{1}{t}$ from $2x$. Hence

$$\begin{aligned} |\phi(x)| &= \frac{1}{2}|\phi(2x)| \leqslant \frac{1}{2}(|\phi(x - x_i)| + |\phi(x_i)|) \\ &\leqslant \frac{1}{2}\|\phi\|\,\|x - x_i\| + \frac{1}{2} < \frac{t}{2t} + \frac{1}{2} = 1. \end{aligned}$$

Letting $N = \{\phi \in X^* : |\phi(x)| < 1, x \in A\}$ we have shown that

$$\{\phi \in tU(X^*) : |\phi(x_i)| < 1, i = 1, \ldots, n(t)\} \subset N \subset A^a,$$

so that A^a contains the bw*-open θ-neighborhood N and is consequently itself a bw*-θ-neighborhood. Note that this argument only uses the total boundedness of the set A.

For the converse let N be a bw*-open θ-neighborhood in X^*. Then by definition of the bw*-topology there is a finite set $F_1 \subset X$ such that $F_1^a \cap U(X^*) \subset N$. Now assume for the sake of an inductive construction that for some integer n we have obtained a finite set $F_n \subset X$ with the property that

$$F_n^a \cap nU(X^*) \subset N. \tag{18.10}$$

We shall show that there is a finite set $H_n \subset \frac{1}{n} U(X)$ such that $(F_n \cup H_n)^a \cap (n + 1)U(X^*) \subset N$. Suppose not. Then the family of sets

$$(F_n \cup H)^a \cap (n + 1)U(X^*) \cap (X \backslash N)$$

where H is a finite subset of $\frac{1}{n} U(X)$ has the finite intersection property. Since $X \backslash N$ is bw*-closed, this family consists of weak*-closed subsets of the weak*-compact set $(n + 1)U(X^*)$. Consequently the intersection of all the sets in this family is non-empty. Any point ϕ in this intersection has the properties that $\phi \in F_n^a \cap (X \backslash N)$ and that $|\phi(x)| \leqslant 1$ whenever $\|x\| \leqslant \frac{1}{n}$. This last property implies that $\|\phi\| \leqslant n$ so that $\phi \in F_n^a \cap nU(X^*) \subset N$, which contradicts $\phi \in X \backslash N$.

Thus if we set $F_{n+1} = F_n \cup H_n$ we have achieved an inductive construction of finite sets F_n with the property (18.10), and such that if the set $\bigcup_n F_n$ is enumerated in any order the resulting sequence $\{x_n\}$ converges to θ in X. It follows that the set $A \equiv \{\theta, \{x_n\}\}$ is compact and that $A^a \subset N$. □

Corollary. *Let X be a normed linear space.*

a) $\{\{x_n\}^a : x_n \in X, \lim_n x_n = \theta\}$ *is a local base for the bw*-topology.*

b) *A net X^* is bw*-convergent if and only if it converges uniformly on each compact subset of X.*

c) *A bounded weak*-convergent net in X^* is bw*-convergent.*

The verification of these assertions follows readily from the lemma; the details are left to exercise 3.35. Parts b) and c) suggest that the bw*-topology is always strictly stronger than the weak*-topology (unless, of course, X is finite dimensional), and this is true as is demonstrated by the following example in particular and exercise 3.36 in general.

Example. Let $X = \ell^p(\aleph_0)$ for $1 \leqslant p < \infty$. Let $\{\phi_n\} \subset X^*$ be defined by $\phi_n = n^{1/p} e_n$, where $e_n \equiv$ the nth-standard unit vector $\equiv \chi_{\{n\}}$. Then, in X^*, θ belongs to the weak*-closure of the sequence $\{\phi_n\}$. To see this, let $\{x_i \equiv (\xi_1^{(i)}, \xi_2^{(i)}, \ldots) : 1 \leqslant i \leqslant k\}$ be a finite subset of X and let $\varepsilon > 0$. Since $\sum_n |\xi_n^{(i)}|^p < \infty$ for each i we have

$$\sum_{n=1}^{\infty} \left(\sum_{i=1}^{k} |\xi_n^{(i)}| \right)^p < \infty.$$

Hence there must be some n for which

$$\sum_{i=1}^{k} |\xi_n^{(i)}| < \varepsilon n^{-1/p}$$

(if not, raise both sides to the pth-power and obtain a contradiction from the divergence of the harmonic series). Choosing such an n we find that

$$|\phi_n(x_i)| \equiv n^{1/p} |\xi_n^{(i)}| < \varepsilon, \qquad i = 1, \ldots, k,$$

thus proving that the sequence $\{\phi_n\}$ intersects every basic weak*-θ-neighborhood.

However, θ does not belong to the bw*-closure of $\{\phi_n\}$. We can see this, for example, by defining the compact set $A \subset X$ to be $\{\theta, n^{-1/p}e_n\}$. Since $\phi_m(n^{-1/p}e_n) = \delta_{mn}$, no subnet of $\{\phi_n\}$ can converge to θ uniformly on A.

Finally we might note that θ also fails to be in the weak*-sequential closure of $\{\phi_n\}$, since no subsequence of $\{\phi_n\}$ can converge on account of the uniform boundedness principle. This observation provides a direct proof that the weak*-topology on $\ell^p(\aleph_0)$ is not metrizable for $1 < p \leqslant \infty$ (compare with exercise 3.29). □

E. We show now that the space of all bw*-continuous linear functionals on X^* can be identified with the completion $\hat{X} \equiv \overline{J_X(X)}$ (**16D**). This fact, a special case of "Grothendieck's completeness theorem", is an important justification of our interest in the bw*-topology. In particular, it leads immediately to the Krein-Šmulian theorem.

Theorem. *Let X be a normed linear space. Then*

(18.11) $$\hat{X} = \{\Phi \in X^{*\prime} : \Phi \text{ is bw*-continuous}\}.$$

Proof. First suppose that $\Phi \in X^{**}\backslash\hat{X}$. Let H be the hyperplane $[\Phi; 1]$ in X^*. We shall show that there is a bounded net in H that weak*-converges to θ. This will imply that H is not bw*-closed, whence Φ cannot be bw*-continuous.

Let $d = d(\Phi, \hat{X}) > 0$ and choose any λ such that $d\lambda > 1$. Let $V \equiv \{x_1, \ldots, x_n\}^a$ be an arbitrary basic weak*-θ-neighborhood in X^*. We shall show that

(18.12) $$\lambda U(X^*) \cap H \cap V \neq \varnothing$$

or, in other words, that θ belongs to the weak*-closure of $\lambda U(X^*) \cap H$. To do so we utilize Helly's condition (**7A**) with $A = (1/d)U(X^*)$. Accordingly, we can prove that there exists $\phi_\delta \in (1 + \delta)A$ for every $\delta > 0$ such that

$$\Phi(\phi_\delta) = 1$$
$$\hat{x}_i(\phi_\delta) = 0, \qquad i = 1, \ldots, n$$

(any such ϕ_δ will certainly belong to the left-hand side of (18.12) if δ is sufficiently small), provided that

(18.13) $$|\alpha| \leqslant \sup\left\{\left|\alpha\Phi(\phi) + \sum_{i=1}^{n} \alpha_i\phi(x_i)\right| : \|\phi\| \leqslant \frac{1}{d}\right\},$$

for arbitrary real $\alpha, \alpha_1, \ldots, \alpha_n$. Now (18.13) is trivial if $\alpha = 0$; if not, we divide both sides by $|\alpha|$ and let $x = \sum_1^n (\alpha_i/\alpha)x_i$. We are then reduced to proving

(18.14) $$1 \leqslant \sup\left\{|\langle\phi, \Phi - \hat{x}\rangle| : \|\phi\| \leqslant \frac{1}{d}\right\}.$$

But (18.14) is certainly valid since $\|\Phi - \hat{x}\| \geqslant d$.

At this point we have shown that every bw*-continuous linear functional on X^* belongs to $\hat{X}$. It remains to select any $\Phi \in \hat{X}$ and to show that Φ is bw*-continuous. We can do this by showing that Φ is bounded on some bw*-θ-neighborhood. Now since $\Phi \in \hat{X} \equiv \overline{J_X(X)}$ there is a sequence $\{x_n\} \subset X$ such that $\lim_n \hat{x}_n = \Phi$. We let $A = \{x_n\}$ and note that since A is totally bounded, A^a is a bw*-θ-neighborhood (**18D**). We can conclude the proof by showing that $|\Phi(\phi)| \leqslant 2$ if $\phi \in A^a$. But

$$\begin{aligned} |\Phi(\phi)| &\leqslant |\phi(x_n)| + |\langle \phi, \Phi - \hat{x}_n \rangle| \\ &\leqslant 1 + \|\phi\| \, \|\Phi - \hat{x}_n\| \end{aligned}$$

for any n and we can certainly choose some n so that $\|\Phi - \hat{x}_n\| \leqslant 1/\|\phi\|$. □

Corollary 1. *Let X be a normed linear space. Then X is complete if and only if every bw*-continuous linear functional on X^* is weak*-continuous.*

This corollary provides a useful method for establishing the weak*-continuity of a given functional $\Phi \in X^{*\prime}$, when X is a Banach space. Indeed, the problem is reduced to verification that $U(X^*) \cap \ker(\Phi)$ is weak*-closed. More generally we can assert that any linear subspace $M \subset X^*$ is weak*-closed if and only if $U(M) \equiv U(X^*) \cap M$ is weak*-closed. This assertion is known as the "Banach-Dieudonné theorem", and is a special case of the next corollary which is the "Krein-Šmulian theorem". This result again emphasizes the important role played by convex sets since it shows that any convex set in X^* has the same closure in both the weak* and the bw*-topologies. We already know that this property is not enjoyed by arbitrary subsets of X^* (unless as usual X is finite dimensional).

Corollary 2. *Let X be a Banach space. A convex subset C of X^* is weak*-closed if and only if it is bw*-closed.*

Proof. Suppose that ϕ does not belong to the bw*-closure of C. Since the bw*-topology is locally convex we can apply the strong separation theorem and separate C from ϕ by a bw*-closed hyperplane H. By the theorem H is weak*-closed so that ϕ cannot belong to the weak*-closure of C. Therefore $\overline{C}^* \subset \overline{C}^{bw^*}$, and the reverse inclusion is trivial since the weak*-topology is weaker than the bw*-topology. □

F. We now give two applications of the Banach-Dieudonné theorem. The first is a new characterization of reflexive spaces. Let X be a Banach space with norm $\|\cdot\|$ and let σ be an equivalent norm on $X: \alpha\|x\| \leqslant \sigma(x) \leqslant \beta\|x\|$ for some $\alpha, \beta > 0$ and all $x \in X$. Then

$$\frac{1}{\beta}\frac{|\phi(x)|}{\|x\|} \leqslant \frac{|\phi(x)|}{\sigma(x)} \leqslant \frac{1}{\alpha}\frac{|\phi(x)|}{\|x\|}, \qquad x \in X, \qquad \phi \in X^*.$$

Thus the norm σ^* defined on X^* by

$$\sigma^*(x) = \sup_{x \neq \theta} \frac{|\phi(x)|}{\sigma(x)}, \qquad \phi \in X^* \tag{18.15}$$

is equivalent to the original norm $\|\cdot\|^*$ on X^*. Any norm on X^* of the form (18.15) derived from an equivalent norm on X is called a *dual norm*. Thus dual norms are equivalent to the original norm on X^*. We shall consider the converse question: for which spaces X is every equivalent norm on X^* necessarily a dual norm?

We first prove a lemma which gives a general topological criterion for a given equivalent norm on X^* to be a dual norm.

Lemma. *Let ρ be a norm on X^* equivalent to the original norm. Then ρ is a dual norm if and only if its unit ball U_ρ is weak*-closed.*

Proof. Suppose that there is an equivalent norm σ on X such that $\sigma^* = \rho$. Then

$$\sigma(x) = \sup_{\phi \neq \theta} \frac{|\phi(x)|}{\sigma^*(\phi)} = \sup_{\phi \neq \theta} \frac{|\phi(x)|}{\rho(\phi)}. \tag{18.16}$$

Thus there is at most one candidate for σ. Let us now define σ according to (18.16). Then if ${}^\circ U_\rho = \{x \in X : |\phi(x)| \leqslant 1, \phi \in U_\rho\}$ is the polar of U_ρ in X, we have ${}^\circ U_\rho = U_\sigma$, whence

$$U_\rho \subset {}^\circ U_\rho^\circ = U_\sigma^\circ = U_{\sigma^*} \equiv \{\phi \in X^* : \sigma^*(\phi) \leqslant 1\}.$$

By **9B** and **12C** it follows that $\rho = \sigma^*$ if and only if $U_\rho = {}^\circ U_\rho^\circ = \bar{U}_\rho^*$. □

Theorem. *Every equivalent norm on X^* is a dual norm if and only if X is reflexive.*

Proof. Assume that X is reflexive and that ρ is an equivalent norm on X^*. Then U_ρ is closed and convex, hence weakly closed, and hence weak*-closed (**16D**). Conversely assume that every equivalent norm on X^* is a dual norm. To show that X is reflexive we must show that every bounded linear functional Φ on X^* is weak*-continuous (**12B**). Since X is complete (by hypothesis) it is sufficient to show that Φ is bw*-continuous, or that $B \equiv \ker(\Phi) \cap U(X^*)$ is weak*-closed (**18E**). But $B = \bigcap_1^\infty B_n$ where $B_n \equiv \left\{\phi \in U(X^*) : |\Phi(\phi)| \leqslant \frac{1}{n}\right\}$, so it suffices to show that each B_n is weak*-closed. However, for each n, B_n is a solid bounded barrel in X^* and so its gauge ρ_n is an equivalent norm on X^*. By hypothesis ρ_n is a dual norm and the lemma then implies that $B_n \equiv U_{\rho_n}$ is weak*-closed. □

There is an extension of this theorem to quasi-reflexive spaces (**16I**): A Banach space X satisfies $\dim(X^{**}/J_X(X)) \leqslant n$ if and only if there is a subspace $M \subset X^*$ of codimension $\leqslant n$ such that for every equivalent norm ρ on X^* there is a dual norm on X^* that agrees with ρ on M. This more general result is known as the "Roth-Williams theorem".

G. As our second application we establish a companion result to the theorem of **16C**. The present result lies deeper than that of **16C** but is correspondingly of greater use. Before stating this result let us make two simple remarks about conjugate operators. Let X and Y be normed linear spaces and $T \in B(X, Y)$.

i) $T^{**} \in B(X^{**}, Y^{**})$ is an extension of T in the sense that $T^{**} \circ J_X = J_Y \circ T$.

ii) If T is an isomorphism of X onto Y then T^* is an isomorphism of Y^* onto X^*.

The proof of i) is completely analogous to the argument required for exercise 1.5b). As for ii), it is easily verified that $(T^{-1})^*$ is a bounded inverse for T^* that is defined on X^*. The converse of ii) is also true and can be established by use of both i) and ii).

Theorem. *Let* $T \in B(X, Y)$.

a) *If* Y *is complete and* T *is surjective, then* T^* *has a bounded inverse.*

b) *If* X *is complete and* T^* *has a bounded inverse, then* T *is surjective (hence* Y *is also complete).*

Proof. The proof of a) is similar to the corresponding argument in **16C** and is left to exercise 3.42. The main difficulty with b) is to prove that $T(X)$ is complete, since it is easy to see that it must be dense in Y (otherwise we could strongly separate some point in Y from $\overline{T(X)}$, and so find a non-zero element in $T(X)^\circ \subset \ker(T^*)$, whence T^* could not have an inverse).

Let $N = \ker(T) \subset X$. We shall prove that $T^*(Y^*) = N^\circ$. Granting this for a moment we can finish the proof by introducing the 1–1 operator $\hat{T}: X/N \to Y$. We find that $\hat{T}^*$ is then (by hypothesis) an isomorphism from Y^* onto $(X/N)^*$ (indeed, $\hat{T}^* = (Q_N^*)^{-1} \circ T^*$ where $Q_N^*: (X/N)^* \to N^\circ$ is an isomorphism (**16E**)). Hence by remark ii) above $\hat{T}^{**}$ is an isomorphism of $(X/N)^{**}$ onto Y^{**}. In particular, $\hat{T}^{**}(J_{X/N}(X/N))$ is complete in Y^{**} and so by remark i) above $T(X) \equiv \hat{T}(X/N)$ is complete in Y.

It remains to prove that $T^*(Y^*) = N^\circ$. We clearly have $T^*(Y^*) \subset N^\circ$. If we knew that $T^*(Y^*)$ were weak*-closed then we could obtain a contradiction by assuming this inclusion to be proper. For in this case there would be a functional $\phi \in N^\circ \backslash T^*(Y^*)$ and hence a weak*-continuous functional $\hat{x}$ that vanishes on $T^*(Y^*)$ but not on ϕ. It would then follow that $\psi(T(x)) = 0$ for all $\psi \in Y^*$ so that $T(x) = \theta$. Thus $x \in N$ but $\phi(x) \neq 0$ for some $\phi \in N^\circ$, a contradiction.

We shall finally prove that $T^*(Y^*)$ is weak*-closed. Let $K = T^*(U(Y^*))$. Then K is absolutely convex and weak*-compact (**12D** and **16C**) and $M \equiv \operatorname{span}(K) = T^*(Y^*)$ is closed in X^* (by hypothesis). Thus M is complete and it follows from **17C** that the gauge ρ_K is a continuous norm on $M: \rho_K(\phi) \leqslant \lambda\|\phi\|$ for all $\phi \in M$ and some $\lambda > 0$. Now if $\{\phi_\delta : \delta \in D\}$ is any net in $U(M)$ with weak*-limit ϕ, then since $\rho_K(\phi_\delta) \leqslant \lambda$ we have by exercise 2.4 that $\{\phi_\delta/\lambda : \delta \in D\} \subset K$. Hence $\phi/\lambda \in K$. This proves that $\phi \in \lambda K \subset M$ and since $U(X^*)$ is weak*-compact, $\phi \in U(M)$. By the Banach-Dieudonné theorem we conclude that M is weak*-closed. □

In this way we obtain a complete "dual" version of the theorem of **16C** under the assumption that both X and Y are Banach spaces.

§19. The Theorem of James

In this section we discuss a final characterization of weakly compact sets in Banach spaces. It is one of the most profound theorems in this book and we shall only completely establish a special case. The result is actually valid in any quasi-complete locally convex space, but we shall not go into this added generality here. The consequences for Banach spaces are already impressive and we shall give several of these.

A. Let A be a weakly compact subset of a real Banach space X. If $\phi \in X^*$ then ϕ is weakly continuous on X and in particular on A. Hence $\sup\{\phi(x): x \in A\}$ is attained. The theorem of James asserts, surprisingly, that this trivial necessary condition is sufficient to guarantee the weak compactness of A under mild restrictions.

Theorem. *Let A be a bounded and weakly closed subset of the real Banach space X. If every continuous linear functional on X attains its supremum on A then A is weakly compact.*

The proof will be given in several steps. The first step is a new characterization of weakly compact sets in Banach spaces known as the "iterated limit condition". This result is in effect another version of the Eberlein-Šmulian theorem (**18B**).

Lemma 1. *Let A be a bounded subset of the Banach space X. Then A is relatively weakly compact if and only if for every pair of sequences $\{x_n\} \subset A$ and $\{\phi_m\} \subset U(X^*)$*

$$\lim_n \lim_m \phi_m(x_n) = \lim_m \lim_n \phi_m(x_n) \tag{19.1}$$

whenever both of the limits exist.

Proof. If A is weakly compact and such a pair $\{x_n\}$, $\{\phi_m\}$ is given, let x_0 be a weak cluster point of $\{x_n\}$ and let ϕ_0 be a weak*-cluster point of $\{\phi_m\}$. Then if either of the limits in (19.1) exists it must be $\phi_0(x_0)$.

For the converse we proceed as in **18B**: since A is bounded $\overline{J_X(A)}^*$ is weak*-compact in X^{**}; if we can prove the inclusion (18.7) we will be done. If $\Phi \in \overline{J_X(A)}^*$ we shall prove that Φ is bw*-continuous; this will establish (18.7) by **18E** and **12B**. If Φ is not bw*-continuous then its restriction to $U(X^*)$ fails to be weak*-continuous at some point ϕ_0. There is thus a $\Phi(\phi_0)$-neighborhood N such that each weak*-ϕ_0-neighborhood contains a point ϕ with $\Phi(\phi) \notin N$.

We now construct a pair of sequences that will violate (19.1). For any $x_1 \in A$ there is a $\phi_1 \in U(X^*)$ such that $|\langle x_1, \phi_0 - \phi_1\rangle| < 1$ and $\Phi(\phi_1) \notin N$. Select $x_2 \in A$ so that $|\langle \phi_0, \hat{x}_2 - \Phi\rangle| < 1$, $|\langle \phi_1, \hat{x}_2 - \Phi\rangle| < 1$. Then select $\phi_2 \in U(X^*)$ so that $|\langle x_1, \phi_0 - \phi_2\rangle| < \frac{1}{2}$, $|\langle x_2, \phi_0 - \phi_2\rangle| < \frac{1}{2}$ and $\Phi(\phi_2) \notin$

N. Proceeding, we inductively construct sequences $\{x_n\} \subset A$, $\{\phi_m\} \subset U(X^*)$ such that

$$\max\{|\langle \phi_j, \hat{x}_n - \Phi\rangle| : 0 \leqslant j \leqslant n-1\} < \frac{1}{n-1},$$

$$\max\{|\langle x_j, \phi_0 - \phi_n\rangle| : 1 \leqslant j \leqslant n\} < \frac{1}{n},$$

$$\Phi(\phi_n) \notin N.$$

It follows that

$$\lim_n \lim_m \phi_m(x_n) = \lim_n \phi_0(x_n) = \Phi(\phi_0)$$

while

$$\lim_n \phi_m(x_n) = \Phi(\phi_m) \notin N.$$

Therefore, if we choose a subsequence of $\{\phi_m\}$ such that $\{\Phi(\phi_m)\}$ converges to a point outside N, the iterated limit condition (19.1) will fail. □

B. Now to begin the proof of James' theorem we suppose that A is not weakly compact. Then the iterated limit condition (19.1) fails for some pair of sequences $\{x_n\} \subset A$, $\{\phi_m\} \subset U(X^*)$. After changing the signs of the ϕ_m (if necessary), and after possibly discarding a finite number of $\{x_n\}$ and/or $\{\phi_m\}$, we can find an $r > 0$ such that for all k

(19.2) $$\phi_k(x_n) - \lim_m \phi_m(x_n) \geqslant r,$$

provided that n is sufficiently large. We shall moreover assume that $\{\phi_m\}$ contains a weak*-convergent subsequence with weak*-limit $\phi_0 \in U(X^*)$. This assumption is certainly warranted if X is separable (**12F**). We relabel the terms of this subsequence as $\{\phi_m\}$ and discard any other terms of the original sequence. Hence (19.2) becomes simply

(19.3) $$\langle x_n, \phi_k - \phi_0\rangle \geqslant r$$

for each k, provided that n is sufficiently large.

For each $n = 1, 2, \ldots$, we let $K_n = \operatorname{co}(\{\phi_m : m \geqslant n\})$ and we let σ_A be the *support function* of A:

$$\sigma_A(\phi) = \sup\{\phi(x) : x \in A\}, \qquad \phi \in X^*.$$

Then we note that for $\phi \in K_1$,

(19.4) $$\begin{aligned}\sigma_A(\phi - \phi_0) &\geqslant \langle x_n, \phi - \phi_0\rangle \\ &= \sum_{j=1}^{p} \lambda_j \langle x_n, \phi_{m_j} - \phi_0\rangle \geqslant r \sum_{j=1}^{p} \lambda_j = r,\end{aligned}$$

where $\lambda_1, \ldots, \lambda_p \geqslant 0$, $\sum_1^p \lambda_j = 1$ exist by virtue of $\phi \in K_1$, and where n is taken sufficiently large.

Lemma 2. *Let $\{\beta_n\}$ be a sequence of positive numbers. Then there exists $\psi_n \in K_n$, $n = 1, 2, \ldots$, such that*

$$\sigma_A\left(\sum_{i=1}^{n} \beta_i(\psi_i - \phi_0)\right) > \frac{1}{2}\beta_n r + \sigma_A\left(\sum_{i=1}^{n-1} \beta_i(\psi_i - \phi_0)\right).$$

Let us grant the validity of this lemma momentarily and see how we can use it to construct a functional $\psi \in X^*$ that fails to attain its supremum on A. Such a construction will of course reduce the proof of James' theorem to that of Lemma 2.

Suppose that the sequence $\{\beta_n\}$ is chosen so that

$$\lim_n \frac{1}{\beta_n} \sum_{i=n+1}^{\infty} \beta_i = 0;$$

for example, $\beta_n = 1/n!$. Define

$$\psi = \sum_{i=1}^{\infty} \beta_i(\psi_i - \phi_0);$$

this series converges absolutely since $\|\psi_i\| \leqslant 1$, and so defines an element of X^* (exercise 3.1). Suppose that ψ attains its supremum on A at $x_0 \in A$: $\sigma_A(\psi) = \psi(x_0)$. Then if $\gamma \equiv \sup\{\sigma_A(\phi) : \phi \in K_1 - \phi_0\}$ we would have

$$\begin{aligned}
\sum_{i=1}^{n} \beta_i\langle x_0, \psi_i - \phi_0\rangle &= \psi(x_0) - \sum_{i=n+1}^{\infty} \beta_i\langle x_0, \psi_i - \phi_0\rangle \\
&\geqslant \psi(x_0) - \gamma \sum_{i=n+1}^{\infty} \beta_i \equiv \sigma_A(\psi) - \gamma \sum_{i=n+1}^{\infty} \beta_i \\
&\geqslant \sigma_A\left(\sum_{i=1}^{n} \beta_i(\psi_i - \phi_0)\right) - \sigma_A\left(\sum_{i=1}^{n} \beta_i(\psi_i - \phi_0) - \psi\right) - \gamma \sum_{i=n+1}^{\infty} \beta_i \\
&\geqslant \sigma_A\left(\sum_{i=1}^{n} \beta_i(\psi_i - \phi_0)\right) - 2\gamma \sum_{i=n+1}^{\infty} \beta_i \\
&> \frac{1}{2}\beta_n r + \sigma_A\left(\sum_{i=1}^{n-1} \beta_i(\psi_i - \phi_0)\right) - 2\gamma \sum_{i=n+1}^{\infty} \beta_i \\
&\geqslant \frac{1}{2}\beta_n r + \sum_{i=1}^{n-1} \beta_i\langle x_0, \psi_i - \phi_0\rangle - 2\gamma \sum_{i=n+1}^{\infty} \beta_i.
\end{aligned}$$

Therefore,

$$\langle x_0, \psi_n - \phi_0\rangle > \frac{1}{2} r - 2\gamma \frac{1}{\beta_n} \sum_{i=n+1}^{\infty} \beta_i,$$

whence

$$\liminf_{n\to\infty} \langle x_0, \psi_n - \phi_0\rangle \geqslant \frac{r}{2}.$$

But this is a contradiction to $\lim_n \psi_n(x_0) = \phi_0(x_0)$ which follows in turn from $\lim_m \phi_m(x_0) = \phi_0(x_0)$ and the fact that $\psi_n \in K_n$.

We shall complete the proof of James' theorem by an inductive construction of the sequence $\{\psi_n\}$ of Lemma 2. This construction uses at each step the following algebraic fact.

Lemma 3. *Let σ be a sublinear function on a linear space X, K a convex subset of X, u a point in X, and α, β, and β' three positive numbers. If $\inf\{\sigma(u + \beta x): x \in K\} > \alpha\beta + \sigma(u)$, then there exists $x_0 \in K$ such that*

$$\inf\{\sigma(u + \beta x_0 + \beta' x): x \in K\} > \alpha\beta' + \sigma(u + \beta x_0).$$

Proof. There is a $\delta > 0$ such that $-\sigma(u) = \alpha\beta - \inf\{\sigma(u + \beta x): x \in K\} + \delta$. For any $x_0, y_1 \in K$, put $z = (\beta x_0 + \beta' y_1)/(\beta + \beta') \in K$. Then $\sigma(u + \beta x_0 + \beta' y_1) \geqslant \sigma((1 + \beta'/\beta)(u + \beta z)) - \sigma(\beta' u/\beta)$. Hence

$$\begin{aligned}
\inf\{\sigma(u + \beta x_0 + \beta' y_1): y \in K\} &\geqslant \left(1 + \frac{\beta'}{\beta}\right)\inf\{\sigma(u + \beta x): x \in K\} - \frac{\beta'}{\beta}\sigma(u) \\
&= \left(1 + \frac{\beta'}{\beta}\right)\inf\{\sigma(u + \beta x): x \in K\} \\
&\quad + \frac{\beta'}{\beta}(\alpha\beta - \inf\{\sigma(u + \beta x): x \in K\}) + \frac{\beta'}{\beta}\delta \\
&= \alpha\beta' + \inf\{\sigma(u + \beta x): x \in K\} + \frac{\beta'}{\beta}\delta.
\end{aligned}$$

We can now achieve the proof of Lemma 3 by selecting $x_0 \in K$ so that $\sigma(u + \beta x_0) < \inf\{\sigma(u + \beta x): x \in K\} + \dfrac{\beta'}{\beta}\delta$. □

Let us finally give the construction of the sequence $\{\psi_n\}$ of Lemma 2. For ψ_1 we apply Lemma 3 with $\sigma = \sigma_A$, $K = K_1 - \phi_0$, $u = \theta$, $\beta = \beta_1$, $\beta' = \beta_2$ and $\alpha = r/2$. The hypothesis of Lemma 3 is an immediate consequence of (19.4) and so there exists $\psi_1 \in K_1$ such that

$$\inf\{\sigma_A(\beta_1(\psi_1 - \phi_0) + \beta_2(\psi - \phi_0)): \psi \in K_1\} > \tfrac{1}{2}\beta_2 r + \sigma_A(\beta_1(\psi_1 - \phi_0)).$$

In general, we obtain ψ_n by applying Lemma 3 with $K = K_n - \phi_0$, $u = \sum_1^{n-1} \beta_i(\psi_i - \phi_0)$, $\beta = \beta_n$, and $\beta' = \beta_{n+1}$. The hypothesis of Lemma 3 in this general case follows from the conclusion of the lemma at the previous step:

$$\begin{aligned}
\inf\{\sigma_A(u + \beta(\psi - \phi_0)): \psi \in K_n\} &\geqslant \inf\{\sigma_A(u + \beta_n(\psi - \phi_0)): \psi \in K_{n-1}\} \\
&> \tfrac{1}{2}\beta_n r + \sigma_A(u).
\end{aligned}$$

This step completes the proof of James' theorem. □

We emphasize that we have only completely proved James' theorem for separable Banach spaces (but see also exercise 3.44). However, as will be noted in the following paragraphs, this restriction is not as serious as it may seem. The reason is that by use of the Eberlein-Šmulian theorem we can always reduce the problem of determining the weak compactness of a specified set A to that of determining the weak compactness of each separable "slice" of A. By this term we mean the intersection of A with a closed separable subspace.

C. We now give two new and striking geometric characterizations of reflexive spaces. First let X be a reflexive real Banach space. As was noted in **16I** every closed convex subset of X is then proximinal. Further, according to exercise 3.10a), any two disjoint, closed, convex subsets of X, one of which is bounded, can be strictly (actually strongly) separated by a (closed) hyperplane. Our next result is that either of these geometric properties is characteristic of reflexivity.

Theorem. *A real Banach space X is reflexive if (and only if)*

a) *every closed convex subset of X is proximinal; or*

b) *each pair of disjoint closed, convex subsets of X, one of which is bounded, can be strictly separated by a hyperplane.*

Proof. Suppose that X is not reflexive.

a) From exercise 3.33 we know that X contains a nonreflexive separable subspace M. The ball $U(M)$ is therefore not weakly compact and so, by James' theorem, there exists a functional $\phi \in M^*$ that fails to attain its norm on $U(M)$. This means (**15B**) that the closed convex set $H \equiv [\phi; \|\phi\|]$ has no minimal element and is consequently not proximinal.

b) We shall show that the disjoint closed convex sets H and $U(M)$ cannot be strictly separated. Suppose otherwise; then there would exist $\psi \in X^*$ and a positive number γ such that $\psi(x) < \gamma < \psi(y)$, for all $x \in U(M)$, $y \in H$. Let us assume that ϕ has been extended to all of X via the Hahn-Banach theorem, and let us call the extension ϕ also. Choose any point $\bar{x} \in X$ such that $\phi(\bar{x}) = \|\phi\|$ and any $z \in X$ for which $\phi(z) = 0$. Then for all $\lambda \in \mathbb{R}$, $\psi(\bar{x} + \lambda z) > \gamma$. Hence $\psi(z) = 0$ and so $\ker(\phi) \subset \ker(\psi)$. This means that the set $\{\phi, \psi\}$ is linearly dependent. We can therefore choose a constant α so that $\psi = (\alpha\gamma/\|\phi\|)\phi$. Now if $y \in H$ then $\gamma < \psi(y) = \alpha\gamma$, so that $1 < \alpha$. On the other hand, if $x \in U(M)$ then $\psi(x) < \gamma$, and so $\phi(x) = (\|\phi\|/\alpha\gamma)\psi(x) < \|\phi\|/\alpha$; since $\alpha > 1$, this contradicts the definition $\|\phi\| = \sup\{\phi(x): x \in U(M)\}$. □

There is another (and stronger) characterization of reflexive spaces by means of a separation property, known as the "Klee-Tukey theorem", which we shall state without proof: a real Banach space is reflexive if (and only if) each pair of disjoint, closed, bounded, convex subsets can be separated by a hyperplane. The proof does not depend on James' theorem but hinges rather on the fact that any non-reflexive space contains a non-reflexive (closed) subspace of infinite codimension. This fact in turn depends on the existence of a bounded sequence with no weakly convergent subsequence (**18B**).

D. In **13E** and **15C** we discussed the notion of a strictly normed linear space. We now define a stronger property. Let us say that a closed convex subset A of a normed linear space X is *uniformly rotund* if there is a non-decreasing function δ on $[0, \infty)$ with $0 = \delta(0) < \delta(t)$, $t > 0$, such that $\frac{1}{2}(x + y) + z \in A$ whenever $x, y \in A$ and $\|z\| \leqslant \delta(\|x - y\|)$. If $U(X)$ satisfies this condition we shall say that X is *uniformly normed.*

The condition that a convex set A be uniformly rotund is a strong geometric constraint on A. It requires that the mid-point of any chord joining two boundary points of A be bounded away from the boundary by a positive quantity that depends only on the length of the chord and not on the location of its end-points. Intuitively the boundary of A cannot come too close to "flattening out" in any region. In particular, any uniformly rotund set is rotund, and on the other hand, any finite dimensional strictly normed linear space is actually uniformly normed.

Our interest in this condition is summarized by the next result known as "Milman's theorem". We see that it is another of those peculiar hybrids wherein a hypothesis of one type (in this case, geometric) leads to a conclusion of a different type (topological). By this nature it reminds us of the Krein-Milman theorem wherein a topological hypothesis implied an algebraic conclusion.

Theorem. *A uniformly normed Banach space X is reflexive.*

Proof. As usual, it is sufficient to prove that $U(M)$ is weakly compact where M is any separable closed subspace of X. If $\phi \in M^*$ let a sequence $\{x_n\} \subset U(M)$ be chosen so that $\lim_n \phi(x_n) = \|\phi\|$. We wish to show that $\{x_n\}$ is a Cauchy sequence. If we do so then its limit will be a point where ϕ attains its norm; hence $U(M)$ will be weakly compact on account of James' theorem. Now as m and n become large, $\phi(x_n) + \phi(x_m) \to 2\|\phi\|$, so that

$$\lim_{m,n} \phi(\tfrac{1}{2}(x_n + x_m)) = \|\phi\|.$$

Hence $\lim_{m,n} \|\frac{1}{2}(x_n + x_m)\| = 1$. Because of the uniform rotundity of $U(M)$ (which follows *a fortiori* from that of $U(X)$), we conclude that $\lim_{m,n} \|x_n - x_m\| = 0$, and so $\{x_n\}$ is indeed a Cauchy sequence. □

Are there any uniformly normed Banach spaces of infinite dimension? The answer is in the affirmative: every $L^p(\Omega, \mu, \mathbb{F})$ space (of infinite dimension) is uniformly normed for $1 < p < \infty$ ("Clarkson's theorem"). This is difficult to prove for general p but not so hard when $p = 2$. For in this case it is easily seen that

$$\|x + y\|_2^2 + \|x - y\|_2^2 = 2(\|x\|_2^2 + \|y\|_2^2),$$

(the "parallelogram law") from which it follows that we may take $\delta(t) = \sqrt{1 + 4t^2} - 1$ in the definition of uniform rotundity of the ball $U(L^2(\Omega, \mu, \mathbb{F}))$.

E. We shall finally give two applications of the theorem of James that do not pertain to reflexivity. The first is "Krein's theorem".

Theorem. *Let A be a weakly compact subset of a real Banach space X. Then $\overline{\text{aco}}(A)$ is weakly compact.*

Proof. As usual, it will suffice to prove that each separable slice of A is weakly compact; we shall therefore assume directly that X is separable. Now let $\phi \in X^*$; we must show that ϕ attains its supremum on $\overline{\text{aco}}(A)$. Let $\alpha = \min\{\phi(x): x \in A\} \leqslant \max\{\phi(x): x \in A\} = \beta$. Since A is weakly compact there exist $u, v \in A$ such that $\phi(u) = \alpha$, $\phi(v) = \beta$. Now the image $\phi(\text{aco}(A))$ is the interval $[-\gamma, \gamma]$ where $\gamma \equiv \max\{|\alpha|, |\beta|\}$, and $\text{aco}(A)$ contains a point (namely $-u$ or v) at which ϕ assumes the value γ. Since $\phi(\overline{\text{aco}}(A))$ also equals $[-\gamma, \gamma]$, we have shown that $\overline{\text{aco}}(A)$ satisfies the hypotheses of James' theorem and consequently is weakly compact. □

Observe that Krein's theorem is only new in the case where X is not reflexive. For we know that all reflexive spaces are weakly quasi-complete (**16F**), and that in any quasi-complete locally convex space $\overline{\text{aco}}(A)$ is compact whenever A is totally bounded (**11C**).

F. Our second application concerns vector measures taking values in a real Banach space X. Let (Ω, Σ) be a measurable space and let $\bar{\mu}: \Sigma \to X$ be a function having the property that for every sequence $\{E_n\} \subset \Sigma$ of mutually disjoint sets

$$\bar{\mu}\left(\bigcup_{n=1}^{\infty} E_n\right) = \sum_{n=1}^{\infty} \bar{\mu}(E_n),$$

where the series on the right is assumed to converge unconditionally in X (that is, to converge regardless of the order of its terms: $\sum_1^{\infty} \bar{\mu}(E_n) = \sum_1^{\infty} \bar{\mu}(E_{p(n)})$, for any permutation p of the positive integers; this is equivalent to absolute convergence of the series when $\dim(X) < \infty$, but otherwise is weaker).

Our aim is to show that the range $\bar{\mu}(\Sigma)$ is relatively weakly compact in X ("theorem of Bartle-Dunford-Schwartz"). The contrast between this situation and that of **15E** is that we are studying measures with values in an arbitrary Banach space rather than in a product space and, more importantly, the measures are not assumed to possess a density.

Theorem. *Under the above assumptions the* range $\bar{\mu}(\Sigma)$ *is relatively weakly compact in* X.

Proof. For any functional $\phi \in X^*$ the composite $\phi \circ \bar{\mu}$ is a finite signed measure on (Ω, Σ), and there is a corresponding Hahn-decomposition of Ω. That is, there is a partition $\Omega = A \cup B$, $A \cap B = \varnothing$, with $\phi \circ \bar{\mu}$ a non-negative measure on A and $-\phi \circ \bar{\mu}$ a non-negative measure on B. Now, to show that the weak closure R of the range $\bar{\mu}(\Sigma)$ is weakly compact, we observe that

$$\begin{aligned}\sup\{\phi(x): x \in R\} &= \sup\{\phi(x): x \in \bar{\mu}(\Sigma)\} \\ &= \sup\{\phi(\bar{\mu}(E)): E \in \Sigma\} = \phi(\bar{\mu}(A)).\end{aligned}$$

This proves that ϕ attains its supremum on R and hence, by James' theorem, that R is weakly compact. (Note that it is implicit in the preceding argument that R is weakly bounded and hence bounded by **17C**.) □

We remark that a stronger conclusion is possible in the special case where X is reflexive and $\bar{\mu}$ is of bounded variation in the sense that

$$\sup_{\pi} \sum_{n} \|\bar{\mu}(E_n)\| < \infty,$$

where the supremum is taken over all partitions $\pi = \{E_1, \ldots, E_m\}$ consisting of a finite collection of disjoint sets in Σ whose union is Ω. Namely, in this case it can be shown that the norm closure of the range $\bar{\mu}(\Sigma)$ is norm-compact, and is also convex provided that $\bar{\mu}$ has no atoms (where this latter term is defined in complete analogy with the scalar case considered in **15E**). The proof of this second assertion utilizes the finite dimensional Lyapunov convexity theorem of **15E**.

§20. Support Points and Smooth Points

In this section we establish the famous theorems of Bishop and Phelps concerning the existence of support points and support functionals for a given convex subset of a Banach space and observe the particular consequence that every Banach space is subreflexive. We also discuss the subdifferentiability of lower semicontinuous convex functions on Banach spaces, a situation which is not covered by the earlier discussion in **14B**. Unlike most of the earlier results in this chapter which admit extensions to certain more general types of locally convex spaces, the present results require both a norm and completeness for their validity, as we see by appropriate examples and exercises. Finally, we resume the discussion begun in **7E–F** of smooth points. The density theorem of Mazur is established for separable spaces, and some applications to the uniqueness of Hahn-Banach extensions of linear functionals are given.

A. Let A be a closed subset of a real Banach space X. We say that $x_0 \in A$ is a *conical support point* of A if there is a closed cone C in X such that

(20.1) $$A \cap (x_0 + C) = \{x_0\}.$$

In terms of the ordering induced on X by C (**5A**) equation (20.1) simply means that x_0 is a maximal element of A (that is, $x_0 \leqslant x \in A$ implies $x_0 = x$). Now in general there is no reason why such points in the set A should exist. However, as we shall now see, their existence can be guaranteed for a certain class of cones. After establishing this technical fact we shall discuss some of its implications.

For $\phi \in X^*$, $\|\phi\| = 1$, and $0 < \gamma < 1$ we define a closed cone $C = C(\phi, \gamma) = \{x \in X : \gamma\|x\| \leqslant \phi(x)\}$. It is easily seen that C is the cone generated by the set $B = B(\phi, \gamma) \equiv U(X) \cap \{x \in X : \gamma \leqslant \phi(x)\}$, that is, $C = [0, \infty)B$. Also, from now on, we shall write $\sup \phi(A)$ in place of $\sup\{\phi(z) : z \in A\}$.

Lemma. *Let A be a closed subset of the Banach space X and suppose that $\phi \in X^*$ ($\|\phi\| = 1$) is bounded above on A. Then for $0 < \gamma < 1$, and any*

$x \in A$, there exists $x_0 \in A$ such that $x_0 \in x + C$ and (20.1) holds for $C = C(\phi, \gamma)$.

Proof. Let $A_n = A \cap (x_n + C)$ where the sequence $\{x_n\}$ is defined inductively as follows: $x_1 = x$ and having obtained $x_1, \ldots, x_n$ we take x_{n+1} to be any point in A_n for which $\sup \phi(A_n) < \phi(x_{n+1}) + 1/n$. Since $x_{n+1} \in A_n \subset x_n + C$, we have $x_{n+1} + C \subset x_n + C$, ahd hence $A_{n+1} \subset A_n$. Now, it $y \in A_{n+1}$, then $\phi(y) \leqslant \sup \phi(A_n)$ and

$$\gamma\|y - x_{n+1}\| \leqslant \phi(y) - \phi(x_{n+1}) \leqslant \sup \phi(A_n) - \phi(x_{n+1}) < \frac{1}{n},$$

whence $\operatorname{diam}(A_{n+1}) \leqslant \dfrac{2\gamma}{n}$. Because A is complete the intersection of the nested sequence $\{A_n\}$ consists of a single point x_0. Since $x_0 \in A_1$ we have $x_0 \in x + C$. Finally, since $x_0 \in A_n = A \cap (x_n + C)$ for all n, we have $A \cap (x_0 + C) \subset A_n$ for all n (because C is a cone), hence $A \cap (x_0 + C) = \{x_0\}$, and so (20.1) is satisfied. □

B. If A is a solid closed convex subset of a real linear topological space then the support theorem **(11E)** assures us that every boundary point of A is a support point. On the other hand, whether or not A is solid, it will in general (exercise 2.18) contain non-support points. For a long time it was unknown whether an arbitrary (not solid, not weakly compact) closed convex set A in a Banach space necessarily contained any support points. We can now see that, in fact, support points must exist in this setting, since any conical support point $x_0 \in A$ (with respect to some solid cone C) is actually a support point of A. This follows from an application of the separation theorem to the convex sets A and $x_0 + C$.

The first theorem of this section will provide a stronger response to the question of the existence of support points by proving their density in the boundary of a given closed convex set A. We will then known that either the support points of A are exactly the boundary points (when A is solid) or else they are dense in A (when A has no interior). For both this theorem and a later one we shall need another technical fact which we shall call the "Phelps-Brondsted-Rockafellar lemma".

Lemma. *Suppose that A is a closed convex subset of the Banach space X, that $\phi \in X^*$ has norm 1, and that $\varepsilon > 0$ and $x \in A$ are such that*

$$\sup \phi(A) \leqslant \phi(x) + \varepsilon.$$

Then for any $\gamma \in (0, 1)$ there exist $\psi \in X^*$ and $x_0 \in A$ such that $\sup \psi(A) = \psi(x_0)$, $\|x_0 - x\| \leqslant \varepsilon/\gamma$, and $\|\phi - \psi\| \leqslant \gamma$.

Proof. By the preceding lemma there is a conical support point $x_0 \in A$ (with respect to the cone $C = C(\phi, \gamma)$) such that $x_0 \in x + C$. We shall obtain the desired functional ψ via a separation argument resembling the one used to obtain subgradients. Let $f(x) = \gamma\|x\| - \phi(x)$ and let $A_1 = \{(z, 0) \in X \times \mathbb{R}^1:$

$x_0 + z \in A\}$. Then $A_1 \cap \text{epi}(f) = \{(\theta, 0)\}$ by (20.1) and so A_1 is disjoint from the interior of epi(f). By the separation theorem we can find a functional $\Phi \in (X \times \mathbb{R}^1)^*$ such that $\sup \Phi(A_1) = 0 = \inf \Phi(\text{epi}(f))$. Now the point $(\theta, 1)$ belongs to int(epi(f)) so that $\Phi(\theta, 1) > 0$. Hence we can write $\Phi(z, t) = \psi(z) + t$ for all $(z, t) \in X \times \mathbb{R}^1$. Now $(z - x_0, 0) \in A_1$ whenever $z \in A$, so that $\Phi(z - x_0, 0) = \psi(z - x_0) \leqslant 0$, whence $\sup \psi(A) = \psi(x_0)$. Also, since $(z, f(z)) \in \text{epi}(f)$ for all $z \in X$, we have $0 \leqslant \Phi(z, f(z)) = \psi(z) + f(z)$, whence $-\psi(z) \leqslant f(z) \equiv \gamma\|z\| - \phi(z)$ and therefore $\|\phi - \psi\| \leqslant \gamma$. Finally, $x_0 - x \in C$ implies $\gamma\|x_0 - x\| \leqslant \phi(x_0 - x) \leqslant \sup \phi(C) - \phi(x) \leqslant \varepsilon$. □

This lemma leads directly to the "first Bishop-Phelps theorem" on the density of support points.

Theorem. *If A is a closed convex subset of a Banach space X, then the support points of A are dense in the boundary of A.*

Proof. Let $x \in \partial(A)$ and $\delta > 0$ be given. Choose $z \in X \setminus A$ so that $\|x - z\| < \delta/2$ and then choose $\phi \in X^*$ such that $\|\phi\| = 1$ and $\sup \phi(A) < \phi(z)$ (**11F**). Then $\phi(z) \leqslant \phi(x) + \|x - z\|$, whence $\sup \phi(A) < \phi(x) + \delta/2$. We now apply the preceding lemma with $\varepsilon \equiv \delta/2$ and $\gamma \equiv 1/2$ to obtain $x_0 \in A$ and $\psi \in X^*$ such that $\sup \psi(A) = \psi(x_0)$, $\|x_0 - x\| \leqslant \delta$, and $\|\phi - \psi\| \leqslant \frac{1}{2}$. This last inequality shows that $\psi \neq \theta$ (since $\|\phi\| = 1$), and thus x_0 is a support point of A within distance δ from x. □

C. In order to show that there is not much hope of extending this theorem beyond the setting of Banach spaces we shall indicate an example of Peck (based on an earlier more specialized example of the same type of phenomenon due to Klee). This example will lead to a bounded closed convex subset of a complete metrizable locally convex space which has no support points at all. The construction serves also as a further application of the theorem of James. Let us say that a linear functional ϕ is a *support functional* of a set A if $\phi \neq \theta$ and ϕ attains its supremum over A: $\sup \phi(A) = \phi(x_0)$, for some $x_0 \in A$.

Example. Let $X = \prod_1^\infty X_k$ be the product of a sequence of non-reflexive real Banach spaces X_k. In its product topology X is locally convex and complete, and this topology is metrizable by exercise 2.4. We are going to construct a closed bounded convex subset A of X such that the projection of A on each $\prod_1^n X_k$ is open, $n = 1, 2, \ldots$. Since any functional $\phi \in X^*$ is bounded on some basic θ-neighborhood in X, ϕ must have the form $\sum_1^n \phi_k \circ \pi_k$, where $\phi_k \in X_k^*$ and $\pi_k: X \to X_k$ is the usual projection. It follows that $\phi(A)$ must be open and hence that ϕ cannot be a support functional of A. Therefore, A can have no support points.

The construction of A is based on an inductive construction of a sequence of closed bounded convex sets $A_n \subset \prod_1^n U(X_k)$. To begin, let $\phi_2 \in X_2^*$ be a norm-one functional that is not a support functional of $U(X_2)$ (**19A**). Let $A_2 = \{(x_1, x_2) \in U(X_1) \times U(X_2) : \|x_1\| \leqslant \phi_2(x_2)\}$. Note that if $\|x_2\| < 1$ and $\phi_2(x_2) > 0$, then (θ, x_2) belongs to the interior of A_2 in $X_1 \times X_2$. For the inductive step we suppose that $n \geqslant 3$ and that a closed bounded convex set $A_{n-1} \subset \prod_1^{n-1} U(X_k)$ has been constructed containing an interior point $y^{(n-1)} \equiv (y_1, \ldots, y_{n-1})$. Choose a norm-one functional $\phi_n \in X_n^*$ which is not a support functional of $U(X_n)$. Let ρ_{n-1} be the gauge of the convex θ-neighborhood $A_{n-1} - y^{(n-1)}$, and define

$$A_n = \{(x_1, \ldots, x_{n-1}, x_n) \in \prod_1^n U(X_k) : (x_1, \ldots, x_{n-1}) \in A_{n-1} \quad \text{and} \quad \rho_{n-1}((x_1, \ldots, x_{n-1}) - y^{(n-1)}) \leqslant \phi_n(x_n)\}.$$

If $y_n \in \operatorname{int}(U(X_n))$ is chosen so that $\phi_n(y_n) > 0$ then $y^{(n)} \equiv (y_1, \ldots, y_{n-1}, y_n)$ is an interior point of A_n.

Having obtained the sets $A_1, \ldots, A_n, \ldots$ we now define

$$A = \{(x_1, x_2, \ldots) \in X : (x_1, \ldots, x_n) \in A_n, n \geqslant 2\}.$$

Clearly A is a closed bounded convex subset of X. To complete the example it will suffice to prove that the projection of A on $\prod_1^n X_k$ is $\operatorname{int}(A_n)$ for $n = 2, 3, \ldots$. Suppose first that $(x_1, \ldots, x_n, x_{n+1}, \ldots) \in A$. Then

$$\rho_n((x_1, \ldots, x_n) - y^{(n)}) \leqslant \phi_{n+1}(x_{n+1}) < 1,$$

whence $(x_1, \ldots, x_n) - y^{(n)} \in \operatorname{int}(A_n - y^{(n)})$, or $(x_1, \ldots, x_n) \in \operatorname{int}(A_n)$. To reverse the inclusion, take $(x_1, \ldots, x_n) \in \operatorname{int}(A_n)$; then $\rho_n((x_1, \ldots, x_n) - y^{(n)}) < 1$. Since $\|\phi_{n+1}\| = 1$, there exists $x_{n+1} \in \operatorname{int}(U(X_{n+1}))$ such that $\rho_n((x_1, \ldots, x_n) - y^{(n)}) < \phi_{n+1}(x_{n+1})$. Therefore, $(x_1, \ldots, x_n, x_{n+1}) \in \operatorname{int}(A_{n+1})$. We can continue this inductive procedure and obtain $(x_1, \ldots, x_n, x_{n+1}, \ldots) \in A$ whose projection on $\prod_1^n X_k$ is $(x_1, \ldots, x_n)$. □

D. Let A be a convex subset of a Banach space X, and let $f \in \operatorname{Conv}(A)$ be lower semi-continuous. Then f is continuous throughout $\operatorname{int}(A)$ (exercise 3.50), and hence subdifferentiable there (**14B**). If A is not solid it is still of interest to inquire about the subdifferentiability of f. Recall (**6D**) that subgradients of f correspond to non-vertical supporting hyperplanes to $\operatorname{epi}(f)$. Making use of arguments analogous to those used in the lemma of **20B**, the following result was established by Brondsted and Rockafellar.

Lemma. *Let $\varepsilon, \gamma > 0$. Suppose that $\phi \in X^*$ satisfies*

$$(f(x) - \varepsilon) + \phi(z - x) \leqslant f(z),$$

for some x and all z in A. Then there exist $x_0 \in A$ and $\psi \in X^*$ such that $\|x_0 - x\| \leqslant \gamma$, $\|\phi - \psi\| \leqslant \varepsilon/\gamma$, and $\psi \in \partial f(x_0)$.

Any such ϕ is called an *ε-approximate subgradient* of f at x and the set of all these is denoted $\partial_\varepsilon f(x)$. The sets $\partial_\varepsilon f(x)$ are non-empty weak*-closed convex subsets of X^* for $\varepsilon > 0$, and they decrease to $\partial f(x)$ as ε decreases to 0. (That $\partial_\varepsilon f(x) \neq \varnothing$ may be seen by strongly separating the point $(x, f(x) - \varepsilon)$ from the closed set epi(f) (**14A**).) How well one of these sets approximates $\partial f(x)$ can be estimated by the above lemma. We shall use this lemma to prove a formula which implies that the points at which f is subdifferentiable constitute a dense subset of A. Let $B = \{z \in X : \partial f(z) \neq \varnothing\}$ and let $\bar{f} = f|B$.

Theorem. *For all $x \in A$,*

(20.2)
$$f(x) = \liminf_{y \to x} \bar{f}(y).$$

Proof. Because f is lower semicontinuous we need only prove that $f(x) \geqslant \liminf \bar{f}(y)$. Given $x \in A$ and $\delta > 0$, put $\varepsilon = \delta/2$ and select $\phi \in \partial_\varepsilon f(x)$. Choose $\gamma > 0$ so small that $\gamma \leqslant \delta$ and $\gamma\|\phi\| \leqslant \delta/2$. Now let $x_0 \in A$ and $\psi \in X^*$ satisfy the conclusions of the lemma. Then

$$\begin{aligned} f(x_0) - f(x) &\leqslant -\psi(x - x_0) \\ &\leqslant \|x - x_0\|\,\|\psi\| \leqslant \gamma\left(\|\phi\| + \frac{\varepsilon}{\gamma}\right) < \frac{\delta}{2} + \varepsilon = \delta. \end{aligned}$$

Thus $x_0 \in B$, $\|x - x_0\| < \delta$ and $f(x_0) < f(x) + \delta$. □

We remark that this result too cannot be extended beyond the confines of Banach spaces. To illustrate, let K be the supportless set constructed in **20C** in the product space X. We choose an arbitrary non-zero $x_0 \in X$ and define a convex function by

(20.3)
$$f(x) = \min\{t \in R : x + tx_0 \in K\},$$

the domain A of f being the set of $x \in X$ for which some such t exists. Because K is closed and bounded in X this function is lower semicontinuous on A. However, if $\partial f(x) \neq \varnothing$ for some $x \in A$ then it can be shown (exercise 3.51) that the set K would have a non-trivial supporting hyperplane at the point $x + f(x)x_0$, and this is a contradiction. Thus (20.3) defines a lower semicontinuous convex function on $A \subset X$ which is nowhere subdifferentiable.

E. We shall now give a second application of the Phelps-Brondsted-Rockafellar lemma. This intended application is motivated by the problem of subreflexivity of Banach spaces, which is in turn motivated by the theorem of James (**19A**). Given a real Banach space X we let $\mathscr{P}(X) = \{\phi \in X^* : \phi$ attains its norm on $U(X)\}$. Then the theorem of James asserts that $\mathscr{P}(X) = X^*$ if and only if X is reflexive. We say that X is *subreflexive* if $\mathscr{P}(X)$ is dense in X^* (in the norm topology). We are thus led to inquire as to which Banach spaces are subreflexive.

For example, if we identify c_0^* with ℓ^1 as in **16H**, then $\mathscr{P}(c_0)$ is that subset of ℓ^1 whose members vanish except on a finite set (the finitely supported elements of ℓ^1). Or, if we identify $L^{1*} \equiv L^1(\Omega, \mu, \mathbb{F})^*$ with $L^\infty \equiv L^\infty(\Omega, \mu, \mathbb{F})$ as in **12C** (here μ is a σ-finite measure), then $\mathscr{P}(L^1) = \{f \in L^\infty : \mu\{t \in \Omega: |f(t)| = \|f\|_\infty\} > 0\}$ (in other words, $\mathscr{P}(L^1)$ is that subset of L^∞ whose members attain their norm on a set of positive measure). In both these cases it is easy to see that the spaces are subreflexive. The next result, the "second Bishop-Phelps theorem", shows that this is not an accident.

Theorem. *Let A be a closed convex set in the real Banach space X, and let $\phi \in X^*$ ($\|\phi\| = 1$) be bounded above on A. Then for any $\delta \in (0, 1)$ there exists a support functional ψ of A with $\|\phi - \psi\| < \delta$.*

Proof. Choose $x \in A$ so that $\sup f(A) \leqslant f(x) + 1$ and apply the lemma of **20B** with $\varepsilon = 1$ and $\gamma = \delta$. We obtain $\psi \in X^*$ such that $\sup \psi(A) = \psi(x_0)$ for some $x_0 \in A$, and $\|\phi - \psi\| \leqslant \delta < 1 \equiv \|\phi\|$. Hence $\psi \neq \theta$ and is therefore a support functional of A. □

This theorem shows that the set of support functionals of A is dense in the space of functionals that are bounded above on A, and leads immediately to the following corollaries.

Corollary 1. *If A is a closed bounded and convex subset of X then the support functionals of A are dense in X^*.*

Corollary 2. *Every Banach space is subreflexive.*

It is interesting to remark that when X is an incomplete normed linear space there is a solid closed bounded convex subset A of X such that the support functionals of A are not dense in X^* (exercise 3.53). Consequently, Corollary 1 is actually a new characterization of Banach spaces within the class of normed spaces. On the other hand, an incomplete normed space may or may not be subreflexive. For example, the space of sequences with only finitely many non-zero terms, normed by the $\ell^p(\aleph_0)$-norm for $1 < p < \infty$, is a subreflexive normed space. But the space of polynomials on $[0, 1]$, normed by the supremum norm, is not reflexive. The proof of this latter assertion depends on the representation of the general continuous linear functional on this space (or, equivalently, on the space $C([0, 1], \mathbb{R})$) as a Stieltjes integral defined by an integrator function of bounded variation (**17I**, **22D**, and exercise 4.9).

F. We consider now a special kind of support point for convex subsets in normed spaces. In general, let A be a solid convex set in a real linear topological space X. A support point of A is called a *smooth point* of A if there is only one (closed) hyperplane supporting A at x. We assume that A has non-empty interior so as to rule out situations where A lies in some hyperplane; in such cases we would not expect A to have any smooth points (except in trivial cases such as the case where A is already a hyperplane.) If every boundary point of A is a smooth point we shall say that A is *smooth.* The set of all smooth points of A is denoted sm(A), so that A is smooth if and only if $\partial(A) = \text{sm}(A)$. If A is the unit ball in the normed space X and if A is smooth, then X will be said to be *smoothly normed.*

Smooth points were introduced in **7F** for the purely linear space situation and in particular we classified the unit vectors of the p-norm unit balls in $\mathbb{R}^n$ as to smoothness ($1 \leqslant p < \infty$). We can also read out of **7F** the following application to smoothness in normed spaces: a unit vector x_0 is a smooth point of $U(X)$ exactly when the norm function has a gradient at x_0. Thus (**7F**) $x_0 \in \mathrm{sm}(U(X))$ exactly when

(20.4) $$g(x_0; x) \equiv \lim_{t \to 0} \frac{\|x_0 + tx\| - \|x_0\|}{t}$$

exists for all x and defines a functional $g(x_0; \cdot)$ in X^*. The functional $g(x_0; \cdot)$ is the gradient of the norm at x_0 and, as an element of X^*, has norm one. When the limit in (20.4) exists we have $g(x_0; x) = \tau_{U(X)}(x_0, x)$, the tangent function of $U(X)$ (**7F**).

We now consider some of the standard normed spaces and determine the smooth points of their unit balls. Most of the details are left as an exercise.

Example. a) Let $L^p(\Omega, \mu, \mathbb{R})$ for $1 < p < \infty$. Then every unit vector is a smooth point of $U(L^p)$. This can be seen from the condition for equality in Holder's inequality, or by directly differentiating under the integral sign in order to compute the limit in (20.4). However, it also follows from the fact that $L^{p*} = L^q$ is strictly normed (**13E**—Ex. 5, and **16G**), and the duality between smoothly normed and strictly normed reflexive spaces (**20G**). For $x_0 \in \partial U(L^p)$ the gradient of the p-norm at x_0 is given by the function $x_0|x_0|^{p-2}/\|x_0\|^{p-1} \in L^q$.

b) Let $L^1 = L^1(\Omega, \mu, \mathbb{R})$. Then $x_0 \in \mathrm{sm}(U(L^1))$ if and only if $\mu(\{t \in \Omega : x_0(t) = 0\}) = 0$. When this condition holds the gradient of the norm at x_0 is the function $\mathrm{sgn}(x_0) \in L^\infty$. (We assume that the measure μ is such that the usual congruence between L^{1*} and L^∞ holds.) In the special case where $\Omega = [0, 1]$, $\mu .=$ Lebesgue measure, it follows that the subspace of polynomials of degree $\leqslant n$, for some integer n, is a smooth subspace of L^1.

c) By contrast, the sequence space $\ell^1(\aleph_0)$ has no smooth subspaces of dimension > 1. This may be seen by taking any two vectors $x = (\xi_1, \xi_2, \ldots)$, $y = (\eta_1, \eta_2, \ldots)$ in ℓ^1, and considering the function $f(t) \equiv \sum_1^\infty |\xi_n + t\eta_n|$. The function f is either a constant (if $y = \theta$) or else fails to be differentiable at $t = -\xi_n/\eta_n$ whenever $\eta_n \neq 0$. But, $f'(t) \equiv g(x + ty; y)$.

d) Let $C = C(\Omega, \mathbb{R})$, where Ω is a compact Hausdorff space. A *peak function* in C is a function x_0 that attains its norm $\|x_0\|_\infty$ at a single point in Ω. Clearly, any unit vector in C that is not a peak function cannot belong to $\mathrm{sm}(U(C))$. On the other hand, a peak function x_0 of unit norm is indeed a smooth point of $U(C)$. To prove this we assume that $x_0(p_0) = 1 = \|x_0\|_\infty$ and verify that $g(x_0; \cdot) = \delta_{p_0}$. For any $x \in C$ we have

$$\|x_0 + tx\|_\infty \equiv |x_0(p_t) + tx(p_t)| \geqslant |x_0(p_0) + tx(p_0)|.$$

Therefore,

$$0 \leqslant 1 - |x_0(p_t)| \leqslant |t|\,|x(p_0)| + |t|\,|x(p_t)| \leqslant 2|t|\,\|x\|_\infty,$$

whence

$$\lim_{t\to 0} |x_0(p_t)| = 1.$$

Since Ω is compact, it follows that $p_t \to p_0$ as $t \to 0$. Now, for sufficiently small $|t|$

$$\begin{aligned} tx(p_0) &= 1 + tx(p_0) - 1 = |1 + tx(p_0)| - 1 \\ &\leqslant |x_0(p_t) + tx(p_t)| - 1 \\ &\equiv \|x_0 + tx\|_\infty - \|x_0\|_\infty \\ &= x_0(p_t) + tx(p_t) - 1 \leqslant tx(p_t). \end{aligned}$$

Now divide through by t and let $t \to 0$. If $x_0(p_0) = -1$, we use the general rule $g(-x_0; \cdot) = -g(x_0; \cdot)$ and so obtain $g(x_0; \cdot) = \operatorname{sgn}(x_0(p_0))\delta_{p_0}$. □

The main result about smooth points is known as the "Mazur density theorem".

Theorem. *Let A be a solid closed convex subset of a separable Banach space X. Then* $\operatorname{sm}(A)$ *is a residual subset of* $\partial(A)$.

Proof. Since A is solid we can assume that $\theta \in \operatorname{int}(A)$. Let $\tau = \tau_{U(X)}$, the tangent function (**7F**) of the unit ball in X. From the properties of τ listed in **7F** we see that

$$\begin{aligned} |\tau(x, y) - \tau(x, z)| &\leqslant \max(\rho_A(y - z), \rho_A(z - y)) \\ &\leqslant \frac{1}{\varepsilon}\|y - z\|, \end{aligned} \tag{20.5}$$

if $\varepsilon > 0$ is chosen small enough that $\varepsilon U(X) \subset A$. From formula (7.8) and the theorem in **7F** we see that a boundary point x_0 of A is a smooth point of A provided that $\tau(x_0, x) = -\tau(x_0, -x)$, for all $x \in X$. Now let $\{y_n\}$ be a countable dense set in $X\backslash\{\theta\}$, and define

$$Z_n = \{x \in X : \tau(x, y_n) = -\tau(x, -y_n)\},$$

$$Z = \bigcap_{n=1}^{\infty} Z_n.$$

Then, because of the continuity of $\tau(x, \cdot)$ as shown by (20.5), a non-zero $x \in Z$ will satisfy $x/\rho_A(x) \in \operatorname{sm}(A)$. Since $\tau(\alpha x, \cdot) = \tau(x, \cdot)$, $\alpha > 0$, we see that the problem is reduced to proving that Z is a residual set in X.

Let

$$Z_{n,i,j} = \left\{x \in X : j\left(\rho_A\left(x + \frac{y_n}{j}\right) - 2\rho_A(x) + \rho_A\left(x - \frac{y_n}{j}\right)\right) < \frac{1}{i}\right\}.$$

These sets are open in X and so $Z_{n,i} \equiv \cup\{Z_{n,i,j} : j = 1, 2, \ldots\}$ is also open. But $Z_n = \cap\{Z_{n,i}: i = 1, 2, \ldots\}$. Thus it remains only to show that each set $Z_{n,i}$ is dense in X. Suppose not; then for some integers i and n there exists a point $x_0 \in X$ and $\delta > 0$ such that $Z_{n,i}$ is disjoint from $x_0 + \delta U(X)$. If we put $g(\lambda) = \rho_A(x_0 + \lambda y_n)$ it follows that g is not differentiable for $|\lambda| < \delta/\|y_n\|$. But g is a Lipschitz continuous (hence absolutely continuous) function of λ, since ρ_A is sublinear, and, as is known from analysis, any such function is differentiable almost everywhere. Thus we arrive at a contradiction by assuming that $Z_{n,i}$ fails to be dense. □

Having obtained such a geometrical fact about convex sets we can, by the usual device of applying the fact to epigraphs, obtain a corresponding conclusion about convex functions.

Corollary. *Let A be a convex subset of the separable Banach space X and $f \in \mathrm{Conv}(A)$. If f is continuous at a relative interior point of A then f has a gradient at each point of a residual subset of* rel-int(A).

Proof. As usual, after passing to $\overline{\mathrm{aff}}(A)$, we can assume that f is continuous throughout int(A). The epigraph of f is then a solid closed convex set in the separable Banach space $X \times \mathbb{R}^1$, and so the smooth points constitute a residual subset of its boundary gr(f). There are therefore open sets $Z_n \subset X \times \mathbb{R}^1$ such that $\mathrm{sm}(\mathrm{epi}(f)) = \bigcap_n (Z_n \cap \mathrm{gr}(f))$ and $Z_n \cap \mathrm{gr}(f)$ is dense in gr(f). Let $P: X \times \mathbb{R}^1 \to X$ be the projection along $\mathbb{R}^1$: $P(x, t) \equiv x$. Then P is a continuous open mapping and $P(\mathrm{gr}(f)) = A$. It follows that the sets $P(Z_n \cap \mathrm{gr}(f))$ are dense and open in A, and hence that their intersection B is a residual subset of int(A). If $x \in B$, then $(x, f(x)) \in \bigcap_n (Z_n \cap \mathrm{gr}(f)) \equiv \mathrm{sm}(\mathrm{epi}(f))$. Thus there is a unique hyperplane of support to epi(f) at the point $(x, f(x))$. By **6D** it follows that there can be at most one subgradient of f at x. But by **14B**, $\partial f(x) \neq \varnothing$. Therefore, there is a unique subgradient of f at x which must, by **14D**—Cor. 1, be the gradient of f at x. This shows that f has a gradient at the points of the residual set B. □

The separability hypothesis in these results is crucial. Lacking this, the conclusion can fail completely. For example, the norm in the Banach spaces $L^\infty([0, 1], \mu, \mathbb{R})$ (μ = Lebesgue measure) and $\ell^1(\aleph)$ ($\aleph > \aleph_0$) is nowhere differentiable. Hence the unit balls of these spaces have no smooth points at all. However, it is known that the conclusions do hold for all reflexive spaces.

G. We now discuss a few miscellaneous topics related to the notion of smoothness. First is the observation that the properties of being strictly normed and smoothly normed are, in a sense, dual to one another.

Theorem. *Let X be a real normed linear space. If X^* is smoothly (resp. strictly) normed then X is strictly (resp. smoothly) normed.*

Proof. The proofs of both assertions are similar, so we shall just prove the first. Suppose that X is not strictly normed. Then the boundary $\partial U(X)$ contains a non-trivial line segment $[u, v]$. Let $\phi \in \partial U(X^*)$ be a support functional to $U(X)$ at the point $(u + v)/2$. Then the functionals $\hat{u}, \hat{v} \in X^{**}$ are both subgradients of the norm at the point ϕ, and so there cannot be a gradient of the norm in X^* at ϕ (**14D**). Hence X^* is not smoothly normed. □

Corollary. *Let X be a real reflexive Banach space. Then X is strictly (resp. smoothly) normed if and only if X^* is smoothly (resp. strictly) normed.*

Let A be an open convex set in a normed linear space X. Suppose that $f \in \operatorname{Conv}(A)$ is continuous and differentiable on A. Thus we have the gradient map $x \to \nabla f(x)$, defined from A into X^*. The following result establishes two basic properties of such a mapping: monotonicity (**3A**) and demicontinuity. In general, a mapping from a subset of X into X^* is *demicontinuous* if it is continuous from the norm topology into the weak*-topology. The kinds of convex functions to which we want to apply this result are those associated with the norm on a smoothly normed space; for example, $f(x) = \|x\|$ or $f(x) = \frac{1}{2}\|x\|^2$. But notice also that even in the finite dimensional case the theorem provides some new information by showing that a differentiable convex function (defined on an open subset of $\mathbb{R}^n$, say) is automatically continuously differentiable.

Theorem. *Let X be a normed linear space and f a continuous and differentiable convex function defined on an open convex set $A \subset X$. Then the gradient map $x \to \nabla f(x)$ is monotone and demicontinuous on A.*

Proof. The monotonicity inequality

$$\langle x - y, \nabla f(x) - \nabla f(y)\rangle \geqslant 0, \qquad x, y \in A,$$

is proved exactly as in **3A**, making use of the subgradient property $\langle u - v, \nabla f(v)\rangle \leqslant f(u) - f(v)$, for $u, v \in A$.

In order to prove that $f(\cdot)$ is demicontinuous at a given point $x_0 \in A$, we first note that there is an x_0-neighborhood V such that the restriction $f|V$ satisfies a uniform Lipschitz condition on V with constant λ (exercise 2.41). It follows that the restriction $\nabla f(\cdot)|V$ is a bounded mapping:

$$\|\nabla f(x)\| \leqslant \lambda, \qquad x \in V. \tag{20.6}$$

Now suppose that $\lim_n x_n = x_0$ for some sequence $\{x_n\} \subset V$. We shall prove that weak*-$\lim_n \nabla f(x_n) = \nabla f(x_0)$ by showing that the sequence $\{\nabla f(x_n)\}$ has the unique weak*-cluster point $\nabla f(x_0)$. Because of the weak*-compactness of $\{\nabla f(x_n)\}$ guaranteed by (20.6), there is some weak*-cluster point $\phi \in X^*$. To see that $\phi = \nabla f(x_0)$ it must be shown that $\phi(y - x_0) \leqslant f(y) - f(x_0)$ for all $y \in A$. Now

$$\begin{aligned}\phi(y - x_0) = \langle y - x_0, \phi - \nabla f(x_n)\rangle + \langle y - x_n, \nabla f(x_n)\rangle \\ + \langle x_n - x_0, \nabla f(x_n)\rangle, \qquad y \in X.\end{aligned} \tag{20.7}$$

Select some $y \in A$ and any $\varepsilon > 0$. Then there is a sequence $\{n_k\}$ of positive integers such that $|\langle y - x_0, \phi - \nabla f(x_{n_k})\rangle| < \varepsilon$, $|f(x_{n_k}) - f(x_0)| < \varepsilon$ and $\lambda\|x_{n_k} - x_0\| < \varepsilon$. Hence, from (20.7),

$$\begin{aligned}\phi(y - x_0) &< \langle y - x_{n_k}, \nabla f(x_{n_k})\rangle + 2\varepsilon \\ &\leqslant f(y) - f(x_{n_k}) + 2\varepsilon < f(y) - f(x_0) + 3\varepsilon.\end{aligned}$$

This proves that ϕ is a subgradient of f at x_0 and hence, by uniqueness of subgradients (**14D**), that $\phi = \nabla f(x_0)$. □

Example. Let X be a smoothly normed space. Then the norm gradient mapping $x \mapsto g(x; \cdot)$ is defined by (20.4) from $\partial U(X)$ into $\partial U(X^*)$. From **7F** it follows that this mapping is positively homogeneous of degree zero:

$$g(\alpha x; \cdot) = g(x; \cdot), \qquad \alpha > 0, \|x\| = 1.$$

Hence we may consider that g is defined on the open set $X \backslash \{\theta\}$, where it is consequently monotone and demicontinuous. The range of the norm gradient consists of certain kinds of extreme points of $U(X^*)$; this idea is developed further in exercise 3.57.

Now let $f(x) = \frac{1}{2}\|x\|^2$. By the chain rule, f is differentiable on all of X and

$$\textbf{(20.8)} \qquad \nabla f(x) = \begin{cases} \|x\| g(x; \cdot), & x \in X, x \neq \theta, \\ \theta & x = \theta. \end{cases}$$

The mapping (20.8) is called the *norm-duality map* and will be denoted by T. Again, T is monotone and demicontinuous. Further, we can assert that $\text{range}(T) = \mathscr{P}(X)$ (**20E**). Hence range(T) is dense in X^* whenever X is complete but it equals X^* only when X is reflexive (James' theorem). Finally, it is easy to see that T is injective exactly when X is strictly normed. Thus, when X is a reflexive Banach space which is both smoothly and strictly normed, it follows that T is a bijection between X and X^*. It may also happen, but not necessarily, that T is a homeomorphism (see exercise 3.58). □

H. Let M be a linear subspace of a normed space X. We know from the Hahn-Banach theorem that any $\phi \in M^*$ has a norm-preserving extension $\bar{\phi}$ in X^*. A most interesting question pertains to the uniqueness of $\bar{\phi}$. We shall study this question briefly, making use of the concepts of rotundity and smoothness. Let us say that M has *property* (U) if every $\phi \in M^*$ admits exactly one norm-preserving extension to all of X. We may as well confine ourselves to closed subspaces in X, since a subspace M has property (U) if and only if $\bar{M}$ has property (U).

Example. The simplest space where property (U) always occurs is Euclidean n-space, that is, $\mathbb{R}^n$ normed by the 2-norm. Any subspace M of $\mathbb{R}^n$ has an orthogonal complementary subspace $M^\perp$ for which $M \oplus M^\perp = \mathbb{R}^n$. Let P_M be the corresponding projection of X on M. Then any $\phi \in M^*$ has the unique norm-preserving extension $\bar{\phi} \equiv \phi \circ P_M$. Indeed, any other such extension, $\bar{\bar{\phi}}$ say, could be represented as an inner product $\bar{\bar{\phi}}(x) = \langle x, z\rangle$, where $z = u + v$, $u \in M$, $v \in M$. By confining x to M we see that $\phi(x) =$

$\langle x, u \rangle$, whence $||\phi|| = ||u||$. But, $||\bar{\bar{\phi}}|| = ||z|| \equiv (||u||^2 + ||v||^2)^{1/2}$ and so $||\bar{\bar{\phi}}|| = ||\phi||$ requires $v = \theta$. That is, $\bar{\bar{\phi}} = \phi \circ P_M \equiv \bar{\phi}$. □

We shall now obtain a substantial generalization of this example. The basic fact here is due to Phelps. Let us say that a subspace N of a normed space is a *Chebyshev subspace* if it is proximinal and every point outside of N has a unique best approximation (**15B**) from N.

Theorem. *The subspace M of the normed linear space X has property (U) if and only if its annihilator M° is a Chebyshev subspace of X^*.*

Proof. If M does not have property (U) there exists some functional $\phi \in M^*$ with two distinct extensions ψ_1 and ψ_2 in X^*. Hence $\psi_1 - \psi_2$ is a non-zero element of M°. We claim that ψ_1 has two distinct best approximations from M°, namely θ and $\psi_1 - \psi_2$. To see this, recall from formula (16.9) that $d(\psi_1, M^\circ) = ||\psi_1|M|| \equiv ||\phi||$. Thus $||\psi_1 - \theta|| = ||\psi_1|| \equiv ||\phi|| = d(\psi_1, M^\circ)$, and $||\psi_1 - (\psi_1 - \psi_2|| = ||\psi_2|| = ||\phi||$, as well. Conversely, suppose that M° is not a Chebyshev subspace. Then there exists a functional $\psi \in X^*$ that has two distinct best approximations from M°, say ψ_1 and ψ_2. (We are implicitly using the fact that M° is proximinal, since it is weak*-closed (exercise 2.52).) After translating by ψ_1 we can assume that $\psi_1 = \theta$. Then ψ and $\psi - \psi_2$ have the same restriction to M (since their difference, ψ_2, belongs to M°), and $||\psi|| = d(\psi, M^\circ) = ||\psi|M|| = ||\psi - \psi_2||$. Therefore, M does not have property (U). □

From **15C** we recall that every convex subset of a strictly normed space contains at most one best approximation to each point. Conversely, it is easy to see that any normed space with this property must be strictly normed (consider one-dimensional convex subsets). With these remarks the theorem is seen to have as an immediate corollary the "Taylor-Foguel theorem".

Corollary. *Every subspace of a normed linear space X has property (U) if and only if X^* is strictly normed.*

I. Finally, we look at the problem of extending a single functional in a unique manner. We shall need the following lemma concerning convex functions, although only for the case of the norm function.

Lemma. *Let A be a solid convex subset of a locally convex space X. Let $f \in \operatorname{Conv}(A)$ be continuous at a point $p \in A$. Then the set*

$$\Gamma(f; p) \equiv \{x \in X : -f'(p; -x) = f'(p; x)\}$$

is a closed linear subspace of X, on which $f'(p; \cdot)$ is a continuous linear functional.

Proof. This follows from **7E** or **14D**, as either of these subsections shows that $x \in \Gamma(f; p)$ if and only if the functional $J_X(x)$ in $X^{*\prime}$ assumes a constant value on $\partial f(p)$. Thus $\Gamma(f; p)$ is the subspace of direction vectors in X for which the function f is differentiable. Furthermore, the directional derivative $f'(p; \cdot)$ being sublinear on X (**7D**), is continuous on X since

$f'(p; x) \leqslant f(p + x) - f(p)$, which is bounded for x in some θ-neighborhood. Therefore, its restriction to $\Gamma(f; p)$, where it is linear, is also continuous. $\square$

We now specialize to the case where X is a normed space and $f(x) = \|x\|$. For $x \neq \theta$ we define

$$\textbf{(20.9)} \qquad \Gamma_x = \left\{ y \in X : \lim_{t \to 0} \frac{\|x + ty\| - \|x\|}{t} \text{ exists} \right\},$$

and we let $\phi_x \in \Gamma_x^*$ be the linear functional whose value at y is the limit in (20.9). Thus for each x in X $(x \neq \theta)$, Γ_x is a smooth subspace of X and ϕ_x is the norm gradient for this subspace. We can now state a sufficient condition for unique extendability of linear functionals.

Theorem. *Let M be a subspace of X and suppose that $\phi \in M^*$ attains its norm at $x \in \partial U(M)$. Then ϕ has a unique norm-preserving extension to the subspace $\overline{M + \Gamma_x}$.*

Proof. Any norm-preserving extension $\bar{\phi}$ in $(\overline{M + \Gamma_x})$ must satisfy $\bar{\phi}|M = \phi$ and $\bar{\phi}|\Gamma_x = \phi_x$, so that $\bar{\phi}$ is uniquely determined on $M + \Gamma_x$ and hence on its closure. $\square$

Corollary. *If $\phi \in M^*$ attains its norm at a point in $\partial U(M)$ which happens to be a smooth point of $U(X)$, then ϕ has a unique norm-preserving extension to all of X.*

From this we can in turn observe that if X is smoothly normed then every reflexive subspace of X has property (U).

§21. Some Further Applications

In this section we discuss a variety of miscellaneous topics, some of which are direct applications of previous developments. All the topics here have been chosen on the basis of their usefulness and intrinsic interest.

A. We begin by establishing the Banach space version of the quotient theorem (**1G**), known as the "Sard quotient theorem". We are given normed linear spaces X, Y, and Z, with X and Y complete, and operators $S \in B(X, Y)$, $T \in B(X, Z)$, with S surjective.

Theorem. *Let the operators S and T satisfy in addition* $\ker(S) \subset \ker(T)$. *Then there exists a uniquely specified operator $R \in B(Y, Z)$ such that $T = R \circ S$.*

Proof. We introduce the 1–1 operators $\hat{S}: X/\ker(s) \to Y$ and $\hat{T}: X/\ker(T) \to Z$ (**16E**). By the inverse mapping theorem (**17F**) the linear mapping $\hat{S}^{-1}$ is bounded. Now to each coset $x + \ker(S)$ we make correspond the coset $x + \ker(T)$. This correspondence is well defined because of our hypothesis that $\ker(S) \subset \ker(T)$, and defines a linear map $P: X/\ker(S) \to X/\ker(T)$. Since $\|x + \ker(T)\| \equiv d(x, \ker(T)) \leqslant d(x, \ker(S)) \equiv \|x + \ker(S)\|$, we have $\|P\| \leqslant 1$. We can thus finally define $R = \hat{T} \circ P \circ \hat{S}^{-1}$ and easily verify that $T = R \circ S$. $\square$

Although a simple enough consequence of the inverse mapping theorem, this result plays an important role in the analysis of certain problems in approximation theory and numerical analysis. We indicate a prototypical application.

Example. Suppose that we have some sort of numerical formula that we wish to apply to functions defined on an interval $[a, b]$ (or perhaps some region of higher dimension). This formula is to be thought of as providing an approximation to a desired quantity. Thus if this desired quantity is the definite integral of a function, the formula may give us a prescribed linear combination of some values of the function (and perhaps of certain of its derivatives) at specified points in $[a, b]$. In this case the formula is usually called a quadrature rule. We have to have some reason for believing that the formula is going to be effective; let us suppose that the formula gives exactly the correct answer when applied to polynomials of degree $\leqslant n - 1$, say. The general problem is then to appraise the error when the formula is applied to functions other than such polynomials.

Let $C^n \equiv C^n([a, b], \mathbb{R})$ be the linear space of n-times continuously differentiable functions on the interval $[a, b]$, normed by

$$||x||^{(n)} \equiv \max\{||x||_\infty, ||x'||_\infty, \ldots, ||x^{(n)}||_\infty\}, \qquad x \in C^n.$$

(Convergence of a sequence in this norm thus means uniform convergence on $[a, b]$ of the functions together with that of their first n derivatives.) The completeness of C^n can be seen either directly or via the observation that C^n is isomorphic to the product space $\mathbb{R}^n \times C \equiv \mathbb{R}^n \times C([a, b], \mathbb{R})$ under the mapping

(21.1) $$S(x) \equiv (x(a), x'(a), \ldots, x^{(n-1)}(a), x^{(n)}), \qquad x \in C^n.$$

Now let $T \in C^{n*}$. Since S is an isomorphism, $\{\theta\} = \ker(S) \subset \ker(T)$ and so the quotient theorem applies. We thus obtain a functional $\Phi \in (\mathbb{R}^n \times C)^*$ such that $T = \Phi \circ S$. If Φ has the form

(21.2) $$\Phi((t_0, t_1, \ldots, t_{n-1}, f)) = \sum_{k=1}^{n-1} c_k t_k + \phi(f)$$

for $(t_0, t_1, \ldots, t_{n-1}) \in \mathbb{R}^n$ and $f \in C$, it follows from (21.1) and (21.2) that

(21.3) $$T(x) = \sum_{k=0}^{n-1} c_k x^{(k)}(a) + \phi(x^{(n)}).$$

Thus we have reduced the problem of determining the form of the continuous linear functionals on C^n to the corresponding problem for C. For the rest of this example we shall grant the validity of the remark made at the end of **20E**, namely that to any $\phi \in C^*$ corresponds a function g of bounded variation on $[a, b]$ such that

(21.4) $$\phi(f) = \int_a^b f(t)dg(t), \qquad f \in C.$$

Returning now to the error analysis problem above, let $T \in C^{n*}$ be the error functional. That is, for each $x \in C^n$, $T(x)$ is the difference between the true (but unknown) value at x and the approximate (but computable) value. We have assumed that $T(x) = 0$ whenever x is a polynomial of degree $\leqslant n - 1$. This condition can be interpreted as the condition that $\ker(T)$ contains the kernel of the linear map $x \mapsto x^{(n)}$ from C^n to C. Applying the quotient theorem we conclude that there exists $\phi \in C^*$ such that $T(x) = \phi(x^{(n)})$, $x \in C^n$. From the representation (21.4) we reach our final conclusion that there exists a function g of bounded variation on $[a, b]$ such that

(21.5) $$T(x) = \int_a^b x^{(n)}(t)dg(t), \qquad x \in C^n.$$

With the achievement of formula (21.5) the contribution of functional analysis to this problem is completed. However, it is clear that for the purposes of numerical analysis the problem is far from solved. The next step is to determine the nature of the integrator function g. For instance, due to additional information that may be available regarding the functional T, we might be able to conclude that g is absolutely continuous. In this case the Stieltjes integral in (21.5) becomes an ordinary Lebesgue integral of the product $x^{(n)}g'$. We could then estimate the error in our approximation by

$$|T(x)| \leqslant ||g'||_1||x^{(n)}||_\infty,$$

or perhaps by

$$|T(x)| \leqslant ||g'||_2||x^{(n)}||_2,$$

if we could be sure that g' is square integrable. If our original goal were to design an optimal approximation formula we might be led, in view of the preceding estimates, to the optimization problem of selecting g so as to minimize either $||g'||_1$ or $||g'||_2$ over a certain class of formulas. However, even the mere computation of these quantities can be difficult, and so we shall leave the problem at this point. □

B. We consider next a special type of optimization problem known as an *abstract spline problem*. Given are two real Banach spaces X and Y, an operator $R \in B(X, Y)$, and a subset K of X. An *R-spline in K* is by definition any solution of the program

(21.6) $$\min\{||R(x)||: x \in K\}.$$

It is usually assumed that K is disjoint from the kernel of R, as otherwise points in the intersection would yield trivial solutions to (21.6).

In all cases of interest the set K is defined by means of a second operator $T \in B(X, Z)$, for some normed space Z and some prescribed subset Γ of Z:

(21.7) $$K = \{x \in X: T(x) \in \Gamma\} \equiv T^{-1}(\Gamma).$$

The most common situations occur when Γ is a singleton, so that K is an affine subspace of X, or when $\Gamma = z_0 + C$ for some cone $C \subset Z$. In this latter case K appears in the form $\{x \in X: T(x) \geqslant z_o\}$, where the inequality refers to the ordering induced on Z by C. As usual, the case $Z = \mathbb{R}^n$ for some n is of special importance. In this case T is defined by a subset $\{\phi_1, \ldots, \phi_n\} \subset X^*$:

$$T(x) = (\phi_1(x), \ldots, \phi_n(x)), \qquad x \in X.$$

Then K is either of the form $\{x \in X : \phi_i(x) = c_i,\ i = 1, \ldots, n\}$ (a finite co-dimensional flat) or $\{x \in X : \phi_i(x) \geqslant c_i,\ i = 1, \ldots, n\}$ (a *polyhedron*).

The operator R is most commonly a linear differential operator of the form

$$\textbf{(21.8)} \qquad R = \sum_{k=0}^{m} a_k(t) \frac{d^k}{dt^k}, \qquad a_m(t) \neq 0, \qquad a_k \in C^k([a, b], \mathbb{R}),$$

on some interval $[a, b]$, and X and Y are accordingly function spaces of such a nature that $R \in B(X, Y)$. At first glance it might seem adequate to take $X = C^m([a, b], \mathbb{R})$ and $Y = C([a, b], \mathbb{R})$. This choice fails to be satisfactory because of the nature of Y in this case: it is not reflexive and in fact not even a conjugate space (**13B**, **13E**, —Ex. 1), hence it will be difficult to guarantee solutions to our basic optimization problem (21.6). Instead we set $Y = L^p \equiv L^p([a, b], \mu, \mathbb{R})$ where μ is Lebesgue measure and $1 < p < \infty$. Then for X we might take the space

$$\textbf{(21.9)} \qquad \begin{aligned} H_p^m \equiv \{ & f \in C^{m-1}([a, b], \mathbb{R}) : f^{(m-1)} \text{ is absolutely continuous} \\ & \text{and } f^{(m)} \in L^p\}. \end{aligned}$$

The spaces H_p^m for $m = 1, 2, \ldots,$ and $1 < p < \infty$ are known as *Sobolev spaces* and are indeed Banach spaces under a variety of norms; for example

$$\|f\|_p^{(m)} \equiv \sum_{i=1}^{m} |f(t_i)| + \|f^{(m)}\|_p$$

where $\{t_1, \ldots, t_m\}$ is a set of distinct points in $[a, b]$.

The classical case occurs where $m = p = 2$, $R = d^2/dt^2$, and n data points $(t_1, c_1), \ldots, (t_n, c_n)$ are given. Here $a \leqslant t_1 < \cdots < t_n \leqslant b$ and the c_i are arbitrary. In this case a solution of the program (21.6) will be a *smoothest interpolant of the data*, that is, a function f satisfying

$$\textbf{(21.10)} \qquad \begin{aligned} & f \in H_2^2, \qquad f(t_i) = c_i, \qquad i = 1, \ldots, n, \\ & \textstyle\int_a^b (f''(t))^2 dt = \min. \end{aligned}$$

This classical problem is only of interest when $n \geqslant 3$, since when $n = 2$ there is a unique polynomial of degree one that interpolates the given data and obviously the minimum in (21.10) is zero in this case. When $n \geqslant 3$ the solution of (21.10) is known to be a *natural cubic spline*, that is, a function of class C^2 whose restriction to each sub-interval (t_i, t_{i+1}) is a cubic polynomial, $i = 1, \ldots, n - 1$, and which in addition reduces to a first degree polynomial in each of the intervals $[a, t_1)$, $(t_n, b]$.

With this background we return to the abstract program (21.6). We shall assume that Y is reflexive and that the range of the operator R is closed in Y. Since any closed subspace of a reflexive space is again reflexive (**16F**) we can replace Y by range(R) and therefore assume that R is surjective. From **16I** we know that every closed convex subset of a reflexive space is proximinal. Hence if the set K is convex and if $R(K)$ is closed in Y we can be sure of the existence of an R-spline in K.

Theorem. *Let Y be reflexive and let K be a closed convex subset of X. Suppose that the operator R has a finite dimensional kernel N which is disjoint from the recession cone of K: $N \cap C_K = \{\theta\}$. Then there exists an R-spline in K.*

Proof. The image $R(K)$ is convex by exercise 1.6 so we must show that it is closed in Y. According to **17H** $R(K)$ is closed exactly when $K + N$ is closed in X. Now N, being a finite dimensional linear subspace of X, is locally compact and equals its own recession cone. We have assumed that $N \cap C_K = \{\theta\}$; hence all the hypotheses of the lemma in **15D** are satisfied and so we may conclude that $K + N$ is indeed closed. □

In the important special case where K is an affine subspace parallel to some linear subspace M, an R-spline in K will always exist provided that $\dim(N) < \infty$. This is because the condition for $R(K)$ to be closed, namely that $M + N$ be closed, is automatically satisfied (**9E**). The hypothesis that the operator R should have a finite dimensional kernel is suggested (and certainly satisfied) by linear differential operators of the form (21.8).

Corollary *Let Y be reflexive and assume that R has finite dimensional kernel N. Suppose that K has the form (21.7) for some surjective operator $T \in B(X, Z)$ and some closed convex set $\Gamma \subset Z$. If $M \cap N = \{\theta\}$, where $M \equiv \ker(T)$, and if $C_\Gamma \cap T(N) = \{\theta\}$, then an R-spline in K exists.*

Proof. We must verify that $C_K \cap N = \{\theta\}$, for then the preceding theorem can be applied. Now, using the surjectivity of T, it is easy to see that $T(C_K) = C_\Gamma$. Then $T(C_K \cap N) \subset C_\Gamma \cap T(N) = \{\theta\}$, whence $C_K \cap N \subset M \cap N = \{\theta\}$. □

We finally consider a class of closed convex sets to which the preceding theorem need not apply. These sets constitute the examples promised in **17H** of non-compact sets whose continuous linear image in a Banach space is always closed.

Example. Let X and Y be real Banach spaces and $R \in B(X, Y)$ a surjective operator. If K is any polyhedron in X then $R(K)$ is closed in Y, and so an R-spline in K exists if Y is reflexive. To prove that $R(K)$ is closed we note first that if $y \in \overline{R(K)}$ then the flat $R^{-1}(y)$ is at zero-distance from K; this observation uses only the fact that the 1–1 operator $\hat{R}$ is an isomorphism and does not depend on the special nature of K. It remains to show that $K \cap R^{-1}(y) \neq \varnothing$.

Let $K = \{x \in X : \phi_i(x) \geqslant c_i,\ i = 1, \ldots, n\}$ for appropriate $\phi_i \in X^*$ and $c_i \in R$. Let $R(\bar{x}) = y$ and $N = \ker(R)$. Setting $c_i' = c_i - \phi_i(\bar{x})$, $i = 1, \ldots, n$, we see that $K \cap R^{-1}(y) \neq \varnothing$ is equivalent to the consistency of the inequality system

(21.11)

$$\begin{aligned} \phi_1(x) &\geqslant c_1', \\ &\vdots \\ \phi_n(x) &\geqslant c_n', \qquad x \in N. \end{aligned}$$

We can prove the consistency of (21.11) by applying Fan's condition (**7B**). Suppose that $\alpha_1, \ldots, \alpha_n$ are non-negative numbers for which $\sum_1^n \alpha_i \phi_i | N = \theta$. Since the sets K and $R^{-1}(y)$ are at zero-distance there are sequences $\{k_j) \subset K$, $\{z_j\} \subset N$, such that $\lim_j \|k_j - \bar{x} - z_j\| = 0$. Then

$$\lim_j |\phi_i(k_j - \bar{x} - z_j)| = 0, \qquad i = 1, \ldots, n,$$

whence

(21.12)
$$\liminf_{j \to \infty} \phi_i(z_j) \geqslant c_i', \qquad i = 1, \ldots, n.$$

Now, given $\varepsilon > 0$, it follows from (21.12) that when j is sufficiently large

$$0 = \sum_{i=1}^n \alpha_i \phi_i(z_j) \geqslant \sum_{i=1}^n \alpha_i (c_i' - \varepsilon);$$

consequently

$$\varepsilon \sum_{i=1}^n \alpha_i \geqslant \sum_{i=1}^n \alpha_i c_i'.$$

Since ε is arbitrary we have shown that $\sum_1^n \alpha_i c_i' \leqslant 0$ and so Fan's condition is satisfied. This in turn proves that the system (21.11) is consistent. □

C. Let Ω and X be sets. A mapping $F: \Omega \to 2^X$ is called a *carrier*. Intuitively a carrier is a "multivalued function" on Ω, assigning a certain subset of X to each point in Ω. In this and the following subsection we are going to study some properties of carriers defined on a certain kind of topological space, and taking values in the family of convex subsets of a Banach space. The theorems we prove concerning the existence of selections and fixed points are extremely powerful and useful, and their range of applicability appears to be limited only by the ingenuity of the user.

Basic to our work is the availability of partitions of unity on normal topological spaces. We recall that a *partition of unity* on a topological space Ω is a family $\{p_\alpha : \alpha \in I\}$ of nonnegative continuous functions on Ω, such that all but a finite number of these functions vanish on some neighborhood of each point in Ω, and

$$\sum_{\alpha \in I} p_\alpha(t) = 1, \qquad t \in \Omega.$$

A partition of unity $\{p_\alpha : \alpha \in I\}$ is *subordinate* to a given covering of Ω if each p_α vanishes outside some member of the covering. We also say that a family $\mathscr{U}$ of subsets of Ω is *locally finite* if each point in Ω has a neighborhood that intersects only finitely many members of $\mathscr{U}$.

Lemma 1. *Let $\{V_\alpha : \alpha \in I\}$ be a locally finite open covering of a normal space Ω. Then there exists a partition of unity $\{p_\alpha : \alpha \in I\}$ which is subordinate to this covering.*

The proof is achieved by shrinking the covering $\{V_\alpha\}$ to obtain a new open covering $\{W_\alpha : \alpha \in I\}$ such that $\overline{W}_\alpha \subset V_\alpha, \alpha \in I$, and then using Urysohn's lemma to obtain continuous functions $q_\alpha : \Omega \to [0, 1]$ such that

$$q_\alpha(t) = \begin{cases} 1, & t \in W_\alpha \\ 0, & t \in \Omega \backslash V_\alpha. \end{cases}$$

Then the definition

$$p_\alpha(t) = \frac{q_\alpha(t)}{\sum_{\alpha \in I} q_\alpha(t)}$$

yields the functions making up the desired partition of unity.

If $\mathscr{U}$ and $\mathscr{V}$ are coverings of a space Ω, $\mathscr{V}$ is said to be a *refinement* of $\mathscr{U}$ if each member of $\mathscr{V}$ is contained in some member of $\mathscr{U}$. Then a Hausdorff space Ω is *paracompact* if every open covering of Ω has an open, locally finite refinement. It is known from topology that metric spaces and compact (Hausdorff) spaces are paracompact, and that every paracompact space is normal.

Let Ω and X be topological spaces. A carrier $F : \Omega \to 2^X$ is *lower semi-continuous* if $\{t \in \Omega : F(t) \cap \mathscr{O} \neq \varnothing\}$ is open in Ω, for every open set $\mathscr{O} \subset X$. When F is an ordinary (single-valued) mapping from Ω to X this definition reduces to the usual requirement of continuity. Observe that if the carrier F is lower semicontinuous and if $\lim\{t_\delta : \delta \in D\} = t$ in Ω, then for each $x \in F(t)$ there exists $x_\delta \in F(t_\delta)$ such that $\lim\{x_\delta : \delta \in D\} = x$.

Before starting our main result we establish a technical lemma concerning convex-valued carriers on paracompact spaces.

Lemma 2. *Let Ω be a paracompact space, X a normed linear space, and $F : \Omega \to 2^X$ a lower semicontinuous carrier whose values are non-empty convex subsets of X. Then if $r > 0$ there exists a continuous map $f : \Omega \to X$ such that $d(f(t), F(t)) < r, t \in \Omega$.*

Proof. For each $x \in X$ let

$$\mathscr{O}_x = \{t \in \Omega : d(x, F(t)) < r\}.$$

These sets $\mathscr{O}_x$ are open in Ω because of the lower semicontinuity of F and therefore $\{\mathscr{O}_x : x \in X\}$ is an open cover of Ω. Hence there exists an open locally finite refinement $\{V_\alpha : \alpha \in I\}$. Let $\{p_\alpha : \alpha \in I\}$ be a partition of unity subordinate to this refinement. Then if for each $\alpha \in I$ we select $x(\alpha) \in X$ so that $V_\alpha \subset \mathcal{O}_{x(\alpha)}$, the desired map f can be defined by

$$f(t) = \sum_{\alpha \in I} p_\alpha(t) x(\alpha), \qquad t \in \Omega.$$

To see that this sum is well-defined and that f is continuous we can note that each $t \in \Omega$ has a neighborhood U which intersects only a finite number of the V_α, and so, on U, f is a finite sum of continuous functions. Thus f is continuous on a neighborhood of each point in Ω, and hence is continuous on all of Ω. Finally, for each $t \in \Omega$, $f(t)$ is a convex combination of points $x(\alpha)$ each of which belongs to the convex set $\{x \in X : d(x, F(t)) < r\}$; it follows that $d(f(t), F(t)) < r$ also. □

Given a carrier $F:\Omega \to 2^X$ a *selection* for F is a mapping $f:\Omega \to X$ such that

$$f(t) \in F(t), \qquad t \in \Omega.$$

The following "Michael selection theorem" asserts the existence of a continuous selection for certain kinds of carriers.

Theorem. *Let Ω be a paracompact space and X a Banach space. If F is a lower semicontinuous carrier on Ω whose values are non-empty closed convex subsets of X, then there is a continuous selection $f:\Omega \to X$ for F.*

Proof. We shall inductively construct a sequence of continuous functions $f_i:\Omega \to X$ such that, for each $t \in \Omega$,

$$\textbf{(21.13)} \qquad \begin{aligned} &\text{a) } \|f_i(t) - f_{i-1}(t)\| < 2^{-i+2}, \qquad i = 2, 3, \ldots, \\ &\text{b) } d(f_i(t), F(t)) < 2^{-i}, \qquad i = 1, 2, \ldots. \end{aligned}$$

This will suffice for the proof, because by a) the sequence is uniformly Cauchy and so converges to a continuous $f:\Omega \to X$; by b) we have $f(t) \in F(t)$ for all $t \in \Omega$. That is, f is a continuous selection for F.

The existence of f_1 satisfying b) follows immediately from Lemma 2. Suppose that $f_1, \ldots, f_n$ have been constructed to satisfy a) and b) for $i = 1, \ldots, n$. We shall construct f_{n+1} so as to also satisfy a) and b).

We define a new carrier F_{n+1} on Ω by

$$F_{n+1}(t) = \{x \in F(t): \|x - f_n(t)\| < 2^{-n}\}, \qquad t \in \Omega.$$

By the induction hypothesis $F_{n+1}(t) \neq \varnothing$, $t \in \Omega$. We claim that F_{n+1} is lower semicontinuous. To see this, let $\mathcal{O}$ be an open set in X and let $U = \{t \in \Omega: F_{n+1}(t) \cap \mathcal{O} \neq \varnothing\}$; we show that each $t_0 \in U$ has a neighborhood contained in U, so that U is open in Ω. Given $t_0 \in U$, select a positive $\lambda < 2^{-n}$ so that $Q \equiv \{x \in X: \|x - f_n(t_0)\| < \lambda\} \neq \varnothing$ Then if

$$\begin{aligned} V_1 &\equiv \{t \in \Omega: F(t) \cap \mathcal{O} \cap Q \neq \varnothing\}, \\ V_2 &\equiv \{t \in \Omega: \|f_n(t) - f_n(t_0)\| < 2^{-n} - \lambda\}, \end{aligned}$$

the set V_1 is open because F is lower semicontinuous, V_2 is open because f_n is continuous, and $t_0 \in (V_1 \cap V_2) \subset U$.

Finally, we apply Lemma 2 to the carrier F_{n+1} and obtain a continuous function $f_{n+1}: \Omega \to X$ such that

$$d(f_{n+1}(t), F_{n+1}(t)) < 2^{-n-1}, \qquad t \in \Omega.$$

But now

$$\|f_{n+1}(t) - f_n(t)\| < 2^{-n-1} + 2^{-n} < 2^{-n+1}$$

which is a), and

$$d(f_{n+1}(t), F(t)) \leqslant d(f_{n+1}(t), F_{n+1}(t)) < 2^{-n-1}$$

which is b). □

It is interesting to remark that the Michael selection theorem admits a converse: if Ω is a Hausdorff space with the property that there exists a continuous selection for every lower semicontinuous carrier on Ω whose values are non-empty closed convex subsets of some Banach space, then Ω is paracompact. The proof is based on the fact (also due to Michael) that Ω is paracompact if to any given open covering of Ω there is a subordinate partition of unity. In terms of the given covering a special Banach space and lower semicontinuous carrier are constructed, and the assumed existence of a continuous selection leads immediately to the desired partition of unity (exercise 3.64).

As an application of the selection theorem we shall establish a useful result known as the "theorem of Bartle and Graves". The setting is a pair of Banach spaces X, Y, and a surjective operator $T \in B(X, Y)$. A *right-inverse* of T is an operator $S \in B(Y, X)$ such that TS is the identity on Y. In this case the operator ST is a projection on X since

$$(ST)^2 = (ST)(ST) = S(TS)T = ST.$$

Since $\ker(ST) = \ker(T)$ it follows that $I - ST$ is a projection of X onto $\ker(T)$. Hence a right-inverse of T can exist only if $\ker(T)$ is a topological direct summand of X. Since an arbitrary (closed) linear subspace of a Banach space need not be a topological direct summand (**22F**), we cannot expect a right inverse to exist in general. What we can always find is a (continuous) *cross-section* of T, that is, a continuous but not necessarily linear map $f: Y \to X$ such that $T(f(y)) = y$, for all $y \in Y$. We can also impose some additional requirements on f, as we see next.

Corollary. *Let X and Y be Banach spaces and let $T \in B(X, Y)$ be surjective. For each $\lambda > 1$ there exists a continuous and homogeneous cross-section f of T such that*

(21.14) $$\|f(y)\| \leqslant \lambda \inf\{\|x\| : T(x) = y\}.$$

Proof. By the open mapping theorem (**17G**) T is open. Hence the carrier F defined on Y by $F(y) \equiv T^{-1}(y)$ is lower semicontinuous. Any continuous selection f of F will therefore be a continuous cross-section of F. In order to obtain a cross-section with the specified additional properties we restrict F to the set $\partial U(Y)$ of all unit vectors in Y; call this restriction F_1.

Let $\gamma(y) = \lambda \inf\{\|x\|: T(x) = y\}$. Since the 1–1 operator $\hat{T}$ is an isomorphism, γ is a continuous function on Y. Define a new carrier F_2 on $\partial U(Y)$ by

$$F_2(y) = \begin{cases} F_1(y) \cap \{x \in X: \|x\| < \gamma(y)\}, & \gamma(y) > 0 \\ \{\theta\}, & \gamma(y) = 0. \end{cases}$$

This carrier is lower semicontinuous and hence so is the carrier F_3 defined by $F_3(y) \equiv \overline{F_2(y)}$. Let g be a continuous selection for F_3 and then define

$$h(y) = \begin{cases} \|y\| g\left(\dfrac{y}{\|y\|}\right), & y \neq \theta \\ \theta, & y = \theta. \end{cases}$$

This function h meets all our requirements except that it is only positively homogeneous: $h(ty) = th(y)$, $t \geqslant 0$.

To satisfy the remaining requirement of homogeneity we distinguish between the cases where the underlying scalar field is real or complex. In the real case we simply define $f(y) = (h(y) - h(-y))/2$; this function meets all our requirements. The complex case is a bit more subtle. We define

$$f(y) = \frac{1}{2\pi} \int_0^{2\pi} e^{-it} h(e^{it} y)\, dt, \qquad y \in Y. \tag{21.15}$$

For each $y \in Y$ the integrand in (21.15) is a continuous 2π-periodic function from $\mathbb{R}$ to X, so that the integral can be defined in the expected manner, namely as the norm-limit of Riemann approximating sums:

$$f(y) = \frac{1}{2\pi} \lim_n \sum_{j=1}^{n} e^{it'_j} h(e^{it'_j})(t_j - t_{j-1}), \tag{21.16}$$

where $0 = t_0 < t_1 < \cdots < t_{j-1} < t'_j < t_j < \cdots < t_n = 2\pi$. The existence of the limit is assured by the completeness of X. Since $\sum_1^n (t_j - t_{j-1})/2\pi = 1$ we see from (21.16) that $f(y)$ belongs to the closed convex hull of the set $\{e^{-it} h(e^{it} y): t \in \mathbb{R}\}$; hence f is a cross-section of T that satisfies (21.14). The continuity of f follows from that of h: if $\lim_n y_n = y_0$ in Y then

$$\lim_n e^{-it} h(e^{it} y_n) = e^{-it} h(e^{it} y_0)$$

uniformly in t, because h is uniformly continuous on the compact set $\{\alpha y_n : n = 0, 1, 2, \ldots, \alpha \in \mathbb{C}, |\alpha| = 1\}$. Finally, because of the periodicity of the integrand we have, for $s \in \mathbb{R}$,

$$\begin{aligned} f(e^{is} y) &\equiv \frac{1}{2\pi} \int_0^{2\pi} e^{-it} h(e^{i(s+t)} y)\, dt \\ &= \frac{e^{is}}{2\pi} \int_0^{2\pi} e^{-i(s+t)} h(e^{i(s+t)} y)\, dt \\ &= \frac{e^{is}}{2\pi} \int_0^{2\pi} e^{-it} h(e^{it} y)\, dt = e^{is} f(y); \end{aligned} \tag{21.17}$$

since f is already positively homogeneous (because h is) we see from (21.17) that f is homogeneous, as desired. □

D. We are now going to establish a fixed point theorem for certain kinds of carriers. If Ω is a set and $F:\Omega \to 2^{\Omega}$ is a carrier, a point $t_0 \in \Omega$ for which $t_0 \in F(t_0)$ is called a *fixed point* of F. This is a natural generalization of the usual notion (**15C**) for single-valued mappings. Fixed point theorems for carriers, or multivalued mappings, have important applications to game theory, mathematical economics, non-linear programming, and to boundary value problems for certain kinds of partial differential equations. As a particular application of the fixed-point theorem below we shall prove the existence of solutions to a so-called "variational inequality". Since many problems can be cast into the form of such an inequality, this result is a useful adjunct to the fixed-point theorem.

If Ω and X are topological spaces a carrier $F:\Omega \to 2^X$ is *upper semicontinuous* if $\{t \in \Omega : F(t) \subset \mathcal{O}\}$ is open in Ω for every open set $\mathcal{O} \subset X$. Again, this definition reduces to that of ordinary continuity when F is a single-valued mapping. We now have the "Fan-Kakutani fixed-point theorem".

Theorem. *Let K be a compact convex subset of a locally convex space X. Let $F:K \to 2^K$ be an upper semicontinuous carrier whose values are non-empty closed convex subsets of K. Then there exists $x_0 \in K$ with $x_0 \in F(x_0)$.*

Proof. Let $\{U_\alpha : \alpha \in I\}$ be a local base in X consisting of absolutely convex open sets. For each index $\alpha \in I$ there exists a finite set $\{x_{\alpha\beta} : \beta \in J(\alpha)\} \subset K$ such that $K \subset \cup\{x_{\alpha\beta} + U_\alpha : \beta \in J(\alpha)\}$. Let $\{p_{\alpha\beta} : \beta \in J(\alpha)\}$ be a partition of unity subordinate to this covering of K (**21C**). Choose $y_{\alpha\beta}$ in $F(x_{\alpha\beta})$ arbitrarily and define the function $f_\alpha : K \to X$ by

$$f_\alpha(x) = \sum_{\beta \in J(\alpha)} p_{\alpha\beta}(x) y_{\alpha\beta}.$$

Now the set $C_\alpha \equiv \text{co}\{y_{\alpha\beta} : \beta \in J(\alpha)\}$ is a finite dimensional compact convex set to which the classical Brouwer fixed-point theorem (**15C**) applies. Hence, since $f_\alpha(C_\alpha) \subset C_\alpha$, there is a point $x_\alpha \in C_\alpha$ with $f_\alpha(x_\alpha) = x_\alpha$.

To produce the desired fixed-point of F we note that the correspondence $U_\alpha \to x_\alpha$ defines a net in K, since the local base $\{U_\alpha : \alpha \in I\}$ is directed (downward) by inclusion. Let $x_0 \in K$ be any cluster point of this net, and suppose that $x_0 \notin F(x_0)$. By the strong separation theorem there is a closed convex neighborhood W of $F(x_0)$ with $x_0 \notin W$. Since F is upper semicontinuous there exists an x_0-neighborhood V such that $F(x) \subset W$ whenever $x \in K \cap V$; clearly we may also assume that $V \cap W = \varnothing$. Choose an index $\gamma \in I$ so that $U_\gamma + U_\gamma \subset V - x_0$. Then, by definition of x_0, there exists $\alpha \in I$ with $U_\alpha \subset U_\gamma$, so that $x_\alpha \in x_0 + U_\gamma$; hence $x_\alpha + U_\alpha \subset V$. Finally, if $p_{\alpha\beta}(x_\alpha) \neq 0$ for any $\beta \in J(\alpha)$ then $x_\alpha \in x_{\alpha\beta} + U_\alpha$, so that $x_{\alpha\beta} \in V$. Hence $y_{\alpha\beta} \in W$ and so

$$x_\alpha = f_\alpha(x_\alpha) = \sum_{\beta \in J(\alpha)} p_{\alpha\beta}(x_\alpha) y_{\alpha\beta} \in W;$$

however, this contradicts $x_\alpha \in V$. We have thus proved that $x_0 \in F(x_0)$. □

Before giving the application to variational inequalities we introduce a new topology on the conjugate space of a given locally convex space X. Let $\mathscr{V}$ be the family of all weak*-closed barrels in X^*. According to **10A** $\mathscr{V}$ is a local base for a unique locally convex topology on X^* called the *strong topology* on X^*. According to exercise 3.67 a net in X^* converges strongly (that is, in the strong topology) only if it converges uniformly on each bounded subset of X. The converse is also true but we shall not need it. Also according to this same exercise, the strong topology on the conjugate of a normed space is just the usual norm topology there.

Corollary. *Let K be a compact convex subset of a real locally convex space X, and let $T:K \to X^*$ be strongly continuous. Then there exists $x_0 \in K$ such that*

(21.18) $$\langle x - x_0, T(x_0)\rangle \geqslant 0, \qquad x \in K.$$

Proof. We define a carrier $F:K \to 2^K$ by

$$F(x) \equiv \{z \in K: \langle z, T(x)\rangle = \min_{y \in K} \langle y, T(x)\rangle\}.$$

The values of F are clearly non-empty closed convex subsets of K, and evidently the fixed points of F (if any) are exactly the points x_0 in K for which (21.18) is valid. Thus, if we show that F is upper semicontinuous we can apply the preceding theorem and complete the proof.

Let $\mathcal{O}$ be a (relatively) open set in K and choose any $y_0 \in K$ such that $F(y_0) \subset \mathcal{O}$; we shall find a y_0-neighborhood N such that $F(y) \subset \mathcal{O}$ for all $y \in N$. Suppose that we can find an $\varepsilon > 0$ such that, with $\phi_0 \equiv T(y_0)$,

(21.19) $$\begin{cases} \sup_{z \in K} |\langle z, \phi - \phi_0\rangle| < \varepsilon \text{ implies} \\ \{x \in K: \phi(x) = \min \phi(K)\} \subset \mathcal{O}. \end{cases}$$

Then we can let

(21.20) $$N = y_0 + T^{-1}(\varepsilon K^a).$$

Since K^a, the absolute polar of K (**18D**), is by definition a strong θ-neighborhood in X^*, and since T is strongly continuous, (21.20) does define a y_0-neighborhood N such that $F(y) \subset \mathcal{O}$ for all $y \in N$. Thus it remains to establish (21.19).

We assert that there is an $\varepsilon_0 > 0$ such that

(21.21) $$x \in K \backslash \mathcal{O} \text{ implies } \phi_0(x) \geqslant \varepsilon_0 + \min \phi_0(K).$$

For otherwise there would be a sequence $\{x_n\} \subset K\backslash\mathcal{O}$ such that $\phi_0(x_n) < \dfrac{1}{n} +$ $\min \phi_0(K)$. Any cluster point of this sequence would be a point in $K\backslash\mathcal{O}$ at which ϕ_0 attains its minimum over K; this, however, is in contradiction to our assumption that $F(y_0) \subset \mathcal{O}$. Finally we show that $\varepsilon \equiv \varepsilon_0/3$ satisfies (21.19). If $\sup\{|\langle z, \phi - \phi_0 \rangle| : z \in K\} < \varepsilon_0/3$, then

$$\phi_0(z) - \frac{\varepsilon_0}{3} \leqslant \phi(z) \leqslant \phi_0(z) + \frac{\varepsilon_0}{3}, \qquad z \in K.$$

Hence, from (21.21)

$$\phi(z) \geqslant \phi_0(z) - \frac{\varepsilon_0}{3} \geqslant \frac{2\varepsilon_0}{3} + \min \phi_0(K)$$

$$\geqslant \frac{\varepsilon_0}{3} + \min \phi(K) > \min \phi(K), \qquad z \in K \backslash \mathcal{O}.$$

This proves the implication (21.19). □

The inequality (21.18) is called a variational inequality because of its interpretation when T is the gradient or "first variation" of a functional f which is to be minimized on the set K. In **14E** it was shown that when $f \in \mathrm{Conv}(K)$ and f has a gradient in X^* then a point $x_0 \in K$ is a solution of the program (K, f) exactly when x_0 is a solution of the program $(K, \nabla f)$. (If f is not convex it is still easy to see that any solution of the program (K, f) must also be a solution of $(K, \nabla f)$, although the converse may fail.) But if we let $T(x) \equiv \nabla f(x)$, $x \in K$, then solutions of this latter program are exactly solutions of (21.18).

Of course, the assumption in the corollary that K is compact does not leave the question of the existence of a minimum in much doubt, unless f is a badly behaved function. There is thus some interest in relaxing the compactness hypothesis; however, this is possible only at the cost of more stringent restrictions on the mapping T. We might mention also that the corollary can be extended in a different direction in that T can be allowed to be an upper semicontinuous carrier on K whose values are compact convex subsets of X^* (all topological statements about T refer to the strong topology on X^*).

E. As an illustration both of the use of the variational inequality and of its extension to certain non-compact sets, we consider a continuous mapping $T: \mathbb{R}^n \to \mathbb{R}^n$. The *complementarity problem* (determined by T) is to find a solution to the system

$$\begin{aligned} y &= T(x) \\ x \geqslant \theta, &\qquad y \geqslant \theta, \\ x \cdot y &= \theta. \end{aligned} \tag{21.22}$$

Since x and y are non-negative vectors in $\mathbb{R}^n$ the bottom line of (21.22) can be interpreted as either the requirement of orthogonality, $\langle x, y \rangle = 0$, or, as indicated, that the componentwise product of x and y is the zero vector. Thus the complementarity problem asks us to find a non-negative vector whose image is also non-negative and such that the two vectors are orthogonal.

The interest in the complementarity problem is that it provides a unified model for certain problems in several different fields such as optimization, game theory, economics, and mechanics. The most important cases lead to a *linear* complementarity problem wherein T is an affine mapping: $T(x) = Ax + b$, for some $n \times n$ matrix A and fixed vector $b \in \mathbb{R}^n$.

Example. Consider the problem of minimizing a convex function f over the positive cone $P \subset \mathbb{R}^n$: $P = \{x \in \mathbb{R}^n : x \geqslant \theta\} \equiv \{x = \{\xi_1, \ldots, \xi_n\}: \xi_1 \geqslant 0, \ldots, \xi_n \geqslant 0\}$. Such a problem would arise in particular if we were interested in approximating from a finite dimensional linear subspace of some normed space X, subject to non-negativity constraints on the coefficients. In this case f would have the form

$$f(\xi_1, \ldots, \xi_n) = \|u - \xi_1 u_1 - \cdots - \xi_n u_n\|,$$

for prescribed $u, u_1, \ldots, u_n \in X$. If f is differentiable then $x_0 \in P$ is a solution of the program (P, f) if and only if $\langle x_0, \nabla f(x_0)\rangle \leqslant \langle x, \nabla f(x_0)\rangle$, for all $x \in P$ (**14E**). It follows that $\nabla f(x_0) \in P$ and that $\langle x_0, \nabla f(x_0)\rangle = 0$. Thus, letting $T = \nabla f(\cdot)$, we are led to a complementarity problem. This problem will be linear exactly when f is a quadratic function: $f(x) = \frac{1}{2}\langle x, Ax\rangle + \langle x, b\rangle + c$, for symmetric A and fixed vectors $b \in \mathbb{R}^n$, $c \in \mathbb{R}$. □

It is easy to see that the complementarity problem (21.22) is equivalent to the variational inequality

$$\textbf{(21.23)} \qquad \langle x - x_0, T(x_0)\rangle \geqslant 0, \qquad x \geqslant \theta, \qquad x_0 \geqslant \theta.$$

That is, $x_0 \geqslant \theta$ solves (21.23) if and only if the pair $(x_0, y_0) \equiv (x_0, T(x_0))$ solves (21.22). However, because of the non-compactness of the cone P it is not so easy to decide on the solvability of either of these problems. Even when $T(x) = Ax + b$ it is only possible to establish the existence of a solution under rather specialized hypotheses on the matrix A. We shall give an existence-uniqueness theorem for the general complementarity problem which will apply in particular to the linear problem when A is positive definite.

Let D be a subset of $\mathbb{R}^n$. A mapping $T: D \to \mathbb{R}^n$ is *strongly monotone* if there exists a constant $\alpha > 0$ such that

$$\textbf{(21.24)} \qquad \langle x - y, T(x) - T(y)\rangle \geqslant \alpha\|x - y\|_2^2. \qquad x, y \in D.$$

This is clearly a strengthening of the concept of monotone mapping introduced in **3A**. In the single variable ($n = 1$) case any function g whose derivative g' satisfies $g'(x) \geqslant \alpha > 0$ for x in some interval D is strongly monotone on D. A generalization to $\mathbb{R}^n$ is given in exercise 3.68.

We know from **3A** that for a smooth function f defined on an open convex set D in $\mathbb{R}^n$, convexity of f is equivalent to the monotonicity of ∇f on D. A function f on D is *strongly convex* if there exists a constant $\alpha > 0$ such that

$$\textbf{(21.25)} \quad f(tx + (1 - t)y) \leqslant tf(x) + (1 - t)f(y) - \alpha t(1 - t)\|x - y\|_2^2,$$

for $x, y \in D$, $0 \leqslant t \leqslant 1$. Intuitively a strongly convex function is a convex function whose graph has positive curvature. Thus, in one variable, $f(x) = x^2$ is strongly convex while $f(x) = x^4$ is not. More generally, of all the p-norms on $\mathbb{R}^n$, only the 2-norm is strongly convex (whereas from **19E** we know that the p-norms are uniform norms on $\mathbb{R}^n$ for $p > 1$). Exercise 3.69 makes the connection between strongly convex functions and strongly monotone mappings: a smooth convex function is strongly convex if and only if its gradient is strongly monotone.

The following result on the solvability of the complementarity problem is due to Karamardian. All norms appearing in the proof are 2-norms.

Theorem. *The complementarity problem (21.22) has a unique solution if the mapping T is continuous and strongly monotone on the positive cone P.*

Proof. In (21.24) we let $y = \theta$:

$$\langle x, T(x)\rangle \geqslant \langle x, T(\theta)\rangle + \alpha\|x\|^2, \qquad x \geqslant \theta.$$

Let K be the compact convex set $\{x \in P : \|x\| \leqslant \|T(\theta)\|/\alpha\}$. Then for any $x \in P\backslash K$ we have

$$\alpha\|x\|^2 > \|x\|\,\|T(\theta)\| \geqslant -\langle x, T(\theta)\rangle,$$

whence

$$\textbf{(21.26)} \qquad \langle x, T(x)\rangle > 0, \qquad x \in P\backslash K.$$

Now for all $u \geqslant \theta$ let $D_u = \{x \in K : \langle u - x, T(x)\rangle \geqslant 0\}$. The sets D_u are closed and we claim that they have the finite intersection property. To verify this assertion select any finite subset $\{u_1, \ldots, u_m\}$ of P, and apply the corollary of **21D** to the compact convex set $D \equiv \operatorname{co}(K \cup \{u_1, \ldots, u_m\})$. The conclusion is that there exists an $x_0 \in D$ for which the variational inequality

$$\textbf{(21.27)} \qquad \langle x - x_0, T(x_0)\rangle \geqslant 0, \qquad x \in D,$$

holds. In particular, $\langle u_i - x_0, T(x_0)\rangle \geqslant 0$ for $i = 1, \ldots, m$. We further have $x_0 \in K$ since otherwise there would result a contradiction to (21.26) (the origin belongs to K, hence to D, so we can take $x = \theta$ in (21.27)). This proves that the family of closed sets $\{D_u : u \geqslant \theta\}$ has the finite intersection property. Since K is compact, $\cap\{D_u : u \geqslant \theta\} \neq \varnothing$; any point in this intersection solves the variational inequality (21.23). Hence the complementarity problem has a solution.

Finally, suppose that we have two solutions to (21.23), say x_1 and x_2. Then, since $0 \leqslant \langle x_1, T(x_2)\rangle$, $0 \leqslant \langle x_2, T(x_1)\rangle$, and $0 = \langle x_1, T(x_1)\rangle = \langle x_2, T(x_2)\rangle$, we have

$$0 \geqslant \langle x_1 - x_2, T(x_1) - T(x_2)\rangle \geqslant \alpha\|x_1 - x_2\|^2,$$

whence $x_1 = x_2$. □

This theorem can be generalized in various ways. First, we can replace the usual positive cone P by an arbitrary closed (convex) cone C in $\mathbb{R}^n$. Under the same hypotheses on T it can be shown that there exists exactly one

$x_0 \in C$ for which $T(x_0) \in C^*$ $(\equiv -C^\circ)$ and $\langle x_0, T(x_0)\rangle = 0$. Second, we can replace $\mathbb{R}^n$ by a real reflexive Banach space X. If C is a closed (convex) cone in X and T is a mapping from C into X^* satisfying the condition of strong monotonicity (21.24), and continuous from the norm to the weak* topology, then we can obtain the same conclusion. The proof proceeds via the observation that any solution $x_0 \in C$ of the variational inequality $\langle x - x_0, T(x_0)\rangle \geqslant 0$, $x \in C$, must satisfy $\|x_0\| \leqslant \|T(\theta)\|$, independently of C. Then, given any finite dimensional linear subspace M of X, the preceding theory is applied to obtain a solution x_M of the finite dimensional problem with cone $C \cap M$. Since $\|x_M\| \leqslant \|T(\theta)\|$, the bounded net $\{x_M : \dim M < \infty\}$ has a weak cluster point in C, and this cluster point turns out to solve the general problem.

Exercises

3.1. Let X be a normed linear space.
   a) Show that X is complete if and only if M and X/M are complete for some, and hence every, closed linear subspace M of X.
   b) A series $\sum_1^\infty x_n$ with $x_n \in X$ is *absolutely convergent* if $\sum_1^\infty \|x_n\| < \infty$. Show that X is complete if and only if every such series converges to an element $z \in X$, in the sense that $\lim_N \sum_1^N x_n = z$.

3.2. Let X be a Banach space and put $B(X) \equiv B(X, X)$.
   a) If $T \in B(X)$ satisfies $\|I - T\| < 1$ (I is the identity operator on X) then T is an automorphism of X, that is, $T^{-1} \in B(X)$. (Consider the series $\sum_1^\infty (I - T)^n$ in $B(X)$.)
   b) Let $\{x_1, \ldots, x_n\}$ be a linearly independent set in X. Show that there is an $\varepsilon > 0$ such that any set $\{y_1, \ldots, y_n\}$ for which $\|x_i - y_i\| < \varepsilon$, $i = 1, \ldots, n$, is also linearly independent. (Construct an automorphism $T \in B(X)$ for which $T(x_i) = y_i$, $i = 1, \ldots, n$.)

3.3. Let $T \in B(X, Y)$ be an operator between normed spaces X and Y which fails to have a bounded inverse. Show that there exists a sequence $\{x_n\} \subset X$ having the properties of (16.6). (Use **16B**.)

3.4. Let X be a real normed linear space and $\phi \in X^*$, $\phi \neq \theta$.
   a) Show that $\inf\{\|x\| : \phi(x) = 1\} = 1/\|\phi\|$.
   b) Show that there is an equivalent norm on X such that the subspace $\ker(\phi)$ is proximinal wrt this new norm.

3.5. Let X be a normed linear space. Show that any weakly-complete subset of X is norm-complete. It follows that a Cauchy sequence is convergent in X if (and only if) it is weakly convergent.

3.6. Let X be a locally convex space and N a linear subspace of X^*. Show that N is weak*-closed in X^* (if and) only if $N = M^\circ$, for some subspace $M \subset X$. (Take $M = {}^\circ N$.)

3.7. Let X be a normed linear space. If $G \subset X^*$ then $({}^\circ G)^\circ = \overline{\text{co}}^*(\{G, \theta\})$ and ${}^\circ(G^\circ) = \overline{\text{co}}(\{G, \theta\})$. Hence $G^\circ = ({}^\circ G)^{\circ\circ}$ if and only if $\overline{\text{co}}(\{G, \theta\})$ is weak*-closed. (The point of this exercise is as follows. Let A be a given convex subset of X with $\theta \in A$. Suppose that we represent A in quasi-linear form as $\{x \in X : \phi(x) \leqslant 1, \phi \in G\}$ for an appropriate set $G \subset X^*$. That is, $A = {}^\circ G$. Then $A^{\circ\circ} = \overline{J_X(A)}^* \subset G^\circ$ and equality holds if and only if $\overline{\text{co}}(\{G, \theta\})$ is weak*-closed in X^*. In other words, whether or not $A^{\circ\circ}$ is "what it should be" in X^{**} depends, in the indicated fashion, on the "richness' of the representing set G.)

3.8. Give an alternative proof of the Goldstine-Weston density lemma in **16F**, proceeding by contradiction and use of the strong separation theorem.

3.9. Let X and Y be Banach spaces over the same field. Show that the congruence $T \mapsto T^*$ from $B(X, Y)$ into $B(Y^*, X^*)$ is surjective if and only if Y is reflexive.

3.10. Let X be a reflexive Banach space.

a) Show that two disjoint closed convex subsets of X can be strongly separated by a closed hyperplane provided that one of the sets is bounded. (Compare with the remark at the end of **19E**.)

b) Show that every closed bounded convex subset of X is equal to the closed convex hull of its extreme points.

3.11. Consider the example in **12G** of a measure space (Ω, Σ, μ) for which the usual congruence of $L^\infty(\Omega, \mu, \mathbb{F})$ into $L^1(\Omega, \mu, \mathbb{F})^*$ is not surjective. Determine a measure space $(\Gamma, \mathcal{T}, \nu)$ for which $L^1(\Gamma, \nu, \mathbb{F}) \cong L^1(\Omega, \mu, \mathbb{F})$, and $L^1(\Gamma, \nu, \mathbb{F})^* = L^\infty(\Gamma, \nu, \mathbb{F})$ via the usual congruence.

3.12. Give the details of the proof of the corollary in **16G**.

3.13. Consider the sequence spaces $\ell^p(\aleph_0)$ for $1 \leqslant p < \infty$. Let e_n be the sequence $(0, \ldots, 0, 1, 0, \ldots)$, where the nth-component is one. Show that this sequence converges weakly to θ if and only if $p > 1$.

3.14. Show that a weakly semi-complete Banach space whose conjugate space is separable must be reflexive.

3.15. Prove the statements made at the end of **16H** pertaining to c_0°. (Under the congruence between $m(\Omega, \mathbb{F})$ and $C(\beta(\Omega, \mathbb{F}))$ the elements of c_0 go into the (continuous) functions that vanish on $\beta(\Omega)\backslash\Omega$.)

3.16. Let X be a Banach space.

a) Show that X is quasi-reflexive if and only if X^* is quasi-reflexive. (This will follow from the observations that, on the one hand

$$J_X(X)^\circ \cong (X^{**}/J_X(X))^*$$

(**16E**) and, on the other hand, that $J_X(X)^\circ$ is isomorphic to $X^{***}/J_{X^*}(X^*)$. This latter isomorphism can be obtained (via exercise 2.2c) from the stronger observation that

$$X^{***} = J_{X^*}(X^*) \oplus J_X(X)^\circ.$$

Indeed, a projection P from X^{***} onto $J_{X^*}(X^*)$ along $J_X(X)^\circ$ can be defined by $P(F) = \hat{\phi}_F$, $F \in X^{***}$, where $\phi_F \in X^*$ is defined by $\phi_F(x) = F(\hat{x})$, $x \in X$.)

b) Let M be a closed linear subspace of X. Then $J_X(X) + M^{\circ\circ}$ is a closed linear subspace of X^{**}. (This follows from

$$J_X(X) + M^{\circ\circ} = Q_M^{**-1} \circ J_{X/M}(X/M),$$

which is in turn a consequence of the formula $J_{X/M} \circ Q_M = Q_M^{**} \circ J_X$.)

c) Show that X is quasi-reflexive if and only if M and X/M are quasi-reflexive for some (hence every) closed subspace $M \subset X$. (For any Banach space X we have in fact that $M^{**}/J_M(M)$ is isomorphic to $(J_X(X) + M^{\circ\circ})/J_X(X)$ and that $(X/M)^{**}/J_{X/M}(X/M)$ is isomorphic to $X^{**}/(J_X(X) + M^{\circ\circ})$. To obtain these isomorphisms let R_M: $X^* \to M^*$ be the restriction map. Then R_M^* is a congruence between M^{**} and $M^{\circ\circ}$, and $R_M^* \circ J_M = J_X|M$. Now for the first isomorphism consider the map $Q_{J_X}(X) \circ R_M^*$, and for the second isomorphism consider

$$Q_{J_{X/M}(X/M)} \circ Q_M^{**}.)$$

3.17. Prove that any locally compact Hausdorff space Ω is a Baire space. (Use the fact that Ω has a basis consisting of relatively compact open sets.)

3.18. Let Ω be a Baire space.

a) Show that each open subset of Ω is again a Baire space.

b) Let f be a lower semicontinuous function on Ω. Then the subset of Ω consisting of points having a neighborhood on which f is bounded above is dense in Ω.

c) Let $\{f_n\}$ be a sequence in $C(\Omega, \mathbb{F})$ that converges pointwise on Ω to a function f. Show that the set of points at which f is continuous is a residual set. ("Osgood's theorem". Without loss of generality, assume $\mathbb{F} = \mathbb{R}$. Observe that if I is any open interval in $\mathbb{R}$ then $f^{-1}(I)$ is an F_σ-set in Ω. Now, given $\varepsilon > 0$, cover $\mathbb{R}$ by a sequence $\{I_n\}$ of open intervals each of length $<\varepsilon$. If $f^{-1}(I_n) = \cup\{C_k^{(n)}$: $k = 1, 2, \ldots\}$ where each $C_k^{(n)}$ is closed, then $\Omega_\varepsilon \equiv \bigcup_n \bigcup_k \operatorname{int}(C_k^{(n)})$ is a dense open set in Ω, at each point of which the oscillation of f is $<\varepsilon$. It follows that f is continuous on the residual set $\bigcap_n \Omega_{1/n}$.)

d) If (Ω, d) is a complete metric space then every non-negative lower semicontinuous function f on Ω is continuous on a residual set in Ω. (Apply c) with $f_n(t) \equiv \inf\{f(x) + nd(t, x) : x \in \Omega\}$.)

3.19. Prove the formula (17.4) for the norm of the Fourier series partial sum functional on $C_{2\pi}$.

3.20. Verify the statements of the lemma in **17F**.

3.21. Let X and Y be Banach spaces and let $T: X \to Y$ be linear.

a) Suppose that T is weakly continuous, that is, T is continuous when both X and Y are given their weak topologies. Show that T must be bounded and hence that $T \in B(X, Y)$.

b) Suppose that $\psi \circ T \in X^*$ for all $\psi \in G \subset Y^*$, where $\overline{\text{span}}(G) = Y^*$. Show that T must be bounded. (In both cases show that T is closed on X. A typical application of b) is to the case where $Y = L^1(\Omega, \mu, \mathbb{F})$ for some σ-finite measure μ and $G \subset L^\infty(\Omega, \mu, \mathbb{F})$ is the set of characteristic functions of chunks.)

3.22. Let $\{a_k\}$ be a sequence of real numbers such that for all $\{b_k\} \in \ell^q(\aleph_0)$, $1 \leqslant q < \infty$, the series $\sum_1^\infty a_k b_k$ is convergent. Prove that $\{a_k\} \in \ell^p(\aleph_0)$ for $p = q/(q-1)$ ($p \equiv \infty$ when $q = 1$.) (Define a linear map T: $\ell^q \to m$ by $T(\{b_k\})_n = \sum_1^n a_k b_k$, and show that T is continuous.)

3.23. Let M be a closed linear subspace of $C \equiv C([0, 1], \mathbb{R})$ consisting of continuously differentiable functions. Show that M must be finite dimensional. (Consider the mapping $x \mapsto x'$ from M into C. In fact, a stronger assertion is true. Namely, M must be finite dimensional if it consists of functions of bounded variation.)

3.24. Let X and Y be Banach spaces with X reflexive. If there exists a surjective operator in $B(X, Y)$ then Y is also reflexive. In other words, the continuous linear image of a reflexive space must also be reflexive.

3.25. In the example in **17H** show that $A + N$ is not closed in X by verifying that $\sum_1^\infty \left(\sin \frac{1}{n}\right) \chi_{\{2n\}}$ belongs to $\overline{A + N} \backslash (A + N)$.

3.26. Let A be a closed convex set in $\mathbb{R}^n$ having no boundary rays nor asymptotes (**17H**). Show that $A + B$ is closed for any closed convex set $B \subset \mathbb{R}^n$. (Proceed via the following steps. Let C be a closed convex set in $\mathbb{R}^n$.

a) Suppose that $p \in C$, $q \in \mathbb{R}^n$ ($q \neq \theta$), and that there are sequences $\{x_k\} \subset C$, $\{t_k\} \subset (0, \infty)$ such that $\lim_k t_k = 0$ and $\lim_k t_k x_k = q$. Then C contains the ray $p + [0, \infty)q$.

b) Suppose that L is a half-line emanating from θ and that x, y are points in $\mathbb{R}^n$ such that $x + L \subset \mathbb{R}^n \backslash C$ and $y + L \subset C$. Then for some $z \in [x, y]$ the half-line $z + L$ is either a boundary ray or an asymptote of C.

c) If $A \cap C = \varnothing$ then $\text{dist}(A, C) > 0$. For suppose that $\text{dist}(A, C) = 0$. Then there exist sequences $\{x_k\} \subset A$, $\{y_k\} \subset C$ such that $\lim_k (x_k - y_k) = \theta$. We get an immediate contradiction unless $\lim_k \|x_k\| = \infty$. We may then assume that $\lim_k x_k/\|x_k\| = q$. Now use a) and b) to get a contradiction to our assumption about A.

d) Finally, let $p \in \overline{A + B}$. Then $\text{dist}(A - p, -B) = 0$.)

3.27. Let A be a subset of a normed linear space X. Show that A is compact if and only if for any sequence $\{\phi_n\} \subset X^*$ with weak*-$\lim_n \phi_n = \theta$, the sequence $\{\phi_n | A\}$ converges uniformly to zero.

3.28. Find the error in the following argument. "Theorem": Let X be a separable normed linear space. Then weak and norm sequential convergence are equivalent. (Recall Schur's lemma in **18C**.) "Proof": Let $\{x_n\}$ be a sequence in X that converges weakly to θ. Then $\{x_n\}$ is bounded and hence the sequence of functionals $\{\hat{x}_n|U(X^*)\}$ is equicontinuous. Since $U(X^*)$ is a compact metric space in its weak*-topology and $\{\hat{x}_n|U(X^*)\}$ converges pointwise to θ, the convergence is uniform. Therefore, $\lim_n \|x_n\| = 0$.

3.29. Let X be a normed linear space of infinite dimension. Then the weak topology on X is not metrizable (**12F**). (If the weak topology on X were metrizable there would be a countable local base for the weak topology and it would follow that X^* would have countable Hamel dimension. But X^* is complete (**16A**) and any infinite dimensional Banach space must have uncountable Hamel dimension.) It follows similarly that if X is complete then the weak*-topology on X^* is not metrizable.

3.30. Let X and Y be normed linear spaces with X complete. Suppose that X and Y are homeomorphic. Then Y is complete. (It is known from topology that any topologically complete subset of a complete metric space is a G_δ set in that space. Applied to the present situation this means that Y is a (dense) G_δ set in its completion $\hat{Y}$, say $Y = \bigcap_n Y_n$ where each Y_n is a dense open set in $\hat{Y}$. Then $\hat{Y}\backslash Y = \bigcup_n (\hat{Y}\backslash Y_n)$ is a set of first category. Finally, if $\hat{Y}\backslash Y \neq \varnothing$ it would follow that Y is a set of first category in $\hat{Y}$, whence $\hat{Y} = Y \cup (\hat{Y}\backslash Y)$ gives a contradiction to the Baire category theorem. Therefore, $\hat{Y}\backslash Y = \varnothing$.)

3.31. In the theorem of **18A**, where is the hypothesis that X is normed used in the proof that e) implies a)? Given an example to show that properties a)–e) are not equivalent for subsets of general locally convex spaces.

3.32. Use the equivalence of weak relative compactness and weak relative countable compactness established in **18B** to prove that each weakly countably compact subset of a normed space is weakly compact, and that each weakly compact subset is weakly sequentially compact (**18A**). (For the second assertion proceed by contradiction, assuming the existence of a sequence with no weakly convergent subsequence.)

3.33. Show that a Banach space X is reflexive if and only if every closed separable linear subspace of X is reflexive.

3.34. Formulate and prove a bipolar theorem for the absolute polar of a subset A of a locally convex space. Use it to show that $({}^aA^a)^a = A^a$. (Observe that ${}^aA^a \equiv {}^a(A^a)$ is a weakly closed absolutely convex set containing A.)

3.35. Prove the assertions of the corollary in **18D**.

3.36. Let X be an infinite dimensional normed linear space and let $\{M_n\}$ be an increasing sequence of n-dimensional subspaces of X^*. For $n = 1, 2, \ldots$, let A_n be a finite $\dfrac{1}{n}$-net in $\partial U(M_n)$, and set $A = \bigcup_n A_n$.

Then θ belongs to the weak*-closure of A but not to the bw*-closure. Conclusion: the bw*-topology on X^* is strictly stronger than the weak*-topology.

3.37. Let M be a closed linear subspace of a Banach space X. Then M is reflexive if and only if $J_X(M)$ is weak*-closed in X^{**}.

3.38. Let X be a real Banach space and $\Phi \in X^{**}$. Suppose that there exists a compact set $A \subset X$ such that $\Phi \leqslant \sigma_A$, where σ_A is the support function of A: $\sigma_A(\phi) = \max\{\phi(x): x \in A\}$, for all $\phi \in X^*$. Prove that $\Phi \in J_X(\overline{\text{co}}(A))$. (If, in fact, $\Phi = \hat{x}$ for some $x \in X$, then necessarily $x \in \overline{\text{co}}(A)$. To prove the existence of some such x, show that σ_A is bw*-continuous.)

3.39. Show that a Banach space X is reflexive if and only if every closed bounded convex subset of X^* is weak*-compact.

3.40. Let X be a Banach space and M a closed linear subspace of X^*. Show that if M, as a Banach space, is reflexive, then M is weak*-closed in X^*.

3.41. Let X be a normed linear space, and let $H(X)$ be the linear space of all real-valued, continuous, positively homogeneous functions on X. Any such function is necessarily bounded on $U(X)$ and so we can norm $H(X)$ by $\|f\| = \sup\{|f(x)|: \|x\| \leqslant 1\}$. Let $N(X)$ be the subset of $H(X)$ consisting of semi-norms, and $E(X)$ the subset of $N(X)$ consisting of equivalent norms. For each $\sigma \in E(X)$ we let $\sigma^* \in E(X^*)$ be the dual norm as defined by (18.15).
a) $H(X)$ is a Banach space and $N(X)$ is a closed cone in $H(X)$;
b) $E(X) = \text{int}(N(X))$ (in particular, $E(X)$ is dense in $N(X)$ (**11A**));
c) $\sigma \mapsto \sigma^*$ is a homeomorphism of $E(X)$ into $E(X^*)$, and is surjective exactly when X is reflexive. (For b), consider the continuous function $\gamma: N(X) \to [0, \infty)$ defined by $\gamma(\sigma) = \inf\{\sigma(x): \|x\| = 1\}$; $\gamma(\sigma)$ is positive if and only if $\sigma \in E(X)$.)

3.42. Give the details of the proof of statement a) of the theorem in **18G**.

3.43. Let X and Y be Banach spaces and $T \in B(X, Y)$ a surjective operator.
a) $\ker(T^{**}) = \ker(T)^{\circ\circ}$;
b) if $\dim(\ker(T)) < \infty$, and if M is a closed linear subspace of X, then $T(M)^{\circ\circ} = T^{**}(M^{\circ\circ})$.

3.44. Let A be a bounded and weakly closed subset of a Banach space X with the property that every functional in X^* attains its supremum on A. Then if Y is any separable Banach space and $T \in B(X, Y)$ is surjective, the image $T(A)$ is weakly compact in Y. Is this statement true if T is not assumed surjective?

3.45. Let A be bounded and weakly closed subset of a Banach space X. Show that the following assertions about A are equivalent.
a) A is weakly compact;
b) if $\{x_n\} \subset A$ and $\lim_n \phi(x_n)$ exists for some $\phi \in X^*$, then there is an $x \in A$ such that $\lim_n \phi(x_n) = \phi(x)$;
c) if B is a weakly closed subset of X which is disjoint from A, then $\text{dist}(A, B) > 0$.

3.46. Let X be a Banach space.

a) Show that X is uniformly normed if and only if any pair of sequences $\{x_n\}$, $\{y_n\}$ of unit vectors in X for which $\lim_n \|x_n + y_n\| = 2$ satisfies $\lim_n \|x_n - y_n\| = 0$.

b) If X is finite dimensional and strictly normed, show that X is uniformly normed.

3.47. Let X be a uniformly normed Banach space.

a) If $\{x_n\}$ is a sequence of unit vectors in X for which weak—$\lim_n x_n \equiv x$ is also a unit vector, then $\lim_n x_n = x$.

b) The metric projection (**15C**) on any closed convex subset of X is single-valued and continuous.

3.48. There exists an incomplete (hence *a fortiori* a non-reflexive) normed linear space X such that every functional $\phi \in X^*$ attains its supremum on $U(X)$. The example is suggested by the observation that if Z is a reflexive Banach space then every $\phi \in Z^*$ attains its supremum on $U(Z)$ at an extreme point of $U(Z)$ (**13B**, **16F**), and that $X \equiv \text{span}(\text{ext}(U(Z)))$ is dense in Z (exercise 3.10b). Consequently, if $X \neq Z$, X will serve as the desired example. (To obtain such an example, let Z_n be the space $\mathbb{R}^n$ normed by the supremum norm, $\|(\xi_1, \ldots, \xi_n)\|_\infty \equiv \max_i(|\xi_i|)$, and let Z be the linear subspace of $\prod_n Z_n$ consisting of those sequences $z = (\xi_1^{(1)}, \xi_1^{(2)}, \xi_2^{(2)}, \xi_1^{(3)}, \xi_2^{(3)}, \xi_3^{(3)}, \ldots)$ for which

$$\|z\|^2 \equiv |\xi_1^{(1)}|^2 + \|(\xi_1^{(2)}, \xi_2^{(2)})\|_\infty^2 + \|(\xi_1^{(3)}, \xi_2^{(3)}, \xi_3^{(3)})\|_\infty^2 + \cdots < \infty.$$

a) So normed, Z is a reflexive Banach space.

b) If $X = \text{span}\{z \in Z: |\xi_1^{(n)}| = |\xi_2^{(n)}| = \cdots = |\xi_n^{(n)}|, n = 1, 2, \ldots\}$ then X is dense in Z (by the above extreme point argument).

c) $X \neq Z$, so that X is not complete (any sequence in Z with distinct terms cannot belong to X).

d) Every $\phi \in Z^*$ has the following form: there is a sequence of numbers $\{\alpha_j^{(n)}\}$ such that

$$\phi(z) = \alpha_1^{(1)}\xi_1^{(1)} + (\alpha_1^{(2)}\xi_1^{(2)} + \alpha_2^{(2)}\xi_2^{(2)}) + (\alpha_1^{(3)}\xi_1^{(3)} + \alpha_2^{(3)}\xi_2^{(3)} + \alpha_3^{(3)}\xi_3^{(3)}) + \cdots.$$

e) Every $\phi \in Z^*$ attains its supremum on $U(Z)$ at a point $z: \|z\| = 1$, $\phi(z) = \|\phi\|$. Taking into account the form of ϕ given in d), the terms of z can be modified so that a point $x \in X$ is obtained with $\|x\| = 1$ and $\phi(x) = \phi(z)$.)

3.49. Let X be a normed linear space and let B be a closed, bounded, convex subset of X with $\theta \notin B$. If $C = [0, \infty)B$, show that C is a closed cone in X.

3.50. Let A be a convex set in a Banach space. If $f \in \text{Conv}(A)$ is lower semicontinuous then f is continuous at each relative interior point of A. (It is sufficient to assume that A is solid. Use **14A** and exercise 3.18b.)

3.51. Prove (by contradiction) that the convex function defined by (20.3) is nowhere subdifferentiable.

3.52. Let M be a subspace of finite codimension in the (real) Banach space X. Suppose that A is a closed convex set in X, that $\varepsilon > 0$, and that $x \in \partial(A) \cap M$. Then there exists a support point x_0 of A such that $x_0 \in M$ and $\|x - x_0\| < \varepsilon$. (If we represent M as $\{x \in X : \phi_1(x) = \cdots = \phi_n(x) = 0\}$ for appropriate functionals $\phi_1, \ldots, \phi_n \in X^*$, then this assertion is a generalization of the first Bishop-Phelps theorem of **20B**: it states that a boundary point of A satisfying a finite number of linear constraints can be approximated by a support point satisfying the same constraints (compare with **9C**). The following lemma is useful for the proof: if x_0 is a support point of the set $A \cap M$ with respect to the subspace M, then x_0 is a support point of A. To prove the lemma, use the support theorem and induction on codim(M).)

3.53. Let X be an incomplete normed linear space. Then there exists a solid, closed, bounded, convex set $A \subset X$ such that the support functionals of A are not dense in X^*. (Embed X in its completion $\hat{X}$ and select a unit vector $x \in \hat{X} \backslash X$. Then select $\phi \in \hat{X}^* \cong X^*$ such that $\|\phi\| = 1 = \phi(x)$. Let $B = U(\hat{X}) \cap \ker(\phi)$, and then put $A_1 = \mathrm{co}(\{x, B\})$. A_1 is a solid closed convex set in $\hat{X}$. Finally, let $A = A_1 \cap X$. Show that any support functional of A must be at distance at least $\frac{1}{2}$ from ϕ.)

3.54. Verify the formulas for $\mathscr{P}(c_0)$ and $\mathscr{P}(L^1)$ given in **20E**, and use these formulas to show directly that c_0 and L^1 are subreflexive.

3.55. Let M be a closed linear subspace of a Banach space X. Let $Y_M = \{\phi \in \mathscr{P}(X) : \phi(x) = \|\phi\|$ for some $x \in M\}$. Suppose that $\bar{Y}_M$ is a linear subspace of X^*. Then $\bar{Y}_M \cong M^*$. (Consider the restriction map from $\bar{Y}_M$ into M^*. The point of the problem is that it gives a condition under which we can identify M^* with a subspace of X^* rather than just a quotient space as in **16E**.)

3.56. Show that the unit ball of $L^\infty([0, 1], \mu, \mathbb{R})$ has no smooth points. (Lebesgue measure is assumed. Let x_0 be a unit vector in L^∞, and suppose that essup $x_0(\cdot) = 1$. Let $\{E_n\}$ be a sequence of pairwise disjoint chunks such that $x_0|E_n \geqslant n/(n + 1)$, and define norm-one functionals ϕ_n on L^∞ by

$$\phi_n(x) = \frac{1}{\mu(E_n)} \int_{E_n} x \, d\mu, \qquad x \in L^\infty.$$

Now the sets $M \equiv \{x \in L^\infty : \lim_n \phi_{2n-1}(x)$ exists$\}$ and $N \equiv \{x \in X : \lim_n \phi_{2n}(x)$ exists$\}$ are subspaces on which the indicated limits define norm-one linear functionals. Let ϕ' and ϕ'' be Hahn-Banach extensions of these functionals to all of L^∞. Then $[\phi'; 1]$ and $[\phi''; 1]$ are distinct hyperplanes of support to $U(L^\infty)$ at x_0.)

3.57. Let X be a real normed linear space. A functional $\phi \in \partial U(X^*)$ is a *regularly exposed point* of $U(X^*)$ if there exists a unit vector $x \in X$

such that $\hat{x}$ attains its supremum $(=\|x\|)$ on $U(X^*)$ only at ϕ. Prove that

a) if $x_0 \in \operatorname{sm}(U(X))$, then the norm gradient $g(x_0;\cdot)$ is a regularly exposed point of $U(X^*)$, and conversely;

b) every regularly exposed point of $U(X^*)$ is an extreme point;

c) if X is a separable Banach space, the regularly exposed points of $U(X^*)$ are weak*-dense in $\operatorname{ext}(U(X^*)$ (use **13B** and **20F**).

3.58. Let $L^p = L^p(\Omega, \mu, \mathbb{F})$ for $1 < p < \infty$. Show that the norm duality map (**20G**) from L^p into L^q is given by

$$x \mapsto \begin{cases} \dfrac{x(\cdot)|x(\cdot)|^{p-2}}{\|x\|^{p-2}}, & x \neq \theta \\ \theta, & x = \theta. \end{cases}$$

Show that this mapping is a homeomorphism from L^p onto L^q (although it is not linear unless $p = 2$).

3.59. Let A be a countable weakly compact subset of an infinite dimensional Banach space. Show that $\overline{\operatorname{co}}(A)$ can have no interior. (Use **17J** and **19E**.)

3.60. Let g be a real-valued function defined on an interval $[a, b]$. If g is of bounded variation then, as is well known, the Riemann-Stieltjes integrals $\int_a^b f(t)dg(t)$ exist for all $f \in C \equiv C([a, b], \mathbb{R})$. Prove the converse: if these integrals exist for every $f \in C$ then g must be of bounded variation. (Otherwise there would exist a sequence $\{\pi_n\}$ of partitions of $[a, b]$, with $\pi_n = \{a \equiv t_1^{(n)}, t_2^{(n)}, \ldots, t_{m(n)}^{(n)} \equiv b\}$, $t_{j-1}^{(n)} < t_j^{(n)}$, such that

$$\lim_n \|\pi_n\| \equiv \lim_n \max\{t_j^{(n)} - t_{j-1}^{(n)} : 1 \leqslant j \leqslant m\} = 0,$$

$$\sum_{j=1}^{m} |g(t_j^{(n)}) - g(t_{j-1}^{(n)})| > n.$$

It follows that each π_n contains a set $\{s_1^{(n)}, \ldots, s_{k(n)}^{(n)}\}$ such that

$$|g(s_2^{(n)}) - g(s_1^{(n)}) + \cdots + g(s_k^{(n)}) - g(s_{k-1}^{(n)})| > \frac{n}{2}.$$

Now define $\phi_n \in C^*$ by

$$\phi_n(f) = \sum_{j=1}^{m(n)} f\left(\frac{t_{j-1}^{(n)} + t_j^{(n)}}{2}\right) [g(t_j^{(n)}) - g(t_{j-1}^{(n)})].$$

Apply **17C** to this sequence of functionals, after noting that $\lim_n \phi_n(f) = \int_a^b f(t)dg(t)$, for each $f \in C$; then make a special choice of f to obtain a contradiction).

3.61. For fixed $t \in [a, b]$ show that the evaluational functional $f \mapsto f(t)$ is a continuous linear functional on the Sobolev space H_p^m of **21B**.

3.62. Let X and Y be Banach spaces and $R \in B(X, Y)$ a surjective operator with kernel N.

a) Let M be a closed linear subspace of X. Then each affine subspace of X parallel to M (that is, each element of X/M) contains a unique R-spline if and only if $R(M)$ is a Chebyshev subspace (**20H**) of Y and $M \cap N = \{\theta\}$.

b) When the conditions of a) hold, the mapping $I - \hat{R}_M \circ T_M \circ R$ assigns to each $x \in X$ the unique R-spline in the flat $x + M$. Here $T_M \equiv$ the metric projection of Y onto $R(M)$ and $R_M \equiv R|M$.

c) Suppose that $\dim(N) < \infty$ and that $T \in B(X, Z)$ for some normed space Z. If Γ is a closed, bounded, and convex set in Z, then $R(K)$ is closed in Y, where K is defined by (21.7). Consequently, if Y is reflexive, an R-spline exists in K.

3.63. Let $F:\Omega \to 2^X$ be a lower semicontinuous carrier, where Ω is a topological space and X is a linear topological space. Prove that the carrier $t \mapsto \overline{\text{co}}(F(t))$ is also lower semicontinuous.

3.64. Prove the converse to the Michael selection theorem: let Ω be a Hausdorff space such that every lower semicontinuous carrier whose values are non-empty closed convex subsets of a Banach space admits a continuous selection; then there is a partition of unity subordinate to any given open covering of Ω, and so Ω is paracompact. (Let $\{\mathcal{O}_\alpha: \alpha \in I\}$ be the given open covering and put $X = \ell^1(I)$. Define a carrier $F:\Omega \to 2^X$ by

$$F(t) = \{x \in X : x_\alpha \geqslant 0, \sum_\alpha x_\alpha = 1, x_\alpha = 0 \text{ if } t \notin \mathcal{O}_\alpha\}.$$

Then F admits a continuous selection $f:\Omega \to X$ and we can put $p_\alpha(t) = (f(t))_\alpha$; $\{p_\alpha : \alpha \in I\}$ is the desired partition of unity.)

3.65. Let X and Y be Banach spaces and $T \in B(X, Y)$ a surjection. Show that T has the *k-covering property*, that is, for every compact set $B \subset Y$ there is a compact set $A \subset X$ such that $B \subset T(A)$.

3.66. Let T, X, Y be as in exercise 3.65. Show that T has a right-inverse if (and only if, by **21C**) $\ker(T)$ is a topological direct summand of X. Show that the set of all such operators is an open subset of $B(X, Y)$. (Suppose that $T_0 \in B(X, Y)$ has a right-inverse $S_0 \in B(Y, X)$. (Choose $T \in B(X, Y)$ to satisfy $\|T - T_0\| < \|S_0\|^{-1}$ and look for a right-inverse of T in the form $S_0 V$, for suitable $V \in B(Y)$; apply exercise 3.2a.)

3.67. Let X be a locally convex space.

a) Show that strong convergence in X^* implies uniform convergence on each bounded subset of X. (Observe that if B is a bounded subset of X then B^a is a weak*-closed barrel in X^*.)

b) Let X be a normed linear space. Then the strong topology on X^* coincides with the norm topology. (X^* is always a Banach space.)

3.68. Let D be an open convex subset of $\mathbb{R}^n$ and let $T:D \to \mathbb{R}^n$ be differentiable. Suppose that there is a constant $\alpha > 0$ such that the spectrum of the symmetric part of the Jacobian matrix J of T lies in the interval $[\alpha, \infty)$, for all points in D. (That is, if λ is an eigenvalue of the matrix $\frac{1}{2}(J + J')$, evaluated at some point in D, then $\lambda \geqslant \alpha$.) Show that T is strongly monotone on D. (Fix $x, y \in D$ and consider the function $\phi(t) \equiv \langle x - y, T(tx + (1 - t)y)\rangle$, for $0 \leqslant t \leqslant 1$.)

3.69. Let D be an open convex subset of $\mathbb{R}^n$ and f a differentiable convex function on D. Prove that f is strongly convex if and only if ∇f is strongly monotone on D. (First show that strong convexity is equivalent to the existence of $\alpha > 0$ such that $\langle y - x, \nabla f(x)\rangle \leqslant f(y) - f(x) - \alpha\|x - y\|_2^2$ for $x, y \in D$.)

3.70. Let X be a linear topological space.
  a) Let A be a convex subset of X. If A is either open or closed, or if X is finite dimensional, then A is ideally convex.
  b) Any intersection of ideally convex subsets of X is again ideally convex.
  c) If $T: X \to Y$ is continuous and linear, and if $A \subset Y$ is ideally convex, then $T^{-1}(A)$ is ideally convex in X.
  Now let X be a Banach space.
  d) If T is as in c) (Y is arbitrary) and if A is a bounded ideally convex subset of X, then $T(A)$ is ideally convex in Y.
  e) The sum of two ideally convex sets in X is again ideally convex, provided that one of them is bounded.

3.71. Let X be an ordered Banach space with closed positive wedge P. P is *non-flat* if there exists a constant $\gamma > 0$ such that to every $x \in X$ corresponds some $y \in P$ with $x \leqslant y$ and $\|y\| \leqslant \gamma\|x\|$.
  a) P is non-flat if and only if P is reproducing. (For the forward implication apply **17E** to the set $P \cap U(X) - P \cap U(X)$.)
  b) If P is reproducing then any positive linear functional ϕ on X is bounded. (That is, $P^+ \subset P^*$. It is enough to show that $\phi|P$ is continuous at θ, and then use the non-flatness of P.)
  c) If P is both reproducing and locally compact then X must be finite dimensional.

Chapter IV

Conjugate Spaces and Universal Spaces

Motivated by the importance of conjugate spaces indicated in earlier sections, we devote the bulk of this final chapter to some further considerations regarding such spaces. We begin with the famous Riesz-Kakutani characterization of $C(\Omega, \mathbb{R})^*$ as the space of regular signed Borel measures on Ω. After giving some applications of this theorem we proceed to some characterizations of general conjugate spaces, and use these to exhibit some new conjugate spaces (spaces of operators and Lipschitz functions). The fact that certain spaces of operators are conjugate spaces has some interesting implications for optimization theory as we shall see. We shall also establish an isomorphism between certain spaces of Lipschitz functions and certain spaces of L^∞ type. A particular consequence of this is an example of a pair of Banach spaces (namely, $\ell^1(\aleph_0)$ and $L^1([0, 1])$) which fail to be isomorphic, yet whose conjugate spaces are isomorphic.

Finally we show that the space of continuous functions defined on an uncountably compact metric space can serve as a "universal" Banach space, in the sense that every (separable) Banach space can be congruently (isometrically) embedded in any such space.

§22. The Conjugate of $C(\Omega, \mathbb{R})$

In this section we identify congruently the space $C(\Omega, \mathbb{R})^*$ with the space $\mathscr{M}_r(\Omega, B, \mathbb{R})$ of regular measures defined on the Borel subsets of the compact space Ω. This is an exceedingly useful representation and we shall indicate a few of its more immediate applications.

22A. As a preliminary to the representation theorem we establish a general fact about the conjugate spaces of certain ordered normed linear spaces. Let X be a real ordered linear space (**5A**) with positive wedge P. The ordering induced by P is *archimedean* (resp. *almost archimedean*) if $x \leqslant ty$ (resp. $-ty \leqslant x \leqslant ty$) for some $y \geqslant \theta$ and all $t > 0$ implies $x = \theta$. An element $e \in P$ is an *order unit* for P if for each $x \in X$ there is some $t > 0$ such that $-te \leqslant x \leqslant te$. If X possesses an order unit e then

(22.1)
$$\|x\| \equiv \inf\{t > 0 : -te \leqslant x \leqslant te\}$$

defines a norm on X exactly when X is almost archimedean ordered. Such a norm is called an *order unit norm.* If X is actually archimedean ordered then

$$U(X) = [-e, e] \equiv \{x \in X : -e \leqslant x \leqslant e\}.$$

The class of order unit normed linear spaces contains (among others) all spaces of bounded continuous functions, all spaces $L^\infty(\Omega, \mu, \mathbb{R})$, and the spaces of hermitian matrices of order n, for any $n \geqslant 2$ (considered as operators on $\mathbb{R}^n$ with the norm (11.4)). The following lemma shows in particular that the conjugate of any such space is positively generated.

Lemma. *Let X be an order unit normed linear space with positive wedge P. Then each $\phi \in X^*$ has a decomposition $\phi = \phi^+ - \phi^-$, where $\phi^+, \phi^- \in P^*$ and $\|\phi\| = \|\phi^+\| + \|\phi^-\|$.*

Proof. Let Y be the product space $X \times X$ with the usual coordinatewise algebraic operations and ordering induced by the wedge $P \times P$. Let e be the order unit defining the norm in X as in (22.1). Then (e, e) is an order unit in Y. Now define a subspace M of Y by

(22.2) $$M = \{y \in Y : y = t(e, e) - (x, -x), t \in \mathbb{R}, x \in X\}.$$

Then given $\phi \in X^*$ define $\psi \in M^*$ by $\psi(y) = t\|\phi\| - \phi(x)$, where t, x are related to y as in (22.2). This functional ψ is positive, since if $y \in M$ is positive then $-te \leqslant x \leqslant te$, whence $\|x\| \leqslant t$ and therefore $\phi(x) \leqslant t\|\phi\|$, that is, $\psi(y) \geqslant 0$.

We can now apply the Krein-Rutman theorem (exercise 2.46) to extend ψ to a positive linear functional $\bar{\psi}$ on all of Y. The hypothesis of this theorem, namely that $\operatorname{int}(P \times P) \cap M \neq \varnothing$, is satisfied by the point (e, e). We now set $\phi^+ = \bar{\psi}(\cdot, \theta)$ and $\phi^- = \bar{\psi}(\theta, \cdot)$. Then ϕ^+ and ϕ^- belong to P^* and $\phi = \phi^+ - \phi^-$. Finally, we observe that

$$\begin{aligned}\|\phi\| &= \bar{\psi}(e, e) = \phi^+(e) + \phi^-(e) \\ &= \|\phi^+\| + \|\phi^-\|.\end{aligned}$$

This computation is justified by the fact that any positive linear functional $\pi \in P^*$ satisfies $\|\pi\| = \pi(e)$. Indeed, $\pi(e) \leqslant \|\pi\| \|e\| \leqslant \|\pi\|$, while if $-te \leqslant x \leqslant te$ then $-t\pi(e) \leqslant \pi(x) \leqslant t\pi(e)$, so that $|\pi(x)| \leqslant t\pi(e)$ and therefore $|\pi(x)| \leqslant \pi(e)\|x\|$, that is, $\|\pi\| \leqslant \pi(e)$. □

We can also note that the dual wedge P^* in the space conjugate to an order unit normed linear space is weak*-locally compact. This is an immediate consequence of exercise 2.34a and the fact that P is solid (since $e \in \operatorname{int}(P)$). From **13C** we then expect that P^* has a weak*-compact base, and indeed such a set is given by $B \equiv \{\phi \in P^* : \phi(e) = 1\}$. Then we can observe the following structure of the unit ball in X^*:

(22.3) $$U(X^*) = \operatorname{co}(B \cup -B).$$

We might finally remark that real Banach spaces having an order-unit are "close" to being spaces of the type $C(\Omega, \mathbb{R})$. Namely, if X is such a Banach space and is in addition a lattice (so that every pair $x, y \in X$ has a supremum in X), then X is congruent to a space $C(\Omega, \mathbb{R})$. In fact, the compact space Ω turns out to be $\operatorname{ext}\{\phi \in P^* : \phi(e) = 1\}$, where P is the positive cone in X and e is the order-unit. (It must first be verified that this set is weak*-closed in

$U(X^*)$. Then the congruence is simply the map $x \mapsto \hat{x}|\Omega$, $x \in X$; this map is also order-preserving. The lattice hypothesis on X is used to guarantee that the range of the congruence is dense in, and hence equal to, $C(\Omega, \mathbb{R})$. The density follows from the order theoretic form of the Stone-Weierstrass theorem, since the range is a linear sublattice of $C(\Omega, \mathbb{R})$ which contains the constant functions and separates the points of Ω.) Any such Banach space is called an *M-space*.

B. Let Ω be a fixed compact Hausdorff space. Recall that the σ-algebra of *Borel* (resp. *Baire*) *sets* in Ω is the σ-algebra generated by the compact (resp. compact G_δ) subsets of Ω. The Baire σ-algebra may alternatively be described as the smallest σ-algebra with respect to which every continuous function on Ω is measureable. A finite signed measure on one of these σ-algebra is naturally called a *Borel* (resp. *Baire*) *measure*. It is known from measure theory that each Baire measure is regular (in the sense that its value at any Baire set A is the supremum of its values on the compact Baire subsets of A), and that every Baire measure can be uniquely extended to a regular Borel measure.

If μ is a Borel (or Baire) measure then as has already been noted in **9C** the mapping

(22.4) $$x \mapsto \int_\Omega x \, d\mu, \qquad x \in C(\Omega, \mathbb{R}),$$

defines an element $\Phi_\mu \in C(\Omega, \mathbb{R})^*$, and $\|\Phi_\mu\| \leqslant \|\mu\|_v \equiv |\mu|(\Omega)$ (**10D**, Ex. 3). Now we claim that if μ is regular then actually $\|\Phi_\mu\| = \|\mu\|_v$. To see this, let $\varepsilon > 0$ and select disjoint Borel sets $A_1, \ldots, A_n$ in Ω such that $\sum_1^n |\mu(A_i)| > |\mu|(\Omega) - \varepsilon$. Let C_i be a compact subset of A_i such that $|\mu|(A_i \backslash C_i) < \frac{\varepsilon}{n}$, and let $\{\mathcal{O}_1, \ldots, \mathcal{O}_n\}$ be a family of disjoint open sets such that $C_i \subset \mathcal{O}_i$, $i = 1, \ldots, n$. Because μ is regular we may assume that $|\mu|(\mathcal{O}_i \backslash C_i) < \frac{\varepsilon}{n}$, $i = 1, \ldots, n$. By Urysohn's lemma there exist $x_1, \ldots, x_n \in C(\Omega, \mathbb{R})$ such that $0 \leqslant x_i(t) \leqslant 1$, $t \in \Omega$, $x_i(t) = 1$, $t \in C_i$, and $x_i(t) = 0$, $t \notin \mathcal{O}_i$. Hence if we put $x_0 = \sum_1^n \operatorname{sgn}(\mu(C_i))x_i$ we find that $\|x_0\|_\infty = 1$ and

$$|\Phi_\mu(x_0) - |\mu|(\Omega)| < \varepsilon.$$

As a final preliminary to the representation theorem we give the following lemma. The proof will be momentarily deferred to **22C**.

Lemma. *Let ϕ be a positive linear functional in $C(\Omega, \mathbb{R})^*$. Then there exists a (unique) positive Baire measure μ such that $\Phi_\mu = \phi$.*

Now let B be the family of Borel sets in Ω and let $\mathcal{M}_r(\Omega, B, \mathbb{R})$ be the linear space of regular Borel measures on Ω, normed by the total variation norm $\|\cdot\|_v$ as in formula (10.3). We then have the "Riesz-Kakutani theorem" (a particular consequence of which is that $\mathscr{M}_r(\Omega, B, \mathbb{R})$ is a Banach space, although this fact can also be proved directly.)

Theorem. *The correspondence $\mu \mapsto \Phi_\mu$ is a congruence between the spaces $\mathscr{M}_r(\Omega, B, \mathbb{R})$ and $C(\Omega, \mathbb{R})^*$.*

Proof. We have already noted that the correspondence $\mu \mapsto \Phi_\mu$, which is clearly linear, is norm-preserving and hence is an isometry. It remains to show that every functional $\phi \in C(\Omega, \mathbb{R})^*$ arises in this fashion.

Now the constantly one function $e \in C(\Omega, \mathbb{R})$ is an order-unit and, indeed, $[-e, e]$ is the unit ball. By **22A**, therefore, any $\phi \in C(\Omega, \mathbb{R})^*$ decomposes as $\phi = \phi^+ - \phi^-$, where ϕ^+, ϕ^- are positive. Let μ^+ and μ^- be the positive Baire measures associated with ϕ^+ and ϕ^- by the lemma. Then $\mu \equiv \mu^+ - \mu^-$ satisfies $\Phi_\mu = \phi$. Finally, μ can be extended to a regular Borel measure as already remarked. □

Since the proof that $\|\Phi_\mu\| = \|\mu\|_v$ applies equally well to complex measures, it is clear that the Riesz-Kakutani theorem is also valid for spaces of complex-valued continuous functions: $C(\Omega, \mathbb{C})^* \cong \mathscr{M}_r(\Omega, B, \mathbb{C})$. We simply apply the real version just proved to the real and imaginary parts of any given functional in $C(\Omega, \mathbb{C})^*$. It is also true that, in the real case, the correspondence $\mu \to \Phi_\mu$ is *bipositive* in the sense that it and its inverse are both order-preserving. In other words, μ is a positive measure if and only if Φ_μ is a positive functional (exercise 4.3).

C. The proof of the lemma in **22B** requires a topological result concerning Stone-Čech compactifications (exercise 2.35) which is of some independent interest. Let us say that a topological space Ω is *extremally disconnected* (a *Stonean space*) if the closure of every open set is again open. Several properties of such spaces are given in exercise 4.5; for example, Ω is extremally disconnected exactly when any two disjoint open subsets of Ω have disjoint closures. The simplest examples of such spaces are the discrete spaces and their Stone-Čech compactifications. We prove this latter assertion now.

Lemma. *The Stone-Čech compactification of a discrete topological space Ω is extremally disconnected.*

Proof. Let $\mathcal{O}_1$ and $\mathcal{O}_2$ be disjoint open subsets of $\beta(\Omega)$, and put $A_i = \mathcal{O}_i \cap \Omega$, $i = 1, 2$. Since Ω is dense in $\beta(\Omega)$ these sets are non-empty (assuming that $\mathcal{O}_i \neq \varnothing$). Since Ω is discrete, the characteristic functions χ_{A_i} are continuous, and so have continuous extensions f_i to $\beta(\Omega)$ (exercise 2.35f)). By continuity each f_i assumes only the values 0 and 1, and we have $f_1 f_2 = \theta$. Since A_i is dense in $\mathcal{O}_i$ it follows that $f_i|\mathcal{O}_i$ is identically 1, $i = 1, 2$, and this shows that $\overline{\mathcal{O}}_1 \cap \overline{\mathcal{O}}_2 = \varnothing$.

We remark that it can be shown that any extremally disconnected compact Hausdorff space is a retract of $\beta(\Omega)$, for some discrete space Ω.

Now, to proceed with the proof of the lemma in **22B**, we let Γ be the space Ω with the discrete topology, and let $f:\beta(\Gamma) \to \Omega$ be the continuous extension of the identity map from Γ to Ω. The formula

$$(Tx)(t) \equiv x(f(t)), \qquad t \in \Omega,$$

defines an isometric embedding T of $C(\Omega, \mathbb{R})$ into $C(\beta(\Gamma), \mathbb{R})$. Hence for any given positive functional $\phi \in C(\Omega, \mathbb{R})^*$ there is, by the Hahn-Banach theorem, a functional $\Phi \in C(\beta(\Gamma), \mathbb{R})^*$ such that $\langle Tx, \Phi \rangle = \phi(x)$, $x \in C(\Omega, \mathbb{R})$, and such that $\|\Phi\| = \|\phi\|$. Since $\Phi(e) = \Phi(Te) = \phi(e) = \|\phi\| = \|\Phi\|$, it follows from exercise 4.2 that Φ is also positive (here we have used e to denote the identically one function on the appropriate space).

Next let $\mathcal{Q}$ be the algebra of open-and-closed sets in $\beta(\Gamma)$. For each $A \in \mathcal{Q}$ the characteristic function χ_A is continuous and so we can define $\nu(A) = \Phi(\chi_A)$. This function ν is a finitely additive measure on $\mathcal{Q}$ and we claim that it is actually countably additive. Indeed, if $\{A_n\}$ is a sequence of disjoint sets in $\mathcal{Q}$ whose union A belongs to $\mathcal{Q}$, then only finitely many A_n can be non-empty, since they are open sets and A is compact. Thus ν is trivially countably additive, and by the usual Carathéodory extension procedure ν can be extended to a measure on the σ-algebra $S(\mathcal{Q})$ generated by $\mathcal{Q}$; let us call the extension ν also.

Since each set in $\mathcal{Q}$ is a G_δ, $S(\mathcal{Q})$ contains only Baire sets. We claim that $S(\mathcal{Q})$ is exactly the σ-algebra of Baire sets; this will follow by showing that each $y \in C(\beta(\Gamma)$ is $S(\mathcal{Q})$-measurable. Given such a y and any real α define $E_n = \{s \in \beta(\Gamma): y(s) < \alpha + 1/n\}$. The sets E_n are open and so, by the lemma, $\bar{E}_n \in \mathcal{Q}$. Hence $\{s: y(s) \leqslant \alpha\} = \bigcap_1^\infty \bar{E}_n \in S(\mathcal{Q})$, whence y is $S(\mathcal{Q})$-measurable. It now also follows that

$$\Phi(y) = \int_{\beta(\Gamma)} y \, d\nu,$$

for each $y \in C(\beta(\Gamma), \mathbb{R})$; we see this by approximating y (in the mean) by simple functions based on sets in $\mathcal{Q}$, and using the definition of ν on such sets.

Finally, if A is any Baire set in Ω, $f^{-1}(A)$ is a Baire set in $\beta(\Gamma)$; define $\mu(A) = \nu(f^{-1}(A))$. Then μ is a Baire measure on Ω and if $x \in C(\Omega, \mathbb{R})$

$$\phi(x) = \langle Tx, \Phi \rangle = \int_{\beta(\Gamma)} Tx \, d\nu = \int_\Omega x \, d\mu.$$

This completes the proof of the lemma and hence of the Riesz-Kakutani theorem. □

D. The original version of the Riesz-Kakutani theorem was given by Riesz and pertains to the spaces $C([a, b], \mathbb{R})$. For such spaces a somewhat more concrete representation of their conjugate spaces is possible. Namely, given any Borel measure μ on $[a, b]$ the definition

(22.5)
$$g(t) \equiv \mu([a, t]), \qquad a < t \leqslant b,$$
$$g(a) \equiv 0,$$

yields a function of bounded variation on $[a, b]$ which is left-continuous at every point in (a, b). Such functions constitute, by definition, the space $NBV([a, b], \mathbb{R})$ of normalized functions of bounded variation. This space can be normed by taking the norm of each such function to be its total variation over the interval $[a, b]$. The correspondence $\mu \mapsto g$ defined by (22.5) is a linear norm-decreasing mapping from the space $\mathscr{M}_r([a, b], B, \mathbb{R})$ into the space $NBV([a, b], \mathbb{R})$ (We recall that any Borel subset of any Euclidean space is automatically a Baire set, so that Borel and Baire measures coincide in this case and, in particular, any Borel measure is regular.)

It turns out that this correspondence is actually a congruence between the spaces $\mathscr{M}_r([a, b], B, \mathbb{R})$ and $NBV([a, b], \mathbb{R})$. One verifies this statement by defining for any $g \in NBV([a, b], \mathbb{R})$ a function μ on certain sub-intervals of $[a, b]$ according to the rules

$$\mu([a, b]) = g(b) - g(a),$$
$$\mu([a, t)) = g(t) - g(a), \qquad a < t \leqslant b.$$

This function μ can then be extended, first to the open sets of $[a, b]$, and then to all Borel sets in a standard fashion, so as to be a (signed) measure. This measure is called the *Borel-Stieltjes measure induced by g*. Since $|\mu|([a, t)) \leqslant$ total variation of g on $[a, t]$, the correspondence $\mu \leftrightarrow g$ is isometric.

In this way we see that every linear functional $\phi \in C([a, b], \mathbb{R})^*$ is given by a Riemann-Stieltjes integral

$$\textbf{(22.6)} \qquad \phi(x) = \int_a^b x(t)dg(t), \qquad x \in C,$$

where $g \in NBV([a, b], \mathbb{R})$ and $\|\phi\| =$ total variation of g on $[a, b]$. (The correctness of (22.6) can be verified by approximating the given continuous function x uniformly by step-functions based on intervals of the form $[c, d)$, $a \leqslant c < d \leqslant b$.) In particular, positive linear functionals on $C([a, b], \mathbb{R})$ are seen to correspond to non-decreasing functions in $NBV([a, b], \mathbb{R})$.

Example. As an illustration of the use of formula (22.6) we establish a result of some interest in probability theory, known as the "Helly selection principle". Recall that a *distribution function* is a bounded, non-decreasing, and left-continuous function defined on $\mathbb{R}$. Typically, such functions arise in connection with random variables: if G is a real-valued random variable defined on some probability space then

$$g(t) \equiv \Pr\{w : G(w) < t\}$$

defines a distribution function g such that

$$g(-\infty) \equiv \lim_{t \to -\infty} g(t) = 0,$$
$$g(+\infty) \equiv \lim_{t \to +\infty} g(t) = 1.$$

Because of its monotonic nature any distribution function is continuous at all but at most countably many points.

Suppose we have a sequence $\{g_n\}$ of distribution functions all of which are concentrated on some interval $[a, b]$, in the sense that $g_n(a) = 0, g_n(b) = 1$ for every n. (Such a sequence might arise in association with a uniformly bounded sequence of random variables.) Then the selection principle asserts the existence of a subsequence $\{g_{n_k}\}$ and a distribution function g such that $\lim_k g_{n_k}(t) = g(t)$ at every point t where g is continuous (in particular, at all but at most countably many points of $[a, b]$).

To prove this assertion we use the fact that we can consider the sequence $\{g_n\}$ to belong to the unit ball of $C([a, b], \mathbb{R})^*$. Since this ball is a compact metric space in its weak*-topology (**12D**, **F**), there is a subsequence $\{g_{n_k}\}$ which converges weak* to some $h \in NBV([a, b], \mathbb{R})$. Suppose that $t_0 \in [a, b]$ is a point of continuity of h. For each $\varepsilon > 0$ define

$$h_\varepsilon(t) = \begin{cases} 1, & t \leqslant t_0 \\ 0, & t_0 + \varepsilon \leqslant t \\ \text{linear} & t_0 < t < t_0 + \varepsilon. \end{cases}$$

Then

$$\lim_{k \to \infty} \int_a^b h_\varepsilon(t) dg_{n_k}(t) = \int_a^b h_\varepsilon(t) dh(t),$$

since h_ε is continuous on $[a, b]$. But,

$$g_{n_k}(t_0) = \int_a^{t_0} h_\varepsilon(t) dg_{n_k}(t) \leqslant \int_a^b h_\varepsilon(t) dg_{n_k}(t),$$

and

$$\int_a^b h_\varepsilon(t) dh(t) \leqslant h(t_0 + \varepsilon).$$

Therefore,

$$\limsup_{k \to \infty} g_{n_k}(t_0) \leqslant h(t_0 + \varepsilon)$$

for each $\varepsilon > 0$, and so $\limsup_k g_{n_k}(t_0) \leqslant h(t_0)$. Similarly, one shows that $\liminf_k g_{n_k}(t_0) \geqslant h(t_0 - \varepsilon)$. In this way we establish the pointwise convergence of $\{g_{n_k}\}$ to h on the continuity set D of h.

We complete the argument by showing that h is a distribution function concentrated on $[a, b]$. Since $h \in NBV([a, b], \mathbb{R})$ we know that h is left-continuous and $h(a) = 0$. Also,

$$\begin{aligned} h(1) &= \int_a^b dh = \lim_{k \to \infty} \int_a^b dg_{n_k} \\ &= \lim_{k \to \infty} (g_{n_k}(b) - g_{n_k}(a)) = \lim_{k \to \infty} 1 = 1. \end{aligned}$$

Therefore, since the total variation of h on $[a, b]$ is at most one, it must be that h is non-decreasing on $[a, b]$, as desired. □

For a more general result see exercise 4.37.

E. As an application of the Riesz-Kakutani theorem we shall give a geometric functional analytic proof of the Stone-Weierstrass theorem. It is convenient to begin by isolating a portion of the argument as a technical lemma. We let Ω be an arbitary compact Hausdorff space.

Lemma. *Let M be a non-dense linear subspace of $C(\Omega, \mathbb{R})$ and let $\mu \in \text{ext}(U(M^\circ))$. Suppose that $g \in L^\infty(\Omega, B, \mathbb{R})$ has the property that $\int_\Omega fg\, d\mu = 0$, for all $f \in M$. Then g is a constant $|\mu|$-almost everywhere.*

Proof. After adding a constant to g and multiplying by a scalar we can assume that $g \geqslant \theta$ and $\int_\Omega g\, d|\mu| = 1$. Now if $||g||_\infty \leqslant 1$ then certainly $g = 1$ $[|\mu|]$. Otherwise $||g||_\infty < 1$. In this case let $\lambda = 1/||g||_\infty$ and define two new (signed) Borel measures μ_1 and μ_2 by

$$\mu_1(E) = \int_E \frac{(1 - \lambda g)}{1 - \lambda}\, d\mu,$$

$$\mu_2(E) = \textstyle\int_E g\, d\mu, \qquad E \in B.$$

By construction, $\mu = (1 - \lambda)\mu_1 + \lambda\mu_2$ and neither μ_1 nor μ_2 equals μ (since $||g||_\infty > 1$ and $\lambda > 0$). Thus we will have obtained a contradiction if we can prove that both $\mu_i \in U(M^\circ)$. Since they clearly belong to M° we need only estimate their norms. Now

$$||\mu_2|| \equiv |\mu_2|(\Omega) = \textstyle\int_\Omega |g| d|\mu| = \int_\Omega g d|\mu| = 1.$$

Also, since $\theta \leqslant \lambda g \leqslant 1$,

$$\textstyle\int_\Omega |1 - \lambda g| d|\mu| = \int_\Omega (1 - \lambda g) d|\mu| = ||\mu|| - \lambda = 1 - \lambda,$$

whence $||\mu_1|| = 1$. □

Let v be a positive regular Borel measure on Ω. Recall that the *support* $\sigma(v)$ of v is the complement of the union of all open subsets $\mathcal{O} \subset \Omega$ for which $|\mu|(\mathcal{O}) = 0$. Consequently, $\sigma(v) = \{t \in \Omega$: every t-neighborhood has positive v-measure$\}$. The support has the important property that $\int_\Omega f\, dv = 0$ for a non-negative $f \in C(\Omega, \mathbb{R})$ if and only if $f|\sigma(v) = \theta$ (exercise 4.7). Now it is not difficult to establish the Stone-Weierstrass theorem.

Theorem. *Let A be a subalgebra of $C(\Omega, \mathbb{R})$ with the properties that for each $t \in \Omega$ there exists $f \in A$ such that $f(t) \neq 0$ and that A separates the points of Ω. Then A is dense in $C(\Omega, \mathbb{R})$.*

Proof. Suppose that A is not dense. Then by the Hahn-Banach, Alaoglu, and Krein-Milman theorems there exists some $\mu \in \text{ext}(U(A^\circ))$. For any $g \in A$ we have $fg \in A$ whenever $f \in A$, so that $\int_\Omega fg\, d\mu = 0$. Hence by the lemma g is constant $|\mu|$-almost everywhere. Since g is continuous g must be constant on $\sigma(|\mu|)$. Since this is true of every $g \in A$ and A is assumed to separate the points of A it follows that $\sigma(|\mu|)$ must consist of a single point, say p. We complete the proof by obtaining the contradiction that $g(p) = 0$, $g \in A$.

Let e be the constantly one function on Ω and choose any $f \in C(\Omega, \mathbb{R})$. Since the function $f - f(p)e$ vanishes on $\sigma(|\mu|)$ we have

$$\left|\textstyle\int_\Omega (f - f(p)e) d\mu\right| \leqslant \int_\Omega |f - f(p)e| d|\mu| = 0,$$

so that $\int_\Omega f\, d\mu = \alpha f(p)$, where $\alpha \equiv \mu(\Omega)$. Because $\mu \neq \theta$ we see that $\alpha \neq 0$.

Therefore,

$$g(p) = \frac{1}{\alpha} \int_\Omega g \, d\mu = 0, \qquad g \in A. \qquad \square$$

The complex version of the Stone-Weierstrass theorem is obtained in the usual manner, by assuming that the algebra A is self-adjoint (**8F**) in addition to the other hypotheses of the theorem. By exercise 1.39 we then have $A = A_R + iA_R$, and the algebra A_R must be dense in $C(\Omega, \mathbb{R})$ by what we have just shown.

F. As our final topic in this section we give an explicit example of a closed linear subspace of a Banach space X which is not the range of any continuous linear projection defined on X. In other words, this subspace is *uncomplemented* in the sense that it has no closed complementary subspace in X. Another example of this phenomenon is given in exercise 4.19 to illustrate a result in §23.

We begin with a simple lemma which provides some information about operators from a general Banach space into a space of continuous functions.

Lemma. *Let X be a Banach space over the field $\mathbb{F}$, let Ω be a compact Hausdorff space, and let $T \in B(X, C(\Omega, \mathbb{F}))$. Then there exists a continuous map $\tau:\Omega \to X^*$ (given the weak*-topology) such that $Tx(t) = \langle x, \tau(t)\rangle$, $t \in \Omega$, $x \in X$, and $\|T\| = \|\tau\|_\infty$.*

Proof. We know that the map $t \mapsto \delta_t$ is a continuous map of Ω onto a weak*-compact subset of $U(C(\Omega, \mathbb{F})^*)$. Hence, if we define $\tau(t) = T^*(\delta_t)$, it follows from **16C** that τ is continuous. Finally,

$$\begin{aligned}\|T\| &\equiv \sup_{\|x\| \leqslant 1} \|T(x)\| = \sup_{\|x\| \leqslant 1} \sup_{t \in \Omega} |Tx(t)| \\ &= \sup_{t \in \Omega} \sup_{\|x\| \leqslant 1} |Tx(t)| \equiv \sup_{t \in \Omega} \sup_{\|x\| \leqslant 1} |\langle x, \tau(t)\rangle| \\ &= \sup_{t \in \Omega} \|\tau(t)\| \equiv \|\tau\|_\infty.\end{aligned}$$

It is clear that, conversely, any such map τ from Ω into X^* defines an operator $T:X \to C(\Omega, \mathbb{R})$ with $\|T\| = \|\tau\|_\infty$. $\square$

Example. Let X be the Banach space of bounded real-valued functions on $[0, 1]$ with the usual sup-norm, and let C be the closed subspace of continuous functions. We shall show that there is no closed complementary subspace for C in X. Indeed, if there were such a subspace then by **17I** there would be a continuous linear projection P from X onto C. Let $\tau:[0, 1] \to X^*$ continuous map associated with P according to the lemma. Thus $Px(t) = \langle x, \tau(t)\rangle$ for $0 \leqslant t \leqslant 1$ and all $x \in X$. In particular, since the restriction of P to C is the identity operator, it follows that $\tau(t)|C = \delta_t$, $0 \leqslant t \leqslant 1$. We claim that $\tau(t)$ is "evaluation at t" as a functional on X; let us call this functional $\bar{\delta}_t$. The claim is valid because δ_t attains its norm on $U(C)$ at peak functions in $U(C)$, for example, at the function x_t defined by $x_t(s) =$

$1 - |s - t|$, $0 \leqslant s \leqslant 1$. The function x_t is a smooth point of $U(C)$ by **20F**, Ex. d) and in fact is also a smooth point of $U(X)$ (exercise 4.10). Consequently, by **20I** δ_t has a unique Hahn-Banach extension to all of X. This proves that $\bar{\delta}_t = \tau(t)$ as claimed.

We can now easily obtain a contradiction. Let x be a discontinuous function in X. To be explicit let us take x to be the characteristic function of $(0, 1]$. Then, for $n = 1, 2, \ldots$,

$$Px(0) = \quad x, \tau(0)\rangle = 0 \neq 1 = \left\langle x, \tau\left(\frac{1}{n}\right)\right\rangle = Px\left(\frac{1}{n}\right),$$

which proves that Px is not continuous at 0, or, equivalently, that τ is not continuous from $[0, 1]$ to the weak*-topology on X^*. □

§23. Properties and Characterizations of Conjugate Spaces

In this section we discuss several special properties of conjugate Banach spaces, some of which are strong enough to characterize such spaces among general Banach spaces. Certain of these properties, such as the Bessaga-Pelczyński necessary condition, pertain specifically to separable spaces. Several examples of these results are also given.

A. Let X be a given Banach space. To say that X is a conjugate space means that there exists a Banach space V such that X is congruent to V^*. We shall begin by presenting a simple condition sufficient to guarantee that such a space V exists. This result will be called the "Dixmier-Ng theorem".

Theorem. *Suppose that there is a (Hausdorff) locally convex topology τ on X such that $U(X)$ is τ-compact. Then X is a conjugate space.*

Proof. Let $V = \{\phi \in X' : \phi | U(X) \text{ is } \tau\text{-continuous}\}$. Then V is a closed linear subspace of X^*, and is therefore a Banach space. (To see that $V \subset X^*$ observe that for any $\phi \in V$ the image $\phi(U(X))$ is a compact hence bounded set of scalars; that is, $\|\phi\|$ is finite and so $\phi \in X^*$. V is closed in X^* because convergence in X^* entails uniform convergence on $U(X)$.) We now bring in the operator $J_{X,V}: X \to V^*$ introduced in **16F**. This operator assigns to each $x \in X$ the functional "evaluation at x" in V^*. We clearly have $\|J_{X,V}\| \leqslant 1$. The proof will be completed by showing that $J_{X,V}$ is a congruence between X and V^*. We do this by showing that $J_{X,V}$ is injective and that it maps $U(X)$ onto $U(V^*)$.

The first assertion follows because V is total. Indeed, V contains the dual space X_τ^* which certainly separates the points of X (**11E**). The second assertion follows from the fact (evident by definition of V) that $J_{X,V}$ is continuous from the τ-topology on X into the weak*-topology on V^*. This means in particular that $J_{X,V}(U(X))$ is w^*-compact in V^*. But, by the Goldstine-Weston density lemma (**16F**), this image is also weak*-dense in $U(V^*)$. □

Let us now illustrate this theorem with a few examples. We first note that it immediately implies the fact (already known from **16E** and exercise 3.6) that any weak*-closed linear subspace X of a conjugate space Y^* is itself a conjugate space. This follows from the observation that $U(X)$ is compact in the (relative) weak*-topology. It also implies the fact (known from **16H**) that the space $m \equiv m(\Omega, \mathbb{F})$ of bounded functions on a set Ω is a conjugate space (exercise 4.12). We now give a new example.

Example. Let $X = \mathrm{Lip}(\Omega, d, \mathbb{F})$ be the space of bounded Lipschitz functions defined on the metric space (Ω, d) and normed by $\|\cdot\|_L \equiv \max\{\|\cdot\|_\infty, \|\cdot\|_d\}$ (**10D**, Ex. 4). Let τ be the topology of pointwise convergence on X (**9D**). Then $U(X)$ is certainly a τ-closed subset of X. In addition, $U(X)$ is contained in the product B^Ω, where $B \equiv \{\lambda \in \mathbb{F}: |\lambda| \leqslant 1\}$. Since B is compact Tychonov's theorem implies that B^Ω is compact in its product topology. Consequently, $U(X)$ is τ-compact and so X is a conjugate space. □

Any space V for which X is (congruent to) V^* is called a *pre-dual* of X. In general, a given conjugate space X can have more than one pre-dual, although it is known, for example, that whenever μ is a σ-finite measure the space $L^1(\mu)$ is the unique pre-dual of $L^\infty(\mu)$ (see also exercise 4.13). In any event, having recognized that a given Banach space is a conjugate space, it is usually of interest to identify a particular pre-dual as a more or less familiar type of space. This problem is considered in exercise 4.14 for the Lipschitz spaces just discussed.

B. We shall now look a little deeper into the question of whether a given Banach space X is a conjugate space. It will also be of interest to raise a companion question: is X isomorphic to a conjugate space? This is definitely a weaker question; for example, it can be shown that every quasi-reflexive space (**16I**) has this latter property. More interesting, perhaps, is the fact that any non-reflexive space can be renormed so as not to be a conjugate space (see **23E**).

Example. As a special case of this last remark let m be the usual space of bounded sequences. Let $A = U(m) + U(m) \cap c_0$. A may be characterized as the set of sequences $(\xi_1, \xi_2, \ldots)$ in m for which $\sup_n |\xi_n| \leqslant 2$ and $\limsup_n |\xi_n| \leqslant 1$. Now $U(m) \subset A \subset 2U(m)$ and so the gauge ρ_A of A is an equivalent norm on m. But the set A has no extreme points, and so ρ_A is not a conjugate space norm. □

We proceed now to reduce the search for answers to either of the above questions to subspaces of X^*. Recall that this is where we found a pre-dual for spaces X satisfying the condition of **23A**. If V is a subspace of X^* let us henceforth write simply J_V for the operator $J_{X,V}$.

Lemma. *Let X and Y be Banach spaces and suppose that $T: X \to Y^*$ is a congruence (resp. an isomorphism) between X and Y^*. Then there exists a subspace V of X^* such that $J_V: X \to V^*$ is a congruence (resp. an isomorphism).*

Proof. Let V be the range of $T^* \circ J_Y$. For any $y \in Y$ set $v = T^*(J_Y(y))$. Then for any $x \in X$

$$\begin{aligned}\langle y, T(x)\rangle &= \langle T(x), J_Y(y)\rangle \equiv \langle x, v\rangle \\ &\equiv \langle v, J_V(x)\rangle \equiv \langle T^*(J_Y(y)), J_V(x)\rangle \\ &= \langle y, (T^* \circ J_Y)^* \circ J_V(x)\rangle.\end{aligned}$$

This proves that $T = (T^* \circ J_Y)^* \circ J_V$ and consequently that

(23.1) $$J_V = (T^* \circ J_Y)^{*^{-1}} \circ T.$$

Since J_Y is always a congruence formula (23.1) exhibits J_V as a composite of congruences (resp. of isomorphisms). Further, range(J_V) is all of V^* because $T^* \circ J_Y : Y \to V$ is surjective, and hence so is $(T^* \circ J_Y)^{*^{-1}} : Y^* \to V^*$. □

This lemma makes it clear that any reflexive space X has a unique pre-dual, namely X^*.

A closed linear subspace V of X^* is said to be *minimal* if it is total and no proper subspace of V is both total and closed. Also, V is said to be *duxial* (or *norm determining*) if $\sup \{|\langle x, v\rangle| : v \in U(V)\} = \|x\|$, $x \in X$. That is, V is duxial exactly when J_V is an isometry. Finally, we say that a closed subspace M of a Banach space X is *constrained* by a subspace N if there exists a norm-one projection $P: X \to M$ such that $\ker(P) = N$.

We now have the following several characterizations of conjugate spaces and their isomorphs. We shall refer to these results collectively as the "Dixmier-Goldberg-Ruston theorem".

Theorem. *Let X be a Banach space.*

a) *X is a conjugate space (resp. is isomorphic to a conjugate space) if and only if there is a total subspace V of X^* such that $U(X)$ is $\sigma(X, V)$-compact (resp. is relatively $\sigma(X, V)$-compact).*

b) *X is isomorphic to a conjugate space (resp. is a conjugate space) if and only if X^* contains a minimal subspace (resp. a duxial minimal subspace).*

c) *X is isomorphic to a conjugate space (resp. is a conjugate space) if and only if $J_X(X)$ has a weak*-closed complementary subspace in X^{**} (resp. is constrained by such a subspace).*

Proof. a) The condition for X to be a conjugate space is an immediate consequence of Alaoglu's theorem and the Dixmier-Ng theorem (**23A**). Suppose that X is isomorphic to a conjugate space. By the lemma we can assume that $J_V : X \to V^*$ is an isomorphism for some (total) subspace V of X^*. Now J_V is also a homeomorphism from the $\sigma(X, V)$-topology into the weak*-topology on V^*. Therefore,

$$J_V\overline{(U(X))}^{\sigma} = \overline{J_V(U(X))}^{*} = U(V^*),$$

where the second equality is a consequence of the Goldstine-Weston density theorem (**16F**). This proves that $\overline{U(X)}^{\sigma}$ is $\sigma(X, V)$-compact.

Suppose conversely that $U(X)$ is $\sigma(X, V)$-relatively compact for some total $V \subset X^*$. Then, as before,

$$J_V(\overline{U(X)}^{\sigma}) = \overline{J_V(U(X))}^{*} \cap J_V(X) = U(V^*) \cap J_V(X).$$

This shows that $U(J_V(X)) = U(V^*) \cap J_V(X)$ is weak*-compact in V^*. Hence $J_V(X)$ is weak*-closed in V^* by the Banach-Dieudonné theorem (**18E**). Since $J_V(X)$ is also (as noted above) weak*-dense in V^*, it follows that J_V is surjective. Consequently, by the inverse mapping theorem (**17F**) J_V is an isomorphism between X and V^*.

b) Suppose that there is a subspace V of X^* such that J_V is an isomorphism (resp. a congruence). Then we claim that $\bar{V}$ is a minimal (resp. a duxial minimal) subspace of X^*. Indeed, if W were a proper closed subspace of $\bar{V}$ there would exist some non-zero $\Phi \in W^\circ \subset \bar{V}^* = V^*$. If $\Phi = J_V(x)$ then $x \in {}^\circ W$ yet $x \neq \theta$, so that W could not be total.

Conversely, suppose that V is a minimal subspace of X^*. We have to prove that J_V is surjective. Select any $\Phi \in V^*$, $\Phi \neq \theta$. Then $\ker(\Phi)$ is a proper closed subspace of V and so cannot be total. Hence there exists a non-zero $x_0 \in X$ such that $\phi(x_0) = 0$ whenever $\Phi(\phi) = 0$. That is, $\ker(\Phi) \subset \ker(J_V(x_0))$, and it follows that $\Phi = \alpha J_V(x_0) = J_V(\alpha x_0)$ for a suitable scalar α. (We have used the fact that $J_V(x_0) \neq \theta$, since V is total and so J_V is injective.) If V is also duxial then we know that J_V is actually a congruence.

c) Suppose that there is a subspace V of X^* such that J_V is an isomorphism. We claim that V° is a closed complementary subspace for $J_X(X)$ in X^{**}. We certainly have $J_X(X) \cap V^\circ = \{\theta\}$, since V is total. Let $R: X^{**} \to V^*$ be the restriction map: $R(\Phi) = \Phi|V$, $\Phi \in X^{**}$. Now define $P: X^{**} \to J_X(X)$ by

$$P = J_X \circ J_V^{-1} \circ R.$$

Then clearly P is a projection of X^{**} onto $J_X(X)$, $\ker(P) = \ker(R) = V^\circ$, and $\|P\| \leqslant \|J_V^{-1}\|$, which is finite by hypothesis. If also J_V is a congruence then $\|P\| \leqslant 1$; hence $\|P\| = 1$ and so $J_X(X)$ is constrained by V°.

Finally, let W be a weak*-closed complementary subspace for $J_X(X)$ in X^{**}. By exercise 3.6 $W = V^\circ$ for some subspace V of X^*. Because $J_X(X) \cap V^\circ = \{\theta\}$, V is total. Now given any $\Phi \in V^*$, let $\bar{\Phi}$ be a Hahn-Banach extension of Φ belonging to X^{**}. Then $\bar{\Phi} = \hat{x} + w$, $w \in V^\circ$, and it follows that

$$\begin{aligned}\langle \phi, \Phi \rangle &= \langle \phi, \hat{x} \rangle + \langle \phi, w \rangle \\ &= \langle x, \phi \rangle \equiv \langle \phi, J_V(x) \rangle, \qquad \phi \in V.\end{aligned}$$

That is, $J_V(x) = \Phi$, so that J_V is surjective and hence is an isomorphism. If $J_X(X)$ is actually constrained by W then we have in addition

$$\|J_V(x)\| \leqslant \|x\| = \|\hat{x}\| \leqslant \|\bar{\Phi}\| = \|\Phi\| = \|J_V(x)\|;$$

this proves that J_V is a congruence. □

Implicit in this proof are some formulas for $\|J_V^{-1}\|$ in the case where J_V is an isomorphism; see exercise 4.15.

If a Banach space is not isomorphic to a conjugate space then it may or may not be complemented in its second conjugate space. Examples are discussed in **23D** and in the exercises. In particular, it happens that $L^1([0,1],\mathbb{R})$ is constrained in its second conjugate, yet is not isomorphic to any conjugate space.

We now indicate a new class of conjugate spaces.

Example. Let X and Y be Banach spaces and consider the space $B \equiv B(X, Y^*)$. We claim that any such operator space is a conjugate space. We can see this by defining for each $x \in X$, $y \in Y$, a functional $x \otimes y \in B^*$ by

$$\langle T, x \otimes y\rangle = \langle y, T(x)\rangle, \qquad T \in B,$$

and letting V be the linear hull of all such functionals. Since

$$\begin{aligned}\sup_{\|x\otimes y\| \leqslant 1} |\langle T, x \otimes y\rangle| &\geqslant \sup_{\substack{\|x\| \leqslant 1 \\ \|y\| \leqslant 1}} |\langle y, T(x)\rangle| \\ &= \sup_{\|x\| \leqslant 1} \|T(x)\| \equiv \|T\|\end{aligned}$$

it follows that V is duxial and hence that J_V is an isometry. To show that J_V is surjective we can either prove that $\bar{V}$ is a minimal subspace of B^* or that $U(B)$ is relatively $\sigma(B, V)$-compact.

It seems more natural to adopt the second course. Accordingly we observe that $U(B)$ is a closed subset of the product space $\Lambda \equiv \Pi\{\|x\|U(Y^*): x \in X\}$. Now, if $U(Y^*)$ is topologized by the weak*-topology, it follows from the theorems of Alaoglu and Tychonov that Λ is compact. Since the product topology on Λ clearly induces the $\sigma(B, V)$-topology on $U(B)$ we see that $U(B)$ is indeed $\sigma(B, V)$-compact. Thus by either part a) or b) of the theorem J_V is a congruence between B and V^*. The topology $\sigma(B, V)$ is called the *weak*-operator topology* on B.

We continue this example by showing an application (see also exercise 4.18). Let I be an index set, let $\{x_\alpha : \alpha \in I\}$ (resp. $\{\psi_\alpha : \alpha \in I\}$) be bounded subsets of X (resp. of Y^*), and let L be a positive number. We consider the operator moment problem: find $T \in B$ such that

(23.2)
$$\begin{aligned} T(x_\alpha) &= \psi_\alpha, \qquad \alpha \in I, \\ \|T\| &\leqslant L. \end{aligned}$$

(When I is finite and $Y = \mathbb{F}^1$ problem (23.2) is known in the literature as the "L-problem of moments", and the smallest value of the parameter L is of special importance. In addition to its purely mathematical interest, the L-problem of moments subsumes many special models of optimization and control.)

Now problem (23.2) may not, as it stands, be consistent. It may not be possible to find an operator interpolating all the data $\{x_\alpha, \psi_\alpha : \alpha \in I\}$ or it may be that no interpolating operator can have norm $\leqslant L$. We intend to prove that there always exists a *Chebyshev* (or *minimax*) solution. That is, we shall show that an operator $\tilde{T}$ exists satisfying $\|\tilde{T}\| \leqslant L$ and

$$\sup_{\alpha} \|\tilde{T}(x_\alpha) - \psi_\alpha\| \leqslant \sup_{\alpha} \|T(x_\alpha) - \psi_\alpha\|,$$

whenever $\|T\| \leqslant L$.

To do this we observe that the functional $f_\alpha : B \to \mathbb{R}$ defined by

$$f_\alpha(T) = \|T(x_\alpha) - \psi_\alpha\|, \qquad T \in B,$$

is weak*-operator lower semicontinuous on B. Indeed, the map $T \mapsto T(x_\alpha) - \psi_\alpha$ is continuous from the weak*-operator topology on B into the weak*-topology on Y^*, and, of course, the norm on Y^* is weak*-lower semicontinuous. Hence the functional $f \equiv \sup\{f_\alpha : \alpha \in I\}$ is also weak*-operator lower semicontinuous on B, and so attains its infimum over the compact set $LU(B)$ (exercise 2.43). Any operator $\tilde{T} \in LU(B)$ at which f attains its minimum is a Chebyshev solution of the moment problem. □

C. We are now going to present a very striking geometric property possessed by all separable conjugate spaces and their isomorphs. This condition is thus necessary for a given separable Banach space to be isomorphic to a conjugate space. We shall see that it follows easily that certain standard separable spaces are not isomorphic to any conjugate space.

The crux of the matter is to establish the following general lemma due to Namioka. Let us agree that if A is a subset of a conjugate space then A_{w*} denotes the set A topologized by the (relative) weak*-topology.

Lemma. *Let Y be a Banach space for which Y^* is separable, and let A be a weak*-compact and convex subset of Y^*. If Z is the set of all points of continuity of the identity map: $A_{w*} \to A$, then $Z \cap \operatorname{ext}(A)$ is weak*-dense in* $\operatorname{ext}(A)$.

Granting momentarily the truth of this lemma we can use it to establish our main result, known as the "Bessaga-Pelczyński theorem".

Theorem. *Let X be a separable Banach space which is isomorphic to a conjugate space. Then every (non-empty) closed, bounded, and convex subset of X is the closed convex hull of its extreme points.*

Proof. According to exercise 2.40 it is sufficient to prove that every such subset has an extreme point. Now if it is known that all separable conjugate spaces have this property then X also has the property, since it is clearly preserved under isomorphism. Hence there is no loss of generality in assuming that X is a conjugate space: $X = Y^*$ for some Y.

Now let B be a closed, bounded, and convex subset of Y^*. Let $A = \bar{B}^*$. Because B is bounded A is weak*-compact, and we may apply the lemma. Let z belong to the set Z of the lemma. Since B is weak*-dense in A there is a net $\{\beta_\alpha\} \subset B$ that converges weak* to z. By definition of Z it follows that $\lim_\alpha \|\beta_\alpha - z\| = 0$. This entails $z \in B$. Consequently $Z \subset B$ and $Z \cap \text{ext}(A) \subset B \cap \text{ext}(A)$. Now the lemma implies that $Z \cap \text{ext}(A)$ is weak*-dense in $\text{ext}(A)$, and of course $\text{ext}(A) \neq \varnothing$ (**13A**). Since $B \subset A$ we have $B \cap \text{ext}(A) \subset \text{ext}(B)$, and so we have shown that in particular $\text{ext}(B) \neq \varnothing$. □

The proof of Namioka's lemma depends heavily on category and is based on two additional lemmas, the first of which is of some independent interest.

Lemma 1. *Let A be a compact convex subset of a locally convex space. If A is also metrizable then* $\text{ext}(A)$ *is a Baire space.*

Proof. Let d be a metric that defines the topology on A, and define $C_n = \left\{\frac{1}{2}(x + y) : x, y \in A \text{ and } d(x, y) \geqslant \frac{1}{n}\right\}$. Then the sets C_n are closed and $\bigcup_n C_n = A \backslash \text{ext}(A)$. This proves that $\text{ext}(A)$ is a G_δ subset of A. Now, since A is compact and metrizable, it is d-complete. From topology it is known that any G_δ subset of a complete metric space is homeomorphic to a complete metric space. Thus, being homeomorphic to a Baire space, $\text{ext}(A)$ is itself a Baire space. □

The conclusion of Lemma 1 remains valid even if A is not metrizable, but the proof is more difficult and the result in this generality will not be needed.

Lemma 2. *Let Y be a Banach space for which Y^* is separable, and let A be a weak*-compact subset of Y^*. Then the set of all points of continuity of the identity map: $A_{w^*} \to A$ is weak*-dense in A.*

Proof. For each $\varepsilon > 0$ let A_ε be the union of all open subsets of (norm) diameter $\leqslant \varepsilon$. Clearly A_ε is open and we shall show that it is dense in A. Since A is separable there is a (norm) dense sequence $\{\phi_n\} \subset A$. Hence $A = \bigcup_n A \cap \left(\phi_n + \frac{\varepsilon}{2} U(X^*)\right)$. From **17A** and exercise 3.17 the union of the interiors of these sets is dense in A, and the union is clearly contained in A_ε. Consequently, since A is a Baire space the set $\bigcap_n A_{1/n}$ is dense in A; but this set is exactly the set of points of continuity of the identity map: $A_{w^*} \to A$. □

Proof of the Lemma. For each $\varepsilon > 0$ let B_ε consist of those points $\phi \in \text{ext}(A)$ for which there exists a weak*-ϕ-neighborhood N such that $\text{diam}(A \cap N) \leqslant \varepsilon$. Clearly B_ε is a weak*-open subset of $\text{ext}(A)$ and it will be shown to also be weak*-dense in $\text{ext}(A)$. Granting this it will follow from Lemma 1 that $\bigcap_n B_{1/n}$ is also weak*-dense in $\text{ext}(A)$. But this set is exactly $Z \cap \text{ext}(A)$. (Notice that in applying Lemma 1 we have tacitly used **12E**, **F** to guarantee that A is bounded and hence weak*-metrizable.) Thus it remains only to show that B_ε is weak*-dense in $\text{ext}(A)$.

Let W be an arbitrary weak*-open set in Y^* such that $\text{ext}(A) \cap W \neq \varnothing$; we must prove that $B_\varepsilon \cap W \neq \varnothing$. Let $D = \overline{\text{ext}(A)}^*$; D is weak*-compact and certainly $D \cap W \neq \varnothing$. By Lemma 2 the set of continuity points of the identity map: $D_{w*} \to D$ is weak*-dense in D. Hence there is a weak*-open set $V \subset Y^*$ such that $\varnothing \neq D \cap V \subset D \cap W$, and $\operatorname{diam}(D \cap V) \leqslant \varepsilon/4$. Let $A_1 = \overline{\text{co}}^*(D \backslash V)$ and $A_2 = \overline{\text{co}}^*(D \cap V)$. Since $\text{ext}(A) \subset A_1 \cup A_2 \subset A$ it follows from **11A** and **13B** that $A = \text{co}(A_1 \cup A_2)$. We can also note that $\operatorname{diam}(A_2) \leqslant \varepsilon/2$ since $D \cap V$ is contained in a ball of radius $\leqslant \varepsilon/4$. Moreover, $A_1 \neq A$ because $\text{ext}(A_1) \subset D \backslash V$ (**13B**) and $D \cap V \neq \varnothing$.

Now let $d = \operatorname{diam}(A)$ and $r = \varepsilon/4d$. We define C to be the image of the set $A_1 \times A_2 \times [r, 1]$ under the map $f(\phi_1, \phi_2, t) \equiv t\phi_1 + (1 - t)\phi_2$. Then C is a weak*-compact and convex subset of A. Further, $C \neq A$ since $\text{ext}(A) \cap C \subset A_1$, and A_1 is a proper subset of A. Now any $\psi \in A \backslash C$ is of the form $\psi = t\phi_1 + (1 - t)\phi_2$, where $\phi_i \in A_i$ and $0 \leqslant t < r$. It follows that $\|\psi - \phi_2\| \leqslant t\|\phi_1 - \phi_2\| \leqslant rd$; consequently $\operatorname{diam}(A \backslash C) \leqslant 2rd + \varepsilon/2 = \varepsilon$, using $\operatorname{diam}(A_2) \leqslant \varepsilon/2$ and the definition of r.

Finally, since $C \neq A$, there exists $\phi \in (A \backslash C) \cap \text{ext}(A)$ (**13B**). Thus $A \backslash C$ is a weak*-ϕ-neighborhood in A of diameter $\leqslant \varepsilon$, so that $\phi \in B_\varepsilon$. And since $D \backslash V \subset A_1 \subset C$, we must have $\phi \in D \cap V \subset D \cap W$. This shows that $\phi \in B_\varepsilon \cap W$. □

D. We shall now give some examples and discussion pertaining to the Bessaga-Pelczynski theorem.

Example 1. Let Ω be a non-compact but locally compact Hausdorff space and let $C_0(\Omega, \mathbb{F})$ be the Banach space of continuous $\mathbb{F}$-valued functions on Ω that vanish at infinity, with the usual sup norm. According to exercise 2.30 the unit ball of this space has no extreme points. Now if Ω is metrizable we know from **15C** (Lemma 3) that $C_0(\Omega, \mathbb{F})$ is separable. Therefore, by **23C** the space $C_0(\Omega, \mathbb{F})$ for metrizable Ω is not isomorphic to any conjugate space. In particular the sequence space c_0 has this property. Hence so does the sequence space c, where c consists by definition of all convergent sequences of scalars. □

It is interesting to remark that c_0 fails to be isomorphic to a conjugate space for another reason: namely, it fails to satisfy criterion c) of the Dixmier-Goldberg-Ruston theorem (**23B**). Indeed, as is to be shown in exercise 4.19, c_0 fails even to be complemented in its second conjugate space m.

Example 2. Again according to exercise 2.30 the Lebesgue space $L^1 \equiv L^1([0, 1], \mathbb{R})$ has the property that its unit ball contains no extreme points. Consequently L^1 fails to be isomorphic to any conjugate space. This result is known as "Gelfand's theorem". The conclusion has been substantially generalized by Pelczyński: a space $L^1(\Omega, \mu, \mathbb{R})$ where μ is a σ-finite measure is isomorphic to a conjugate space (if and) only if μ is purely atomic. This last condition means that every chunk A in Ω differs only by a null set from the union of all the atoms contained in A.

By contrast with Example 1 the space L^1 is complemented (in fact, constrained) in its second conjugate space. To prove this assertion we must define a norm-one projection from $L^{1**} \cong L^{\infty*}$ onto $J_{L^1}(L^1)$. Given $\Phi \in L^{\infty*}$ let $\phi = \Phi | C([0, 1], \mathbb{R})$. Let $g \in NBV([0, 1], \mathbb{R})$ correspond to ϕ according to formula (22.6). Then the derivative g' belongs to L^1 and we can define the desired projection P by $P(\Phi) = J_{L^1}(g')$. Since

$$\begin{aligned}\|P(\Phi)\| &= \|g'\|_1 \equiv \textstyle\int_0^1 |g'(t)|\, dt \\ &\leqslant \operatorname{var}(g) = \|\phi\| \leqslant \|\Phi\|, \qquad \Phi \in L^{\infty *},\end{aligned}$$

we see that $\|P\| \leqslant 1$. Further, if $\Phi = J_{L^1}(f)$ for some $f \in L^1$ then g is an indefinite integral of f, and so $P(\Phi) \equiv J_{L^1}(g') = J_{L^1}(f) \equiv \Phi$. This proves that P is a norm-one projection of L^{1**} onto $J_{L^1}(L^1)$. (In exercise 4.20 it is to be shown that $\ker(P)$ is weak*-dense in L^{1**}.) □

Although more difficult to prove in complete generality it is true that any $L^1(\Omega, \mu, \mathbb{R})$ space is constrained in its second conjugate.

Either by direct proof or as a consequence of results in the next section it follows that $C([0, 1], \mathbb{R})$ is also not isomorphic to a conjugate space.

Next, let us note that the separability hypothesis in the Bessaga-Pelczynski theorem is really necessary. For example, consider the sequence space m; m is a non-separable conjugate space as we know (**16H** and exercise 2.24a). However, m certainly contains closed, bounded, convex subsets having no extreme points; for example, $U(c_0)$ and the set A appearing in the example in **23B**.

Finally, it should be remarked that there is no converse to the Bessaga-Pelczynski theorem. This disappointment has been rather dramatically illustrated by an example of Lindenstrauss: there exists a closed linear subspace of $\ell^1(\aleph_0)$ which fails to be isomorphic to any conjugate space. Such a subspace appears in the following manner. Given any separable normed linear space X there exists an operator $T \in B(\ell^1, X)$ such that T^* is an isometry. Namely, letting $\{x_n\}$ be a sequence dense in $\partial U(X)$, we define

$$T(\xi) = \sum_{n=1}^{\infty} \xi_n x_n, \qquad \xi = (\xi_1, \xi_2, \ldots) \in \ell^1.$$

Clearly, $\|T\| \leqslant 1$ and

$$\begin{aligned}\|T^*(\phi)\|_\infty &= \sup_n |\langle e_n, T^*(\phi)\rangle| \\ &= \sup_n |\phi(x_n)| = \sup_{\|x\|=1} |\phi(x)| = \|\phi\|, \qquad \phi \in X^*.\end{aligned}$$

By **18G**, T is surjective and X is necessarily complete. (It also follows that $\hat{T}: \ell^1/\ker(T) \to X$ is a congruence, proving that any separable Banach space is (congruent to) a quotient space of ℓ^1; however, this fact is not essential to the example.) Lindenstrauss' example is now obtained by choosing $X = L^1([0, 1], \mathbb{R})$; the kernel M of the associated operator T is shown to have the property that it is not complemented in any conjugate space, whence by part c) of the Dixmier-Goldberg-Ruston theorem M cannot be isomorphic to a conjugate space.

E. We now establish the result alluded to in **23B** concerning non-reflexive spaces. This theorem, formulated by Davis and Johnson, provides a new characterization of reflexive spaces: a Banach space is reflexive if and only if every isomorph of X is a conjugate space. (The necessity of this condition is clear since any isomorph of a reflexive space is actually reflexive, as we know.)

We shall prepare for the theorem with another renorming lemma, due originally to Kadec and Klee.

Lemma. *Let X be a Banach space and V a closed separable subspace of X^*. Then there is an equivalent norm σ on X such that if $\{\phi_\delta:\delta \in D\}$ is any net in X^* that converges weak* to $\phi \in V$, and if $\lim_\delta \sigma^*(\phi_\delta) = \sigma^*(\phi)$, then $\lim_\delta ||\phi_\delta - \phi|| = 0$.*

Proof. Since V is separable there is an increasing sequence $\{G_n\}$ of finite dimensional subspaces of V such that $\bigcup_n G_n$ is dense in V. Define

$$\rho(\psi) = ||\psi|| + \sum 2^{-n} d(\psi, G_n), \qquad \psi \in X^*.$$

Then ρ is an equivalent norm on X^* and we claim that its unit ball U_ρ is weak*-closed in X^*. This follows from the observation that a semi-norm $\psi \mapsto d(\psi, G)$ is lower semicontinuous whenever the subspace G is weak*-closed (which in turn follows from formula (16.9) and exercise 3.6). Since U_ρ is weak*-closed we know from **18F** that ρ is a dual norm on X^*, say $\rho = \sigma^*$ for some equivalent norm σ on X.

Now, by weak*-lower semicontinuity we have $\underline{\lim}_\delta ||\phi_\delta|| \geqslant ||\phi||$ and $\underline{\lim}_\delta d(\phi_\delta, G_n) \geqslant d(\phi, G_n)$, for every n. In conjunction with the relation $\lim_\delta \rho(\phi_\delta) = \rho(\phi)$ we have $\lim_\delta d(\phi_\delta, G_n) = d(\phi, G_n)$. But $\lim_n d(\phi, G_n) = 0$ since $\phi \in V$. This fact plus the compactness of balls in the subspaces G_n enables us to show that the net $\{\phi_\delta:\delta \in D\}$ is relatively compact in X^*. (The details of the argument needed here are routine but tedious; we therefore defer them to exercise 4.21.) But any cluster point of this net must be ϕ, since weak*-$\lim_\delta \phi_\delta = \phi$. Consequently, $\lim_\delta ||\phi_\delta - \phi|| = 0$. □

Theorem. *Let X be a non-reflexive Banach space. Then there is an equivalent norm σ on X such that X so normed is not (congruent to) a conjugate space.*

Proof. By **18B** $U(X^*)$ is not countably compact: there exists a sequence $\{\phi_n\} \subset U(X^*)$ with no weak cluster point. Let $\{\phi_\delta:\delta \in D\}$ be a subnet of $\{\phi_n\}$ that converges weak* to some $\phi \in U(X^*)$, and set $V = \text{span}\ \{\phi, \phi_n:n = 1, 2, \ldots\}$. Letting $J_V = J_{X,V}$ as in **23B**, we have

$$\begin{aligned} \lim_\delta \langle \phi_\delta, J_V(x)\rangle &\equiv \lim_\delta \phi_\delta(x) = \phi(x) \\ &= \phi(x) \equiv \langle \phi, J_V(x)\rangle, \qquad x \in X. \end{aligned}$$

On the other hand there exists some $\Phi \in V^*$ such that $\langle \phi_\delta, \Phi\rangle$ fails to converge to $\langle \phi, \Phi\rangle$. This argument shows that J_V is not surjective.

Let σ be the equivalent norm on X defined by V according to the lemma. Suppose that X so normed is (congruent to) some conjugate space Y^*.

For each $\phi \in V \subset Y^{**}$ there is a net $\{y_\delta : \delta \in D\} \subset Y$ such that $\sigma^*(y_\delta) \leqslant \sigma^*(\phi)$ and weak*-$\lim_\delta J_Y(y_\delta) = \phi$ (Goldstine-Weston density lemma, **16F**). The weak*-convergence entails $\underline{\lim}_\delta\, \sigma^*(J_Y(y_\delta)) \geqslant \sigma^*(\phi)$ and hence $\lim_\delta \sigma^*(J_Y(y_\delta)) = \sigma^*(\phi)$. By the lemma it follows that $\lim_\delta J_Y(y_\delta) = \phi$, whence $\phi \in J_Y(Y)$. But now any $\Phi \in V^*$ can be viewed as defined on a subspace of Y. Any Hahn-Banach extension of Φ belonging to Y^* corresponds to some $x \in X$, and we see that $\langle \phi, J_V(x) \rangle \equiv \phi(x) = \langle \phi, \Phi \rangle$, $\phi \in V$. Thus we have arrived at the contradiction that J_V is surjective. □

§24. Isomorphism of Certain Conjugate Spaces

In this section we prove that the L^∞ spaces defined on a separable measure space are all isomorphic to the space of Lipschitz functions on $[0, 1]$. In addition to its intrinsic interest this result stands in contrast with the corresponding negative fact for the L^p spaces with $1 \leqslant p < \infty$, $p \neq 2$. For such values of p the spaces $\ell^p(\aleph_0)$ and $L^p([0, 1], \mathbb{F})$ fail to be isomorphic.

A. We shall first work to establish the existence of an isomorphism between the spaces m and $L^\infty([0, 1], \mathbb{F})$; the general isomorphism theorem will then follow with little difficulty. There is clearly no loss of generality in assuming the scalars to be real: $\mathbb{F} = \mathbb{R}$. We shall need two preliminary lemmas which provide some interesting information about general spaces of L^∞ type. Let us say that an ordered linear space is *boundedly complete* if every subset which has an upper bound has a least upper bound.

Lemma 1. *Let (Ω, Σ, μ) be a σ-finite measure space. Then $L^\infty \equiv L^\infty(\Omega, \mu, \mathbb{R})$ is boundedly complete.*

Proof. Let A be a subset of L^∞ which is bounded above: there exists $g \in L^\infty$ such that $f \leqslant g$ (that is, $\theta \leqslant g - f[\mu]$) for all $f \in A$. After replacing A by the suprema of its finite subsets (if necessary) we may suppose that A is directed by $\leqslant$, and hence that $A = \{f_\alpha : \alpha \in A\}$ is a non-decreasing net in L^∞ with $f_\alpha \leqslant g$, $\alpha \in A$. Now for any non-negative $x \in L^1$ we have $\langle x, f_\alpha \rangle \leqslant \langle x, g \rangle$, $\alpha \in A$, and so $\lim_\alpha \langle x, f_\alpha \rangle$ exists and defines $\phi(x) \in \mathbb{R}$. Next, for any non-negative x, $y \in L^1$ we define $\phi(x - y) = \phi(x) - \phi(y)$, and obtain a linear functional: $\phi \in (L^1)'$. Further, ϕ is continuous since $\langle x, f_\alpha \rangle \leqslant \phi(x) \leqslant \langle x, g \rangle$ for $x \geqslant \theta$ and $\alpha \in A$, so that there exists $\lambda > 0$ with $|\phi(x)| \leqslant \lambda \|x\|_1$; then decomposing any $x \in L^1$ as $x = x^+ - x^-$, $x^\pm \geqslant \theta$, we see that

$$
\begin{aligned}
|\phi(x)| \equiv |\phi(x^+ - x^-)| &\leqslant |\phi(x^+)| + |\phi(x^-)| \\
&\leqslant \lambda \|x^+\|_1 + \lambda \|x^-\|_1 = \lambda \|x\|_1.
\end{aligned}
$$

Let $h \in L^\infty$ correspond to ϕ under the usual congruence between $(L^1)^*$ and L^∞. Since $\langle x, h - f_\alpha \rangle \geqslant 0$, $x \geqslant \theta$, we have $h - f_\alpha \geqslant \theta$, and so h is an upper bound for A. Finally, if $k \in L^\infty$ is any upper bound for A, then $\langle x, f_\alpha \rangle \leqslant \langle x, k \rangle, x \geqslant \theta$, whence $\phi(x) \equiv \langle x, h \rangle \leqslant \langle x, k \rangle, x \geqslant \theta$. This proves that $h \leqslant k$ and therefore that $h = \sup\{f_\alpha : \alpha \in A\} = \sup A$. □

It may be noted that the only use of the σ-finiteness hypothesis was to guarantee $(L^1)^* \cong L^\infty$, so that a slightly more general theorem is true via the same argument.

For our second lemma we need the concept of an injective Banach space. A Banach space X is *injective* if given any congruence between X and a subspace of a Banach space Y, its image has a complementary subspace in Y. In other words, X is injective if it is complemented in every Banach space containing it. If there is always a projection of norm $\leqslant \lambda$ onto X from every space containing it then X is a *P_λ-space*. For example, every finite dimensional space X is injective (cf. exercise 2.2d) and in fact is a P_λ-space with $\lambda \leqslant \sqrt{n}$. Further, it is known that for $X = (\mathbb{R}^n, \|\cdot\|_2)$ we have for the smallest possible value of λ,

$$\lambda = \frac{n\Gamma\left(\frac{n}{2}\right)}{\sqrt{\pi}\Gamma\left(\frac{n+1}{2}\right)} \sim \sqrt{\frac{2n}{\pi}}, \qquad (n \to \infty),$$

so that the general estimate $\lambda \leqslant \sqrt{n}$ is close to optimal, in an asymptotic sense. It will now be shown that the L^∞ spaces are injective (actually P_1-spaces).

Lemma 2. *Let L^∞ be as in Lemma* 1 *and suppose that L^∞ is (congruent with) a subspace of a Banach space X. Then L^∞ is constrained in X.*

Proof. Let e be the identically one function in L^∞ and define a sublinear mapping $g: X \to L^\infty$ by $g(x) = \|x\|e$. Let f be the identity map on L^∞. Then $f \leqslant g|M$, and any extension of f to an operator $P: X \to L^\infty$ for which $P \leqslant g$ will be a norm-one projection of X onto L^∞. The proof that such an extension exists is essentially a copy of a standard proof of the Hahn-Banach extension theorem **(6A)**, except that L^∞ plays the role of range space instead of $\mathbb{R}$.

To proceed with this proof we consider the family of all linear extensions F of f to some subspace M with $L^\infty \subset M \subset X$ such that $F \leqslant g|M$. This family can be partially ordered by saying that $F_1 \leqslant F_2$ if F_2 is an extension of F_1, and use of Zorn's lemma yields a maximal extension $\tilde{F}$ of f such that $\tilde{F} \leqslant g|\tilde{M}$, where $\tilde{M}$ is the domain of $\tilde{F}$. It remains only to see that $\tilde{M} = X$.

If not, there exists $x_0 \in X\backslash\tilde{M}$ and we can define $\tilde{F}_0$ on $\tilde{M}_0 \equiv \operatorname{span}\{x_0, \tilde{M}\}$ by $\tilde{F}_0(x + tx_0) = \tilde{F}(x) + t\beta$, for every $t \in \mathbb{R}$. Then $\tilde{F}_0$ is a proper extension of $\tilde{F}$ and we complete the proof by showing that a choice of β exists for which $\tilde{F}_0 \leqslant g|\tilde{M}_0$. For any $x, y \in \tilde{M}$ we have

$$\tilde{F}(y) - \tilde{F}(x) = \tilde{F}(y - x) \leqslant g(y - x) \leqslant g(y + x_0) + g(-x_0 - x),$$

whence

$$-g(-x_0 - x) - \tilde{F}(x) \leqslant g(y + x_0) - \tilde{F}(y).$$

Since the left side of this inequality is independent of y, and the right side is independent of x, we can apply Lemma 1 to obtain the existence of $\beta \in L^\infty$ for which

$$-g(-x_0 - x) - \tilde{F}(x) \leqslant \beta \leqslant g(x + x_0) - \tilde{F}(x), \qquad x \in \tilde{M}.$$

With this definition of β the inequality $\tilde{F}_0 \leqslant g|\tilde{M}_0$ is easily seen to hold, and we can therefore conclude that $\tilde{F}_0 \geqslant \tilde{F}$, $\tilde{F}_0 \neq \tilde{F}$, which contradicts the maximality of $\tilde{F}$. □

It is evident from an inspection of the proof that the Banach space structure of X and L^∞ was not used at all, and so it would have been possible to formulate a purely algebraic version of this extension theorem for ordered linear spaces with the bounded completeness property. However, the prime examples of such spaces (among Banach spaces) are the L^∞ spaces so that the extra generality has not seemed worthwhile.

Throughout this section we shall use the notation $X \sim Y$ to indicate that the Banach spaces X and Y are isomorphic. Let $L^\infty = L^\infty([0, 1], \mathbb{R})$. Then it is easy to verify that $L^\infty \sim L^\infty \times L^\infty$ and that $m \sim m \times m$ (these product spaces may each be normed by $\|(x, y)\| = \max(\|x\|_\infty, \|y\|_\infty)$. We now have our basic result due to Pelczyński.

Theorem. $m \sim L^\infty$.

Proof. From **23D** we know that L^∞, being dual to the separable space L^1, is congruent with a subspace of m. Since L^∞ is injective by Lemma 2 there is a complementary subspace X in m, and so we have $m \sim X \times L^\infty$ (**9D**). On the other hand, we can directly embed m into L^∞; for example, by choosing a pairwise disjoint sequence $\{E_n\}$ of chunks in $[0, 1]$ and defining $T: m \to L^\infty$ by

$$T(x) = \sum_{n=1}^{\infty} \xi_n \chi_{E_n}, \qquad x = (\xi_1, \xi_2, \ldots) \in m.$$

(The sum is pointwise convergent but does not, of course, converge in the L^∞ topology.) Hence there is a complementary subspace Y for $T(m)$ because m is injective (the existence of Y is easily seen directly in this case), and so $L^\infty \sim m \times Y$. We now have

$$\begin{aligned} L^\infty &\sim m \times Y \sim (m \times m) \times Y \sim m \times (m \times Y) \\ &\sim m \times L^\infty \sim (X \times L^\infty) \times L^\infty \sim X \times (L^\infty \times L^\infty) \\ &\sim X \times L^\infty \sim m. \end{aligned}$$

□

It will be noted that this is quite a non-constructive proof and, in fact, an explicit isomorphism between m and L^∞ is not known.

It is clear that we can similarly embed ℓ^p isometrically in L^p $(1 \leqslant p < \infty)$. However, it was shown by Banach that L^p cannot be even isomorphically embedded in ℓ^p (unless $p = 2$).

Some of the interest in this theorem derives from the fact that the respective pre-dual spaces ℓ^1 and L^1 are not isomorphic (although by **23D** L^1 is a quotient space of ℓ^1). Indeed, L^1 cannot be isomorphic to any conjugate space (**23D**, Ex. 2). Thus we see that although isomorphic spaces must have isomorphic conjugates, the converse is not true. An even more surprising possibility along these lines is pointed out in **25E**.

B. We are now in position to prove our general isomorphism theorem. Let (Ω, Σ, μ) be a measure space for which the spaces $L^p(\Omega, \mu, \mathbb{F})$ are separable for $1 \leqslant p < \infty$ (exercise 2.24). Note that the measure μ is necessarily σ-finite. Also we consider the space $\mathrm{Lip}([0, 1], \mathbb{F})$ of all Lipschitz continuous $\mathbb{F}$-valued functions on $[0, 1]$ with the norm $\|\cdot\|_L$ defined by (10.5).

Theorem. $L^\infty(\Omega, \mu, \mathbb{F}) \sim \mathrm{Lip}([0, 1], \mathbb{F})$.

Proof. As usual, we can restrict our attention to the case of real scalars: $\mathbb{F} = \mathbb{R}$. Let M be the subspace of $\mathrm{Lip}([0, 1], \mathbb{R})$ consisting of those functions which vanish at 0. The map that sends $f \in M$ into $f' \in L^\infty([0, 1], \mathbb{R})$ is clearly an isomorphism (actually a congruence). By **24A** there is an isomorphism $T: M \to m$. Now consider the map $S: \mathrm{Lip}([0, 1], \mathbb{R}) \to \mathbb{R}^1 \oplus m$ defined by $S(f) = (f(0), T(f - f(0)e))$, where e is the identically one function on $[0, 1]$. Then S is clearly an isomorphism, and since $m \sim \mathbb{R}^1 \oplus m$ trivially, we have shown that $\mathrm{Lip}([0, 1], \mathbb{R}) \sim m$. It remains to see that all $L^\infty(\Omega, \mu, \mathbb{R})$ spaces are isomorphic. This is a consequence of **24A** along with some measure theory.

Consider first the case where μ is non-atomic (**15E**). If $\mu(\Omega) = 1$ then there is an isometry between its associated metric space (exercise 2.24b) and that of Lebesgue measure on $[0, 1]$. This is a consequence of the isomorphism theorem from measure theory which states that any separable, non-atomic, normalized measure algebra is isomorphic to the measure algebra of the unit interval. Given this correspondence between measurable sets we obviously can obtain a correspondence between characteristic functions defined by subsets of Σ and those defined by the measurable subsets of $[0, 1]$. This correspondence extends by linearity to the simple functions where it becomes isometric, and thence to an isometry from $L^\infty(\Omega, \mu, \mathbb{R})$ onto $L^\infty([0, 1], \mathbb{R})$. If $\mu(\Omega)$ is finite we still clearly can obtain an isometry from $L^\infty(\Omega, \mu, \mathbb{R})$ onto $L^\infty([0, 1], \mathbb{R})$. Finally, if $\mu(\Omega)$ is infinite we can in this manner obtain a congruence between $L^\infty(\Omega, \mu, \mathbb{R})$ and $L^\infty(\mathbb{R}, \mathbb{R})$. However, $L^\infty(\mathbb{R}, \mathbb{R})$ is congruent to $L^\infty([0\ 1], \mathbb{R})$; for example, via the map $f \mapsto f \circ g$, where $g(t) = \tan \pi(t - \frac{1}{2})$, $0 < t < 1$.

Now the most general σ-finite measure space (Ω, Σ, μ) will contain some atoms, but at most countably many. Thus we can obtain a partition $\Omega = A \cup B$ where A is the union of the atoms in Σ and $B \equiv \Omega \backslash A$ is either a null set or else a subset of Σ on which μ defines by restriction a non-atomic measure. Clearly $L^\infty(\Omega, \mu, \mathbb{R}) \cong L^\infty(A, \mu_A, \mathbb{R}) \oplus L^\infty(B, \mu_B, \mathbb{R})$ where μ_A (resp. μ_B) is the restriction of μ to the measurable subsets of A (resp. of B). The first of these summands is isomorphic to either $(\mathbb{R}^n, \|\cdot\|_\infty)$ if $n \equiv \mathrm{card}(A)$ is finite, or else to m; the second summand is isomorphic to $L^\infty([0, 1], \mathbb{R})$ or else

$\mu(B) = 0$. Making use of **24A** it is now clear that we have $L^\infty(\Omega, \mu, \mathbb{R}) \sim m$ in all cases. □

It is interesting to remark that some other function spaces on $[0, 1]$ are also known to be isomorphic to m (and hence to all of the preceding spaces). Namely, the spaces $H_\alpha \equiv H_\alpha([0, 1], \mathbb{R})$ of all Holder-continuous functions on $[0, 1]$ of order α are each isomorphic to m, if the norm on H_α is defined as $\max\{\|f\|_\infty, \|f\|_\alpha\}$, where

$$\|f\|_\alpha \equiv \inf\{\lambda > 0 : |f(s) - f(t)| \leqslant \lambda|s - t|^\alpha, 0 \leqslant s, t \leqslant 1\} \qquad 0 \leqslant \alpha < 1.$$

Furthermore, in contrast with the preceding case ($\alpha = 1$), Ciesielski has shown that an isomorphism between H_α and m can be effectively written down in terms of a standard family of step functions on $[0, 1]$ known as the Haar functions.

C. There are two general reasons why isomorphism theorems are of interest. The first and more obvious reason is that we may thereby easily gain some new information about particular spaces. For example, based on our knowledge of L^∞ spaces the preceding theorem allows us to conclude that $\mathrm{Lip}([0, 1], \mathbb{F})$ is a non-separable P_λ-space. A second and more basic reason for the importance of such theorems depends on their interpretation as providing equivalent norms that have more pleasant or useful geometric properties than a given norm. This technique is frequently employed in non-linear functional analysis wherein non-convex sets (for example, manifolds) and non-linear mappings between them are studied. In many cases only topological properties of the sets and mappings are of interest, and these of course are unchanged by a renorming. However, the new norm may facilitate certain constructions. We have seen some uses of this technique in **15C** and a further example is mentioned in **25F**.

§25. Universal Spaces

In this final section we discuss the concept of a universal Banach space and give some examples. The interest in such spaces is discussed in **25F**.

A. A Banach space Y is *universal* for a class $\mathscr{C}$ of Banach spaces if every $X \in \mathscr{C}$ is congruent to a subspace of Y. The simplest example of a universal space is the sequence space m.

Example 1. Let $\mathscr{C}$ consist of all separable Banach spaces and their conjugate spaces. Then m is universal for $\mathscr{C}$. Indeed, let X be a separable Banach space. Then $U(X^*)$ is a compact metric space in the weak*-topology, and in particular there is a sequence $\{\phi_n\}$ that is weak*-dense in $U(X^*)$. Define $T: X \to m$ by $T(x) = (\phi_n(x))$, $x \in X$. Then T is linear and

$$\|T(x)\|_\infty = \sup_n |\phi_n(x)| = \|x\|,$$

so that T is a congruence of X with some subspace of m. On the other hand, the argument given at the end of **23D** shows how to construct a congruence between X^* and a subspace of m. □

Thus m is, in a sense, a "macrocosm" of all the Banach spaces of much real interest. It can further be shown that any separable metric space can be isometrically embedded in m. Clearly, if a Banach space Y contains a subspace congruent to m then Y is itself universal for the class of separable spaces and their conjugates. Examples of such spaces Y are the spaces $L^\infty(\Omega, \mu, \mathbb{F})$ of **24B**.

Let Ω be a topological space. The *density character* dens(Ω) of Ω is the smallest cardinal number of a dense subset of Ω. Thus, for instance, Ω is separable if and only if $\text{dens}(\Omega) \leqslant \aleph_0$. We note that the density character of a topological space is always well-defined, since any set of cardinal numbers when ordered according to size is well ordered. Some further examples and properties of dens($\cdot$) occur in the exercises.

Example 2. Let X be a Banach space. Then there exists a compact Hausdorff space Ω with $\text{dens}(\Omega) \leqslant \text{dens}(X)$ such that X is congruent to a subspace of $C(\Omega, \mathbb{F})$. To see this we let $\Omega = U(X^*)$ given the (relative) weak*-topology. The map $x \mapsto \hat{x}|\Omega$, $x \in X$, is clearly a linear isometry of X into $C(\Omega, \mathbb{F})$, since

$$\|x\| = \sup\{|\phi(x)|:\|\phi\| \leqslant 1\} \equiv \sup\{|\langle\phi, \hat{x}\rangle|:\|\phi\| \leqslant 1\}.$$

To estimate dens(Ω) we let $\{x_\alpha:\alpha \in I\}$ be a dense set in $\partial U(X)$ and for each α choose $\phi_\alpha \in \partial U(X^*)$ with $\phi_\alpha(x_\alpha) = 1$. Then, for each $x \in X$, $\|x\| = \sup\{|\langle x, \phi_\alpha\rangle|:\alpha \in I\}$, so that by the extended Krein-Milman theorem (**13B**) we have $\overline{\text{co}}^*(\{\phi_\alpha:\alpha \in I\}) = U(X^*) \equiv \Omega$. In particular, rational convex combinations of the ϕ_α are dense in Ω, whence $\text{dens}(\Omega) \leqslant \aleph_0 \,\text{card}(I) = \text{card}(I)$, and therefore $\text{dens}(\Omega) \leqslant \text{dens}(\partial U(X)) = \text{dens}(X)$. □

Thus any Banach space can be isometrically embedded in some space $C(\Omega, \mathbb{F})$ where Ω is "not too large". In particular, every separable space can be embedded in some space $C(\Omega, \mathbb{F})$ where Ω is compact metric, that is, in a separable space of type $C(\Omega, \mathbb{F})$. It is now natural to inquire whether there is a fixed compact metrizable space Ω such that $C(\Omega, \mathbb{F})$ is universal for the class of all separable Banach spaces (over the field $\mathbb{F}$). In exercise 4.30 it is to be shown that a necessary condition for Ω to serve this purpose is that Ω be uncountably infinite. We are now going to prove that this condition on the cardinality of Ω is also sufficient.

B. We begin with the most famous case of this result due to Banach: the space $C \equiv C([0, 1], \mathbb{F})$ is universal for all separable Banach spaces. The most expedient proof of this result depends on the result from topology which states that any compact metric space is the continuous image of the Cantor set.

Theorem. *Any separable Banach space X is congruent with a subspace of C.*

Proof. Let h be a continuous mapping from the Cantor set K onto $U(X^*)$. Now the complementary set $[0, 1]\backslash K$ is a union of open intervals $I_n \equiv (s_n, t_n)$ ("middle thirds"). The map h can be extended over each interval I_n via the definition $h(\lambda s_n + (1 - \lambda)t_n) = \lambda h(s_n) + (1 - \lambda)h(t_n)$, $0 < \lambda < 1$. The extended map, which we continue to call h, is clearly a continuous map of $[0, 1]$ onto $U(X^*)$. Define $T: C(U(X^*), \mathbb{F}) \to C$ by

$$T(g)(t) = g(h(t)), \qquad 0 \leqslant t \leqslant 1.$$

Clearly T is linear and isometric. Composing T with a congruence from X into $C(U(X^*), \mathbb{F})$ yields the desired congruence between X and a subspace of C. □

C. A closed subset A of a topological space Ω is *perfect* if it has no isolated points, that is, if every point of A is an accumulation point of A. If Ω contains no perfect subsets it is said to be *dispersed* (*scattered*). Thus in a dispersed space the isolated points are dense. Also, any dispersed space must be totally disconnected in the sense that it contains no non-trivial connected components.

We shall be interested in dispersed compact metric spaces. The simplest example of such a space is the one-point compactification of a countable discrete metric space. We can also consider *ordinal sections*. Let α be an ordinal number and set $\Gamma_\alpha = \{\xi : \xi \leqslant \alpha\}$. With the usual order topology Γ_α is compact, and is metrizable exactly when it is countable, that is, when α is less than the first uncountable ordinal ω_1. The spaces Γ_α are dispersed, and it is known that conversely any dispersed compact metric space is homeomorphic to some space Γ_α, where $\alpha < \omega_1$.

Lemma 1. *A real Banach space X is congruent with a subspace of $C \equiv C(\Omega, \mathbb{R})$, where Ω is a dispersed compact Hausdorff space, (if and) only if $\overline{\operatorname{ext}(U(X^*))}^*$ is dispersed.*

Proof. Suppose that T is a congruence of X with a subspace of C. Let $\phi \in \operatorname{ext}(U(X^*))$. Then $\phi \circ T^{-1} \in \operatorname{ext}(U(T(X)^*))$ and so extends to a functional $\psi \in \operatorname{ext}(U(C^*))$ (exercise 4.32). It follows that $T^*(\psi) = \phi$ and we may conclude that $\operatorname{ext}(U(X^*)) \subset T^*(\operatorname{ext}(U(C^*)))$. Since T^* is weak*-continuous (**16C**) and $\operatorname{ext}(U(C^*))$ is weak*-compact (**13E**, Ex. 4, and exercise 2.35), we have

$$\overline{\operatorname{ext}(U(X^*))}^* \subset T^*(\operatorname{ext}(U(C^*))). \tag{25.1}$$

Now T^* is continuous and surjective (**16C**) and hence is an open mapping (**17G**). It is easy to check that the continuous open image of a dispersed space is again dispersed. Therefore, if Ω is dispersed it follows that $\operatorname{ext}(U(C^*))$ is dispersed, and from this that $T^*(\operatorname{ext}(U(C^*))$ is dispersed. Hence from (25.1) we conclude that $\overline{\operatorname{ext}(U(X^*))}^*$ is also dispersed.

The converse is a consequence of the fact that a Banach space X can always be isometrically embedded into $C(\Omega, \mathbb{F})$ where $\Omega = \overline{\operatorname{ext}(U(X^*))}^*$ (**13B**). □

From this lemma we can already see that not all infinite compact metric spaces Ω have that property that $C(\Omega, \mathbb{R})$ is universal for real separable Banach spaces. For example, the Euclidean spaces $X = (\mathbb{R}^n, \|\cdot\|_2)$ certainly have $\mathrm{ext}(U(X^*))$ not dispersed, and therefore cannot be congruent with a subspace of $C(\Omega, \mathbb{R})$, for Ω dispersed. In a moment we shall see conversely that if $C(\Omega, \mathbb{R})$ does contain a copy of any Euclidean space (of dimension > 1) then it must be universal, and Ω must not be dispersed.

Lemma 2. *Let Ω be a compact metric space which is not dispersed. Then there exists a continuous mapping from Ω onto $[0, 1]$.*

Proof. Let P be a perfect subset of Ω. We distinguish two cases.

a) P is totally disconnected. In this case P is known to be homeomorphic to the Cantor set. Since any compact metric space (in particular, $[0, 1]$) is the continuous image of the Cantor set, there exists a continuous map of P on $[0, 1]$. This map can now be extended to all of Ω by the Tietze extension theorem (**15C**, Ex. 2).

b) If P is not totally disconnected then it contains a non-trivial compact connected subset Q which is necessarily infinite. By virtue of being second countable and completely regular Q can be homeomorphically embedded in the "cube" $[0, 1]^{\aleph_0}$. Now considering Q as a subset of this cube project it onto the various factors. Not all of these projections can consist of a single point or else Q would be a singleton. Therefore, some projection is a non-trivial compact connected subset of $[0, 1]$, and hence is an interval homeomorphic to $[0, 1]$. In this way we obtain a continuous map from Q onto $[0, 1]$, and this map can be extended to all of Ω as usual. □

We now have the following theorem due to Lacey and Morris.

Theorem. *Let Ω be a compact metric space. The following assertions are equivalent.*

a) *$C(\Omega, \mathbb{R})$ is universal for the class of separable real Banach spaces.*

b) *Some Euclidean space $(\mathbb{R}^n, \|\cdot\|_2)$ for $n \geqslant 2$ is congruent with a subspace of $C(\Omega, \mathbb{R})$.*

c) *$C(\Omega, \mathbb{R})$ contains a smooth subspace of dimension $\geqslant 2$.*

d) *$C(\Omega, \mathbb{R})$ contains a subspace of dimension $\geqslant 2$ with a strictly normed conjugate space.*

e) *Ω is not dispersed.*

Proof. It is clear that a) implies b) implies c). If X is a smooth subspace of $C(\Omega, \mathbb{R})$ let M be a subspace of X with $1 < \dim(M) < \infty$. Then M^* is strictly normed by **20G** and so d) holds. Next let X be a subspace satisfying the condition of d). Suppose that Ω is dispersed. Then by Lemma 1 so is $\overline{\mathrm{ext}(U(X^*))}^*$. But this set is just $\overline{\partial U(X^*)}^*$ (**13E**, Ex. 5). Now either $\overline{\partial U(X^*)}^*$ equals $\partial U(X^*)$ (if X is finite dimensional) or else it equals $U(X^*)$ (exercise 2.23C). Certainly $U(X^*)$ is not dispersed and since $\dim(X) \geqslant 1$ neither is $\partial U(X^*)$. Therefore, Ω cannot be dispersed.

Finally, if e) holds let $h: \Omega \to [0, 1]$ be the continuous surjection guaranteed by Lemma 2. Then the map $g \to g \circ h$ defines a linear isometric embedding of $C([0, 1], \mathbb{R})$ into $C(\Omega, \mathbb{R})$. It now follows from **25B** that $C(\Omega, \mathbb{R})$ is universal. □

D. The final step in our program can now be taken. This involves proving that an uncountable compact metric space is not dispersed. This fact is in turn an immediate consequence of the following classical topological result, known as the "Cantor-Bendixson lemma". Let us recall that a point p in a topological space Ω is a *condensation point* if every p-neighborhood contains uncountable many points of Ω.

Lemma. *Any separable metric space Ω can be partitioned into the union of a perfect set and a countable dispersed set.*

Proof. Let Ω_1 be the union of all perfect subsets of Ω and set $\Omega_2 = \Omega \backslash \Omega_1$. Then Ω_1 is closed, hence perfect, while Ω_2 is by definition dispersed. Let Ω_c be the set of all condensation points of Ω. Then Ω_c is a perfect set and so $\Omega_c \subset \Omega_1$. The proof can now be completed by showing that $\Omega \backslash \Omega_c$ is countable, since we have $\Omega_2 \equiv \Omega \backslash \Omega_1 \subset \Omega \backslash \Omega_c$.

Since Ω is separable it is 2nd countable and there is a countable basis $\{V_1, V_2, \ldots\}$ for the topology. For each $p \in \Omega \backslash \Omega_c$ there is a p-neighborhood W such that W is countable, and there is an integer $n(p)$ such that $V_{n(p)} \subset W$, whence $V_{n(p)}$ is countable. Now the set $A = \cup\{V_{n(p)} : p \in \Omega \backslash \Omega_c\}$ is countable and contains $\Omega \backslash \Omega_c$; this proves that $\Omega \backslash \Omega_c$ and hence Ω_2 is countable. □

The main result on universal spaces is now at hand. Notice that in contrast with **25C** there is no restriction on the underlying scalar field.

Corollary. *Let Ω be a compact metric space. Then $C(\Omega, \mathbb{F})$ is universal for the class of separable Banach spaces (over the field $\mathbb{F}$) if and only if Ω is uncountable.*

Proof. The necessity is a consequence of exercise 4.30. For the converse it is sufficient to find a continuous map h from Ω onto $[0, 1]$, as this will show that $C([0, 1], \mathbb{F})$ is congruent with a subspace of $C(\Omega, \mathbb{F})$ and **25B** can be applied. By Lemma 2 the map h can be constructed provided that Ω is not dispersed. But this is a consequence of the Cantor-Bendixson lemma and the assumption that Ω is uncountable. □

E. Let Ω be a compact metric space. It is seen from the Lacey-Morris theorem that the decisive geometric criterion for determining the universality of the space $C(\Omega, \mathbb{R})$ is whether or not it contains a non-trivial smooth subspace, or, equivalently, whether or not it contains a non-trivial subspace with a strictly normed conjugate. It is interesting to consider what kinds of continuous functions on Ω can compose such a subspace. In general such functions exhibit a somewhat pathological behavior. For example, let $\Omega = [0, 1]$, and suppose that X is a subspace of $C([0, 1], \mathbb{R})$ with a strictly normed conjugate and satisfying $3 \leqslant \dim(X) < \infty$. Let $\{x_1, \ldots, x_n\}$ be a linearly independent set in X with $n < \dim(X)$. Then, as has been observed by Donoghue and Smith, the curve $t \mapsto (x_1(t), \ldots, x_n(t))$ is a space-filling curve in $\mathbb{R}^n$, that is, it covers some open set in $\mathbb{R}^n$.

F. Let us make a few further and final remarks about the material in this section. The theory of universal spaces is of interest for several reasons. First, it is a conceptual aid in thinking about Banach spaces in general to be able to encapsulate the most important classes into a single space, such as m or $C([0, 1], \mathbb{F})$. Second, and more importantly, it is frequently possible to establish some property for a class of Banach spaces by establishing it first for a particular universal space and then verifying that the property is hereditary, in the sense that it is possessed by all subspaces of the universal space. It then follows that all Banach spaces in the class under consideration have the property.

An outstanding example of this method is the problem of renorming certain Banach spaces so that the new norm has some desirable property not enjoyed by the original norm. For example, to prove that all separable Banach spaces have an equivalent strict norm (**15C**) it suffices simply to note that the norm $\|\cdot\|_\infty + \|\cdot\|_2$ is an equivalent strict norm on $C([0, 1], \mathbb{F})$. Of greater import is the fact that it is possible to prove that $C([0, 1], \mathbb{R})$ admits an equivalent *locally uniform norm*, that is, a norm ρ with the property that whenever $\rho(x)$, $\rho(x_n) \leqslant 1$ and $\lim_n \rho\left(\dfrac{x + x_n}{2}\right) = 1$ then $\lim_n x_n = x$. This notion is evidently mid-way between the notions of strict norm and uniform norm that we have encountered earlier. Once this has been done it follows that all separable real Banach spaces admit equivalent locally uniform norms. The existence of such equivalent norms was one ingredient in the proof of the famous theorem of Kadec to the effect that all separable infinite dimensional Banach spaces are homeomorphic.

A second remark concerns the spaces $C_\alpha \equiv C(\Gamma_\alpha, \mathbb{R})$, where Γ_α is the ordinal section introduced in **25C**. Assuming that $\alpha < \omega_1$ the spaces Γ_α are countable, and hence C_α^* is congruent to $\ell^1(\aleph_0)$. On the other hand, the spaces C_α for infinite ordinals $\alpha < \omega_1$ are not all isomorphic. In fact, Bessaga and Pelczyński have shown that for $\alpha \leqslant \beta < \omega_1$, $C_\alpha \sim C_\beta$ if and only if $\beta < \alpha^\omega$, where ω is the first infinite ordinal. In particular, the spaces C_ω and C_{ω^ω} are not isomorphic yet their conjugate spaces are congruent (!) This surprising example answered a long standing question posed originally by Banach, and may be contrasted with the earlier example of the non-isomorphic spaces $\ell^1(\aleph_0)$ and $L^1([0, 1], \mathbb{R})$ whose conjugate spaces are isomorphic (but not congruent).

A third remark pertains to some further work of Pelczyński, who has shown that while there are other kinds of separable universal Banach spaces besides those of **25D**, the space $C(K, \mathbb{C})$ is, in a sense, the smallest possible such space. (Here K is the Cantor set.) More precisely, let Ω be an uncountable compact metric space, and let A be a *function algebra* on Ω; that is, let A be a closed subalgebra of $C(\Omega, \mathbb{C})$ which contains the constant functions and separates the points of Ω. An example of such an algebra is the space $A(\Omega)$ where Ω is a closed disc in the complex plane (**10D**, Ex. 5). Then any function algebra is universal for the class of separable Banach spaces. However, it has also been shown that any separable universal Banach space contains a constrained subspace which is congruent to $C(K, \mathbb{C})$.

Finally, we remark that while no (separable) reflexive space can be universal for all separable Banach spaces (as a consequence of exercise 4.33), it is possible for such a space to be universal for the class of all finite dimensional spaces (but *not* for the class of all separable reflexive spaces). Indeed, there is an example due to Szankowski of a separable reflexive space X such that every finite dimensional Banach space is congruent with a constrained subspace of X.

Exercises

4.1. Prove formula (22.3). (To prove the inclusion from left to right consider first the case where $\phi \in U(X^*)$ has norm one.)

4.2. Let X be an order unit normed linear space with order unit e. If $\phi \in X^*$ satisfies $\|\phi\| = \phi(e)$ then ϕ is a positive linear functional.

4.3. Show that the correspondence $\mu \mapsto \Phi_\mu$ of **22B** is bipositive in the sense that μ is a positive Borel measure if (and only if) $\int_\Omega x\, d\mu \geqslant 0$ for all non-negative $x \in C(\Omega, \mathbb{R})$.

4.4. Let Ω be a compact Hausdorff space. Suppose that $\{x_n\}$ is a bounded sequence in $C(\Omega, \mathbb{F})$ that is pointwise convergent to θ: $\lim_n x_n(t) = 0$, $t \in \Omega$. Show that x_n converges weakly to θ. Show by example that this conclusion may fail if $\{x_n\}$ is replaced by a bounded pointwise convergent net in $C(\Omega, \mathbb{F})$.

4.5. Let Ω be an extremally disconnected topological space.
a) Any two disjoint open subsets of Ω have disjoint closures.
b) If Ω is metrizable (more generally, first countable) then Ω is discrete.
c) No sequence in Ω can converge unless it is eventually constant.

4.6. a) Use the Riesz-Kakutani theorem to give a new proof of the fact that a Banach space X is reflexive if $U(X)$ is weakly compact (**16F**). (Given $\Phi \in U(X^{**})$, define a Borel measure μ on $U(X)$ by $\Phi(\phi) = \int_{U(X)} \phi|U(X) d\mu$, $\phi \in X^*$. Then $|\mu|(U(X)) = \|\mu\|_V \leqslant 1$. By **13B**, **E**, μ is the weak*-limit of a net of atomic measures of the form $\sum_i c_i^{(\alpha)} \delta_{x_i^{(\alpha)}}$, where $\{x_i^{(\alpha)}\}$ is, for each α, a finite subset of $U(X)$, and $c_i^{(\alpha)} \geqslant 0$, $\sum_i c_i^{(\alpha)} = 1$. Now consider any weak cluster point in $U(X)$ of the net $\{\sum c_i^{(\alpha)} x_i^{(\alpha)}\}$.)
b) Use the fact that reflexivity of a Banach space is equivalent to the weak compactness of its unit ball to give a new proof of the reflexivity of all closed subspaces and quotient spaces of a reflexive space. (For the quotient space argument use **15B** and **16I**.)

4.7. Let ν be a positive regular Borel measure on a compact Hausdorff space Ω. Prove the two assertions made about the support of ν in **22E**.

4.8. For any real Banach space X the set $\mathscr{P}(X)$ was defined in **20E**. For any compact Hausdorff space Ω show that $\mathscr{P}(C(\Omega, \mathbb{R}))$ can be identified with the set of measures $\mu \in \mathscr{M}_r(\Omega, B, \mathbb{R})$ such that $\sigma(\mu^+) \cap \sigma(\mu^-) = \varnothing$. ($\mu^+$ and μ^- were defined in **22B**.)

4.9. Prove that the space of all polynomial functions on $[0, 1]$ normed by the uniform norm is not subreflexive.

4.10. Let x be a normalized peak function in $m \equiv m(\Omega, \mathbb{R})$, so that $|x(t)| = \|x\|_\infty = 1$ for a single $t \in \Omega$. Show that $x \in sm(U(m))$. (This can be done in two ways: either directly by use of the representation of m^* (**16H**), or by use of the congruence $m \to C(\beta(\Omega), \mathbb{R})$, and the result of **20F**, Ex. d.)

4.11. Generalize the example of **22F** to compact spaces other than $[0, 1]$. What must be assumed about such spaces for that proof to still apply?

4.12. Use the Dixmier-Ng theorem to show that the spaces $m(\Omega, \mathbb{F})$ and $\ell^1(\Omega, \mathbb{F})$ are conjugate spaces.

4.13. Show that any reflexive space has a unique pre-dual.

4.14. Determine a pre-dual of the Lipschitz space Lip$(\Omega, d, \mathbb{F})$. (Consider the linear span of the evaluation functionals $\{\delta_t : t \in \Omega\}$ in Lip$(\Omega, d, \mathbb{F})$. This space can in turn be identified with the free vector space generated by Ω.)

4.15. Let X be a Banach space and V a subspace of X^*.

a) Suppose that $J_V \equiv J_{X,V}$ has a bounded inverse. Prove

$$\|J_V^{-1}\| = \sup\{\|x\| : x \in \overline{U(X)}^{\sigma}\},$$

where $\sigma \equiv \sigma(X, V)$.

b) Suppose that $X^{**} = J_X(X) \oplus V^\circ$ and let $P: X^{**} \to J_X(X)$ be the associated projection. Prove that $\|J_V^{-1}\| = \|P\|$.

4.16. Let X be a Banach space.

a) Show that X is reflexive if and only if X^* contains no proper total closed linear subspace.

b) Assume that X is separable. Show that X is reflexive if and only if every total sequence in X^* is fundamental (**9F**).

4.17. Let X be a separable Banach space.

a) Show that X^* contains a separable duxial subspace.

b) Let X be the Lebesgue space $L^1([0, 1], \mathbb{F})$. Show that the space $C([0, 1], \mathbb{F})$ as a subspace of $L^\infty([0, 1], \mathbb{F})$ is a (separable) duxial subspace of X^*.

4.18. Let M be a (closed) complemented linear space of a Banach space X, and suppose that M is a conjugate space. Then there exists a *minimal projection* on M, that is, a projection: $X \to M$ whose norm is $\leqslant$ that of any other projection of X on M. (Use the method of the example in **23B**.)

4.19. Show that the sequence space c_0 is not complemented in the space m, thereby proving anew that c_0 is not isomorphic to any conjugate space. (A simple proof can be constructed along the following lines. Suppose that Z is a complementary subspace for c_0 in $m: c_0 \oplus Z = m$. Then Z is isomorphic to m/c_0 (exercise 2.2). Now there exists a countable total set in m^*, hence there is such a set in Z^*, and therefore also in $(m/c_0)^*$. This last assertion leads to a contradiction. To obtain it, we make use of a fact about any countable set N: there exists an uncountable family $\{U_\alpha\}$ of infinite subsets of N such that $U_\alpha \cap U_\beta$ is finite whenever $\alpha \neq \beta$. Applying this fact to the case where $N = \{1, 2, \ldots\}$ we let f_α be the coset in m/c_0 that contains the characteristic function of the set U_α. Show that for any $\phi \in (m/c_0)^*$ the set $\{f_\alpha: \phi(f_\alpha) \neq 0\}$ is countable. From this it follows that if $\{\phi_k\}$ is any sequence in $(m/c_0)^*$ then $\{f_\alpha: \phi_k(f_\alpha) \neq 0$ for some $k\}$ is countable, whence $\{\phi_k\}$ cannot be a total subset of $(m/c_0)^*$.)

4.20. Let $P: L^{\infty *} \to L^1$ be the projection constructed in **23D**, Ex. 2. Show that $^\circ\ker(P) = \{\theta\} \subset L^\infty$. (One way to proceed is to select any $f \in L^\infty$ ($f \neq \theta$) and show that there is some $\phi \in \ker(P)$ such that $\langle f, \phi\rangle \neq 0$. Consider separately the cases where f is or is not continuous on $[0, 1]$.)

4.21. Fill in the details of the proof of the lemma in **23E**. (The problem is to show that the net $\{\phi_\delta: \delta \in D\}$ is a relatively compact subset of X^*. For every $\varepsilon > 0$ show that there exists $\delta_\varepsilon \in D$ such that the tail $\{\phi_\delta: \delta \geqslant \delta_\varepsilon\}$ has a finite ε-net. This result can then be used to show that any subnet of $\{\phi_\delta: \delta \in D\}$ has a Cauchy subnet.)

4.22. Consider the subspace of $\mathscr{M}([0, 1], B, \mathbb{R})$ consisting of those measures that are absolutely continuous with respect to Lebesgue measure. Is this subspace weak*-closed in $C([0, 1], \mathbb{R})^*$?

4.23. Give a direct proof that any space $m(\Omega, \mathbb{R})$ is a P_1-space.

4.24. In §24 it is shown that $m \sim L^\infty([0, 1], \mathbb{F})$. Are these spaces in fact congruent?

4.25. Show that the spaces $(\mathbb{R}^n, \|\cdot\|_p)$ are P_λ-spaces with $\lambda \leqslant n^{1/p}$ $(1 \leqslant p < \infty)$.

4.26. Show that any separable metric space can be isometrically embedded in m, and consequently can be so embedded in $C \equiv C([0, 1], \mathbb{R})$. (Thus C is universal for all separable metric spaces.)

4.27. Show that the density character of a metric space is equal to the largest cardinal of a discrete (or isolated) subset.

4.28. Let X be a normed linear space. Prove that dens$(X) \leqslant$ dens(X^*), and give examples where equality (resp. strict inequality) holds.

4.29. Compute the density character of a space $m(\Omega, \mathbb{R})$. (Answer: $2^{\operatorname{card} \Omega}$.)

4.30. Let Ω be a countable compact metric space. Show that $C(\Omega, \mathbb{F})$ cannot be universal for the class of separable Banach spaces.

4.31. a) Prove that a dispersed topological space is totally disconnected. b) Prove that the oridinal sections Γ_α $(\alpha < \omega_1)$ are dispersed compact metric spaces in the order topology.

4.32. Let M be a subspace of a normed linear space X. Suppose that $\phi \in \operatorname{ext}(U(M^*))$. Show that there exists an extremal extension of ϕ, that is, a functional $\bar{\phi} \in \operatorname{ext}(U(X^*))$ such that $\bar{\phi}|M = \phi$. (Consider the set of all norm-preserving extensions of ϕ in X^*).

4.33. Let X be a separable Banach space, universal for the class of all separable Banach spaces. Show that X cannot be isomorphic to a conjugate space.

4.34. Show that the sequence space c (space of all convergent scalar sequences), and hence its subspace c_0, has no infinite dimensional reflexive subspace. (Let M be an infinite dimensional subspace of c. Apply **25C** and **17J**. Compare with the example in **18C**.)

4.35. Prove that any separable Banach space X can be smoothly renormed, that is, admits an equivalent smooth norm. (Construct an equivalent strict norm on X^* via **15C**, Lemma 2, which is weak*-lower semicontinuous; then apply **18F** and **20G**.)

4.36. Determine a congruent representation of the conjugate of the space $C_0(\Omega, \mathbb{F})$ of exercise 2.30 as a space of Borel measures on Ω.

4.37. Let $NBV \equiv NBV([a, b], \mathbb{R})$ be considered as the conjugate space of $C([a, b], \mathbb{R})$ as in **22D**. Show that a bounded sequence $\{g_n\} \subset NBV$ converges weak* to $g \in NBV$ if and only if $\lim_n g_n(t) = g(t)$ at each point t of continuity of g.

References

The purpose of the following remarks is to suggest collateral reading to supplement the material in this book. The references given below have been chosen, for the most part, not to be redundant with the present material, but rather to indicate further developments of topics studied above or else, in a few cases, to serve as introductions to material that has not been discussed above, but which is felt to constitute an important aspect of functional analysis and its applications.

For general introductory treatments of functional analysis the texts by Taylor (1959), Goffman-Pedrick (1965), Brown-Page (1970), Larsen (1973), and Rudin (1973) are recommended. More compendious treatments are the Edwards volume (1965) and the massive Dunford-Schwartz trilogy (1958–71). An overview of Banach space theory (in a somewhat compressed format) is given by Day (1973), and the classical Banach spaces are studied in the recent monographs of Lindenstrauss-Tzafriri (1973) and Lacey (1974). The theory of general linear topological spaces is well covered by Kelley, Namioka, et al. (1963) and Schaefer (1971), and the specialized theory of Hilbert spaces is treated by Halmos (1951, 1967) and Maurin (1967). Operators on general Banach spaces are discussed by Goldberg (1966) and Kato (1966), as well as by Dunford-Schwartz, and on Hilbert spaces by Riesz-Nagy (1955), Gohberg-Krein (1969), Beals (1971), as well as by Halmos and Dunford-Schwartz, Part II.

Functional analysis provides (as we hope has already been demonstrated) a powerful and unified approach to problems of optimization. Detailed developments of this theme are given by Luenberger (1969), Balakrishnan (1971), Pshenichnii (1971), Girsanov (1972), Holmes (1972a), and Laurent (1972). Applications to engineering are given by Porter (1966) and Naylor-Sell (1971), to optimal control by Hermes-LaSalle (1969), and to mathematical economics by Telser-Graves (1972). Functional analytic treatments of partial differential equations have been given by Treves (1967) and Carroll (1969), as well as by Dunford-Schwartz, Part II. Applications to approximation theory are covered by Singer (1974), as well as by Holmes and Laurent.

We now indicate some specific references to accompany particular sections or sub-sections.

§1. Jacobson (1953).
§2. Valentine (1964), Stoer-Witzgall (1970).
§3. Rockafellar (1970), Stoer-Witzgall, Holmes.
§4. Klee (1969).
§5. Jameson (1970).
§7. **7B–C**: Fan (1956).
§8. **8B–E**: Rockafellar; **8F**: Phelps (1963).
§9. Kelly, Namioka, et al., Schaefer; **9C**: Deutsch (1966).

§10. Kelly, Namioka, et al., Schaefer.

§11. Kelly, Namioka, et al.

§12. Kelly, Namioka, et al.

§13. **13C**: Klee (1957); **13D**: Buck (1965); **13E**, Ex. 2: Hoffman (1962); Ex. 3: Roy (1968); Ex. 6: Bohnenblust-Karlin (1955).

§14. **14B–E**: Moreau (1967), Pshenichnii; **14F–H**: Tuy (1972).

§15. **15A**: Fan (1956, 1968); **15C**, Ex. 1: Bonsall (1962); Ex. 2: Michael-Pelczyński (1967); **15E**: Kingman-Robertson (1968), Hermes-LaSalle.

§16. **16I**: Civin-Yood (1957).

§17. **17H**: Gale-Klee (1959); **17J**: Lindenstrauss-Phelps (1968).

§18. **18A–E**: Day.

§19. **19A–C**: James (1964); **19F**: Bartle-Dunford-Schwartz (1955).

§20. **20A–B**: Phelps (1974); **20C**: Peck (1971); **20D**: Brondsted-Rockafellar (1965); **20H**: Phelps (1960), Holmes (1971).

§21. **21A**: Sard (1963); **21B**: Laurent, Holmes (1972b); **21C**: Michael (1956), Parthasarathy (1972); **21D**: Browder (1968); **21E**: Bazaraa et al. (1972), Karamardian (1972).

§22. **22F**: Lindenstrauss-Tzafriri (1971).

§23. **23C**: Namioka (1967).

§24. **24B**: Ciesielski (1960), Lacey-Bernau (1974).

§25. **25C**: Pelczyński-Semadeni (1959); **25D**: Lacey-Morris (1968); **25F**: Kadec (1967), Semadeni (1963), Pelczyński (1967), Szankowski (1972).

Bibliography

A. Balakrishnan. 1971. Introduction to Optimization Theory in a Hilbert Space. Berlin-Heidelberg-New York: Springer-Verlag.

R. Bartle, N. Dunford, and J. Schwartz. 1955. Weak compactness and vector measures. Can. J. Math. **7**, 289–305.

M. Bazaraa, J. Goode, and M. Nashed. 1972. A nonlinear complementarity problem in mathematical programming in Banach space. Proc. AMS **35**, 165–170.

R. Beals. 1971. Topics in Operator Theory. Chicago: Univ. of Chicago Press.

H. Bohnenblust and S. Karlin. 1955. Geometrical properties of the unit sphere of Banach algebras. Ann. Math. **62**, 217–229.

F. Bonsall. 1962. On Some Fixed Point Theorems of Functional Analysis. Bombay: Tata Institute of Fundamental Research.

A. Brondsted and R. T. Rockafellar. 1965. On the sub-differentiability of convex functions. Proc. AMS **16**, 605–611.

F. Browder. 1968. The fixed point theory of multi-valued mappings in topological vector spaces. Math. Ann. **177**, 283–301.

A. Brown and A. Page. 1970. Elements of Functional Analysis. London-New York: Van Nostrand-Reinhold.

R. C. Buck. 1965. Applications of duality in approximation theory. In: Approximation of Functions, H. Garabedian, Ed., pp. 27–42. Amsterdam: Elsevier.

R. Carroll. 1969. Abstract Methods in Partial Differential Equations. New York: Harper and Row.

Z. Ciesielski. 1960. On the isomorphisms of the spaces H_α and m. Bull. Acad. Polon. Sci. **8**, 217–222.

P. Civin and B. Yood. 1957. Quasi-reflexive spaces. Proc. AMS **8**, 906–911.

M. Day. 1973. Normed Linear Spaces, 2nd Ed. Berlin-Heidelberg-New York: Springer-Verlag.

F. Deutsch. 1966. Simultaneous interpolation and approximation in topological linear spaces. SIAM J. Appl. Math. **14**, 1180–1190.

N. Dunford and J. Schwartz. 1958. Linear Operators, Part I. New York: Interscience.

——— ———. 1963. Linear Operators, Part II. New York: Interscience.

——— ———. 1971. Linear Operators, Part III. New York: Interscience.

R. Edwards. 1965. Functional Analysis. New York: Holt, Rinehart and Winston, Inc.

K. Fan. 1956. On systems of linear inequalities. In: Linear Inequalities and Related Systems, H. Kuhn and A. Tucker, Ed's., pp. 99–156. Princeton: Princeton Univ. Press.

———. 1968. On infinite systems of linear inequalities. J. Math. Anal. Appl. **21**, 475–478.

D. Gale and V. Klee. 1959. Continuous convex sets. Math. Scand. **7**, 379–391.

I. Girsanov. 1972. Lectures on the Mathematical Theory of Extremum Problems. Berlin-Heidelberg-New York: Springer-Verlag.

C. Goffman and G. Pedrick. 1965. First Course in Functional Analysis. Englewood Cliffs: Prentice-Hall.

I. Gohberg and M. Krein. 1969. Introduction to the Theory of Linear Nonselfadjoint Operators. Providence: Amer. Math. Society.

S. Goldberg. 1966. Unbounded Linear Operators. New York: McGraw-Hill.

P. Halmos. 1957. Introduction to Hilbert Space. New York: Chelsea.

———. 1967. A Hilbert Space Problem Book, Princeton: D. Van Nostrand Co., Inc.

H. Hermes and J. LaSalle. 1969. Functional Analysis and Time Optimal Control. New York: Academic Press.

K. Hoffman. 1962. Banach Spaces of Analytic Functions. Englewood Cliffs: Prentice-Hall.

R. Holmes. 1971. Smoothness indices for convex functions and the unique Hahn-Banach extension problem. Math. Z. **119**, 95–110.

———. 1972a. A Course on Optimization and Best Approximation. Berlin-Heidelberg-New York: Springer-Verlag.

———. 1972b. R-splines in Banach spaces: I. Interpolation of Linear Manifolds. J. Math. Anal. Applic. **40**, 574–593.

N. Jacboson. 1953. Lectures in Abstract Algebra, Vol. II. Princeton: D. Van Nostrand Co., Inc.

R. James. 1964. Weak compactness and reflexivity. Israel J. Math. **2**, 101–119.

G. Jameson. 1970. Ordered Linear Spaces. Berlin-Heidelberg-New York: Springer-Verlag.

M. Kadec. 1967. Proof of the topological equivalence of all separable infinite dimensional Banach spaces. Functional Anal. Appl. **1**, 53–62.

S. Karamardian. 1972. The complementarity problem. Math. Prog. **2**, 107–129.

T. Kato. 1966. Perturbation Theory for Linear Operators. New York: Springer-Verlag.

J. Kelley, I. Namioka, et al. 1963. Linear Topological Spaces. Princeton: D. Van Nostrand Co., Inc.

J. Kingman and A. Robertson. 1968. On a theorem of Lyapunov. J. London Math. Soc. **43**, 347–351.

V. Klee. 1969. Separation and support properties of convex sets—a survey. In: Control Theory and the Calculus of Variations, A Balakrishnan, Ed., pp. 235–304. New York: Academic Press.

———. 1957. Extremal structure of convex sets. Arch. Math. **8**, 234–240.

H. Lacey. 1974. The Isometric Theory of Classical Banach Spaces. New York: Springer-Verlag.

——— and S. Bernau. 1974. Characterizations and classifications of some classical Banach spaces. Adv. Math. **12**, 367–401.

——— and P. Morris. 1968. On universal spaces of the type $C(X)$. Proc. AMS **19**, 350–353.

R. Larsen. 1973. Functional Analysis: an Introduction. New York: Dekker.

P. Laurent. 1972. Approximation et Optimisation. Paris: Hermann.

J. Lindenstrauss and R. Phelps. 1968. Extreme point properties of convex bodies in reflexive Banach spaces. Israel J. Math. **6**, 39–48.

——— and L. Tzafriri. 1971. On the complemented subspaces problem. Israel J. Math. **9**, 263–269.

———. 1973. Classical Banach Spaces. Berlin-Heidelberg-New York: Springer-Verlag.

D. Luenberger. 1969. Optimization by Vector Space Methods. New York: Wiley.

K. Maurin. 1967. Methods of Hilbert Spaces. Warsaw: Polish Scientific Publishers.

E. Michael. 1956. Continuous Selections, I. Ann. Math. **63**, 361–382.

——— and A. Pelczyński. 1967. A linear extension theorem. Ill. J. Math. **11**, 563–579.

J. J. Moreau. 1967. Sous-différentiabilité. In: Proc. Colloquim on Convexity, pp. 185–201, Copenhagen, 1965.

I. Namioka. 1967. Neighborhoods of extreme points. Israel J. Math. **5**, 145–152.

A. Naylor and G. Sell. 1971. Linear Operator Theory in Engineering and Science. New York: Holt, Rinehart and Winston, Inc.

T. Parthasarathy. 1972. Selection Theorems and their Applications. Berlin-Heidelberg-New York: Springer-Verlag.

N. Peck. 1971. Support points in locally convex spaces. Duke Math. J. **38**, 271–278.

A. Pelczyński. 1967. Some linear topological properties of separable function algebras. Proc. AMS **18**, 652–660.

——— and Z. Semadeni. 1959. Spaces of continuous functions (III). Studia Math. **18**, 211–222.

R. Phelps. 1960. Uniqueness of Hahn-Banach extensions and unique best approximation. Trans. AMS **95**, 238–255.

———. 1963. Extreme positive operators and homomorphisms. Trans. AMS **108**, 265–274.
———. 1974. Support cones in Banach spaces and their applications. Adv. Math. **13**, 1–19.
B. Pshenichnii. 1971. Necessary Conditions for an Extremum. New York: Dekker.
F. Riesz and B. Sz.-Nagy. 1955. Functional Analysis, New York: Ungar.
R. T. Rockafellar. 1970. Convex Analysis. Princeton: Princeton Univ. Press.
A. Roy. 1968. Extreme points and linear isometries of the Banach space of Lipschitz functions. Can. J. Math. **20**, 1150–1164.
W. Rudin. 1973. Functional Analysis. New York: McGraw-Hill.
A. Sard. 1963. Linear Approximation. Providence: Amer. Math. Society.
H. Schaefer. 1971. Topological Vector Spaces. New York: Springer-Verlag.
Z. Semadeni. 1963. Isomorphic properties of Banach spaces of continuous functions. Studia Math., Seria Spec. **1**, 93–108.
I. Singer. 1974. Theory of Best Approximation and Functional Analysis. CBMS Lecture Series, SIAM.
J. Stoer and C. Witzgall. 1970. Convexity and Optimization in Finite Dimensions I. Berlin-Heidelberg-New York: Springer-Verlag.
A. Szankowski. 1972. An example of a universal Banach space. Israel J. Math. **11**, 292–296.
A. Taylor. 1959. Introduction to Functional Analysis. New York: Wiley.
L. Telser and R. Graves. 1972. Functional Analysis in Mathematical Economics. Chicago: Univ. of Chicago Press.
F. Treves. 1967. Locally Convex Spaces and Partial Differential Equations. New York: Springer-Verlag.
H. Tuy. 1972. Convex inequalities and the Hahn-Banach theorem. Diss. Math. **47**.
F. Valentine. 1964. Convex Sets. New York: McGraw-Hill.